ENCYCLOPEDIA OF MATHEMATICS AND ITS APPLICATIONS
Volume 26

Theory of Matroids

ENCYCLOPEDIA OF MATHEMATICS
and Its Applications

GIAN-CARLO ROTA, Editor
Massachusetts Institute of Technology

Editorial Board

For other books in this series see page 317

Contents

CAMBRIDGE UNIVERSITY PRESS
Cambridge, New York, Melbourne, Madrid, Cape Town, Singapore, São Paulo, Delhi

Cambridge University Press
The Edinburgh Building, Cambridge CB2 8RU, UK

Published in the United States of America by Cambridge University Press, New York

www.cambridge.org
Information on this title: www.cambridge.org/9780521309370

First published 1986
This digitally printed version 2008

A catalogue record for this publication is available from the British Library

Library of Congress Cataloguing in Publication data
Main entry under title:
Theory of matroids.
(Encyclopedia of mathematics and its applications; v. 26)
Bibliography: p.
1. Matroids. I. White, Neil. II. Series.
QA166.6.T44   1986   511'.6   85–6682

ISBN 978-0-521-30937-0 hardback
ISBN 978-0-521-09202-9 paperback

THEORY OF MATROIDS

Edited by

Neil White

University of Florida

The right of the
University of Cambridge
to print and sell
all manner of books
was granted by
Henry VIII in 1534.
The University has printed
and published continuously
since 1584.

CAMBRIDGE UNIVERSITY PRESS

Cambridge

London   New York   New Rochelle

Melbourne   Sydney

Contributors

Thomas Brylawski Department of Mathematics
University of North Carolina
Chapel Hill, NC 27514

Henry Crapo Institut National de Recherche en
Informatique et en Automatique, B. P. 105
78153 le Chesnay Cedex, France

Ulrich Faigle Institut für Ökonometrie und
Operations Research, Universität Bonn
Nassestrasse 2, D-5300 Bonn 1
Federal Republic of Germany

Joseph P. S. Kung Department of Mathematics
North Texas State University
Denton, TX 76203

Hien Q. Nguyen Department of Mathematics
University of Montana
Missoula, MT 59812

Giorgio Nicoletti Istituto di Geometria
Universita Bologna
Piazza di Porta S. Donato
Bologna, Italy

James Oxley   Department of Mathematics
                   Louisiana State University
                   Baton Rouge, LA 70803

Neil White   Department of Mathematics
                   University of Florida
                   Gainesville, FL 32611

Series Editor's Statement

A large body of mathematics consists of facts that can be presented and described much like any other natural phenomenon. These facts, at times explicitly brought out as theorems, at other times concealed within a proof, make up most of the applications of mathematics, and are the most likely to survive change of style and of interest.

This ENCYCLOPEDIA will attempt to present the factual body of all mathematics. Clarity of exposition, accessibility to the nonspecialist, and a thorough bibliography are required of each author. Volumes will appear in no particular order, but will be organized into sections, each one comprising a recognizable branch of present-day mathematics. Numbers of volumes and sections will be reconsidered as times and needs change.

It is hoped that this enterprise will make mathematics more widely used where it is needed, and more accessible in fields in which it can be applied but where it has not yet penetrated because of insufficient information.

Gian-Carlo Rota

Foreword

It is a rare event that mathematicians, cozily ensconced in the world of established theories, should extract, by dint of pioneering work, some new gem that later generations will spend decades polishing and refining. The hard-won theory of matroids is one such instance. Rich in connections with mathematics, pure and applied, deeply rooted in the utmost reaches of combinatorial thinking, strongly motivated by the toughest combinatorial problems of our day, this theory has emerged as the proving ground of the idea that combinatorics, too, can yield to the power of systematic thinking.

A superficial look at the theory of matroids might lead to the conclusion that it is largely an abstraction of linear algebra. It was noticed quite some time ago that the elementary theory of linear dependence can be developed from the MacLane-Steinitz exchange axiom. In fact, this abstraction was first exploited in the theory of transcendence degrees of fields. But such a conclusion would be unwarranted. What the theory of matroids provides is a variety of cryptomorphic axiomatic approaches, each of which corresponds to a genuinely new way of looking at linear algebra. Linear algebraists might not easily be led to these axiomatic approaches. The axiomatization of matroids by the notion of a minimal dependent set, for example, leads to the deeper matching theorems for sets of vectors. The axiomatization by the notion of rank leads to the classification of projectively invariant constructions of new matroids from old. New axiomatizations are still appearing. Matroid theory is unique in mathematics in the number and variety of its equivalent axiom systems; this accounts in part for the versatility and applicability of the subject.

The idea of geometric lattice arose from the combinatorial theory of matroids and now provides a framework for the invariant theory of sets of points. The abstraction of the coloring problem of graphs to arbitrary geometric lattices has provided a remarkable unification, now called the critical problem, of a variety of deep extremal set-theoretic problems. One of the seldom-stated motivations for much current work on matroid theory is in fact the idea that the critical problem may eventually be solved by joint action of invariant theory, extremal set theory, and finite geometry, perhaps with a touch of homological algebra. The closely related and now well developed theory of coordinatization of matroids is spearheading these developments.

Surely some of the most beautiful results of contemporary combinatorics are Tutte's theorems relating the coordinatizability of matroids to the absence of certain forbidden minors (called obstructions) in its geometric lattice. Tutte's theory culminates in his characterization of unimodular matroids, namely matroids coordinatizable over every field, by the absence of the Fano plane.

Matroids have proved to be an essential concept in discrete optimization. The greedy algorithm is the optimization-theoretic analog of the MacLane-Steinitz exchange property. From this elementary beginning, interest in matroids in this field has exploded: polymatroids, oriented matroids, greedoids, and submodular functions now abound in the literature of combinatorial optimization.

What will happen in the long run to the theory of matroids? We predict that it will soon feed back profoundly upon linear and multi-linear algebra, and, most of all, upon homological algebra, in at least two ways: firstly, by providing a rich problematique which these fields can test themselves upon, and secondly, by feeding its own techniques directly back onto homological algebra. Matroid theory reflects in an exemplary way the mathematical preoccupations of our day, such as the meeting of the cross-currents of pure and applied mathematics and the cutting across party lines of separate fields, while not losing sight of the concrete objectives of solving some of the longstanding problems of contemporary mathematics.

Gian-Carlo Rota

Preface

This book had its beginnings over a decade ago, as a simple rewriting of Crapo and Rota's preliminary edition of *Combinatorial Geometries*, to be accomplished by Crapo, Rota, and White. We soon realized that the subject had grown enough, even then, that a more comprehensive compendium would be of greater benefit. This led, in turn, to the idea of soliciting contributions from many of the workers in matroid theory. Consequently, this work has grown too lengthy to be contained in a single volume. This is but the first of a projected three-volume series, although we are giving separate titles to each of the volumes. We are planning to call the remaining volumes *Combinatorial Geometries*, and *Advances in Matroid Theory*.

This first volume is a primer in the basic axioms and constructions of matroids. It will prove useful as a text because exposition has been kept a prime consideration throughout. Proofs of theorems are often omitted, with references given to the original works, and exercises are included. This volume will also be useful as a reference work for matroid theorists, especially Brylawski's encyclopedic chapter "Constructions" and his cryptomorphism appendix.

The volume starts with Crapo's chapter "Examples and Basic Concepts." This chapter is a very informal introduction to matroids, with lots of examples, that provides an overview of the subject. The next chapter is "Axiom Systems," by Nicoletti and White. This gets into the necessary work of proving the equivalence of some of the major axiom systems, a chore made easier by keeping in mind the familiar analogous concepts from linear algebra. The presentation in this chapter is based on Nicoletti's interesting self-dual metasystem of the axiom systems. In the third chapter, Faigle presents the lattice-theoretic approach to matroids. Originated by Birkhoff, this was the cornerstone of the original Crapo-Rota volume, although it is less heavily relied on in the current work. Then Kung provides more depth on one family of related axiom systems, the basis axioms. In the fifth chapter, Crapo explains the fundamental duality concept, which generalizes duality of planar graphs (a prime motivation in Whitney's creation of the concept of matroid) and orthogonality of vector subspaces. Next comes Oxley's exposition of one of the most important and most elementary classes of examples of matroids, those that arise from a graph. Included are a complete characterization of when two graphs produce the same matroid and a description of an important class of graphs that are characterized matroidally, the series-parallel networks. The seventh chapter is Brylawski's comprehensive compendium of matroid constructions, together with a marvelous separate index for this chapter only. This chapter includes new material never before published, most notably the idea of matroid bracing. Kung's "Strong Maps" is the first of two chapters on mappings between matroids. Strong maps have the beautiful characterization of always being factorable into an embedding followed by a contraction. Kung and Nguyen then present a much more general type of mapping, the weak map, which formalizes the intuitive notion that one matroid is "more dependent" than another. Nguyen then presents the theory of semimodular functions, another approach to matroids using their rank (or dimension) functions. Finally comes Brylawski's "Appendix of Matroid Cryptomorphisms," a detailed listing of all of the known equivalent axiomatizations of matroids, the cryptomorphisms that relate these axiomatizations, and their applications to the most important classes of examples of matroids. A previous version of this listing has been available to matroid theorists for some years and has proved to be a useful reference work.

     Exercises marked with an asterisk tend to be more difficult, and those with two asterisks are unsolved. In order to make it easier to distinguish in the figures between affine diagrams of matroids and other representations, such as lattice diagrams and graphs, we have adopted a new convention of using large dots for elements in affine diagrams, and small dots for lattice elements and vertices of graphs.

I would like to thank all the contributors to this project for their continuing support, despite the delays in the appearance of this volume. I would also like to thank Henry Crapo and Gian-Carlo Rota for involving me in this project and for their support. Thanks are due as well to several outside referees, who must remain unnamed. I also thank Rhodes Peele for working many of the exercises before their inclusion. Finally, Tom Brylawski wishes to thank the National Science Foundation for their partial support of his work under grant MCS 7801149, and Hazeline Lewis and Daniela Calvetti for their assistance in preparing his manuscript.

Neil L. White

University of Florida

CHAPTER 1

Examples and Basic Concepts

Henry Crapo

1.1. EXAMPLES FROM LINEAR ALGEBRA AND PROJECTIVE GEOMETRY

As an introduction to the concepts of *combinatorial geometry* and *matroid*, we wish to emphasize those features of the theory that have given it a unifying role in other branches of mathematics, that have permitted it to be fruitfully applied in disparate domains of science, and that continue to arouse broader interest in the subject. It is our intention in this chapter to clarify the basic concepts by showing how they appear and are interrelated in a list of significant examples. This will give the reader a general orientation with respect to the basic concepts, prior to their axiomatic treatment in Chapter 2. Most of these examples will be dealt with in full detail in subsequent chapters.

The concept of *combinatorial geometry* arose from work in projective geometry and linear algebra. The focus of this work was to understand the basic properties of two relations:

(1) the *incidence* between points, lines, planes, and so on (which in general we call *flats*) in geometries and geometric configurations
(2) the *linear dependence* of sets of vectors.

The task to characterize (axiomatize) these relations of incidence and linear dependence seems in retrospect both urgent and feasible in the light of far-reaching applications, both to new geometries and to more general algebraic and combinatorial structures.

Research in combinatorial geometry has been concentrated on

(1) *synthetic* (*combinatorial*) *methods*, involving only *incidence relations* between flats, and the fundamental operations of *projection* and *intersection*
(2) *intrinsic properties* of configurations, internal properties that configurations possess independent of the way in which they may be represented or constructed within some conventional space.

This emphasis on synthetic methods and on configurations has permitted the theory and its applications to develop without undue reliance on coordinate systems, vector algebra, determinants, and the like. In the resulting combinatorial theory, theorems are often phrased in terms of the existence of certain geometric incidences or subconfigurations. Premises and conclusions of this nature are often more practical to use in applications, and are then a greater support to intuition, than their equivalent algebraic formulations. But there is a price to pay: There exist combinatorial geometries whose points cannot be coordinatized in any satisfactory manner.

1.1.A. Subsets of Projective Spaces

The most fundamental example of a combinatorial geometry is the structure of an arbitrary set of points in a finite-dimensional projective space. To begin with, the projective space has a structure of *flats*, namely its points, lines, planes, and so on, and the relation of incidence between them. Each flat has a *rank* equal to its projective dimension plus 1; points have rank 1, lines have rank 2, and so forth. The rank of any flat is thus equal to the number of points needed to determine that flat.

Our immediate task is to see how this structure of flats can be induced on a *subset* of the set of points of the space. Let $E \subseteq S$ be a subset of the set S of points in some projective space. Each projective flat itself consists of a set of points, whose intersection with E we call a *flat* of the combinatorial geometry $G(E)$ induced on the set E. How should we assign a rank to such a flat? Say a subset $A \subseteq E$ is the intersection with E of some projective flat. Then there is a *least* such projective flat whose intersection with E is A, and it has a rank that we assign as the rank $r(A)$ of the flat A in the geometry $G(E)$.

In Figure 1.1a we sketch a combinatorial geometry with six points. The flats and their relations of incidence are indicated in the lattice diagram of Figure 1.1b. Straight lines in the sketch indicate those lines of the geometry that contain more than two points, but pairs like dg are also lines of the combinatorial geometry, and have rank 2.

This geometry is easily shown to be the geometry induced on a set of six points in the real projective plane. In Figure 1.1c we indicate one such choice of six points. Here, for instance, the collinearity of points g, l, and f

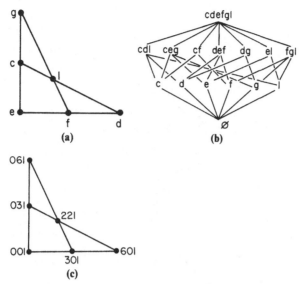

FIGURE 1.1. A six-point geometry, presented in three ways.

is proved by the computation

$$-g + 3l = -1(0, 6, 1) + 3(2, 2, 1) = (6, 0, 2) \simeq (3, 0, 1) = f,$$

because coordinates differing only by a common scalar multiple refer to the same projective point.

The lattice of flats of a combinatorial geometry, such as that in Figure 1.1b, is a *geometric lattice*. The *join* of two flats is the least flat containing both of them; the *meet* of two flats is their intersection. The main properties of geometric lattices are summed up in the following statement: If a flat A is contained in a flat B, then they are in consecutive ranks if and only if there is a point $p \notin A$ such that $A \vee p = B$. Put another way, the set of points *not* on any fixed flat A is partitioned by inclusion of those points in the flats of the next higher rank, containing A. Thus, in a combinatorial geometry of rank ≥ 3, the points not on a given line are partitioned by inclusion in the planes passing through that line. For instance, in the projective cube of Figure 1.2a, the set *abcefh* of points not on the line *dg* is partitioned $(af)(b)(ch)(e)$ by inclusion in the planes *adfg*, *bdg*, *cdgh*, and *deg*, respectively.

We wish to hold off axiomatic treatment of combinatorial geometries until the next chapter; here we shall deal further with examples and sketch the general outlines of the basic theory. For the present, let us agree that a combinatorial geometry consists of *flats* with a well-defined *rank*, each flat being a set of *points*, with the following property:

(1) The set of points *not* on any fixed flat A is partitioned by inclusion in the flats of the next higher rank, containing A.

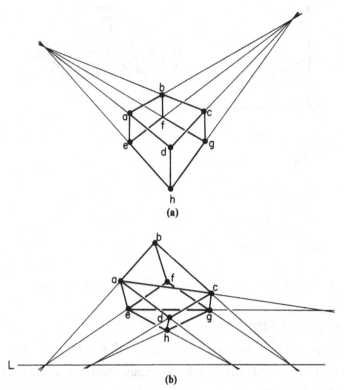

FIGURE 1.2. (a) A projective cube. (b) A noncoordinatizable geometry.

That statement, together with the assertions that

    (2) any intersection of flats is a flat,
    (3) rank agrees with height in the lattice,
    (4) the overall geometry has finite rank, and
    (5) the empty set and single points are flats,

already constitutes more than enough axioms to characterize combinatorial
geometries (see Exercise 2.7).

    There are significant differences between combinatorial geometries in
general and projective geometries in particular. In any combinatorial geometry, concurrent lines must be coplanar, and planes meeting in a line must
be cospatial. In a projective geometry, we also know that coplanar lines are
concurrent, and cospatial planes must meet in a line. This is not true for
combinatorial geometries in general. For instance, the coplanar lines *cf* and
el do not meet in the combinatorial geometry of Figure 1.1a. If Figure 1.2a
is construed as a combinatorial geometry, the cospatial planes *aceg* and *bdfh*
do not meet at all!

A slight modification of the geometry of a set of points in projective space can produce a geometry that cannot be realized in any projective space. Consider Figure 1.2b, which is intended to suggest a projective cube with six flat faces, in which the diagonal set of four points *aceg* is coplanar, but the diagonal set *bdfh* is skew. If eight points lying in this way on seven four-point planes were to be found in a projective geometry, the following contradiction would arise: The planes *bafe*, *aceg*, and *cbgf* would meet in some point *p*, a point that would thus lie on the three lines *ae*, *cg*, and *bf* of intersection of those planes. Similarly, the lines *ae*, *cg*, and *dh* would be concurrent and could meet only at the same point *p*. Thus, *bf* and *dh* would be concurrent at *p* and would *have to be coplanar*, rather than skew. The drawing is designed to make it seem reasonable that the lines *bf* and *dh* could be skew. Actually, in any three-dimensional realization of this particular plane drawing, the lines *ac* and *eg* are also skew, and neither of the diagonal sets *aceg* and *bdfh* is coplanar. If the set *aceg* is to be coplanar, the lines *ac* and *eg* must meet on the line *L*, to provide a point of intersection for the plane *aceg* with the top and bottom planes *abcd* and *efgh* of the projective cube.

From the foregoing discussion we can see that it is a theorem of projective geometry that if one of these diagonal sets is coplanar, so is the other. However, a set of eight points in rank 4, no three points collinear, forming 7 four-point planes *abcd*, *abef*, *aceg*, *adeh*, *bcfg*, *cdgh*, and *efgh* and 28 three-point planes *does form a combinatorial geometry*. We say the geometry is *not projectively coordinatizable*.

It would not be correct to imagine that all finite-point sets in two- or three-dimensional projective space can be illustrated by drawings such as those of Figures 1.1 and 1.2. When points are chosen from projective spaces over finite fields, we are often forced to represent the major lines and planes as *curved*. Figure 1.3b shows the entire projective plane over the field $GF(3) = \{0, 1, 2\}$ mod 3. To prove that this geometry is the projective plane over GF(3), assign coordinates to each of the points in a basis, say $a = (0, 0, 1)$, $b = (0, 1, 0)$, $d = (1, 0, 0)$, $e = (1, 1, 1)$. Because the point *i* is on the line *ae*, and every line has exactly four points over GF(3), the coordinates of the point *i* must be either $(1, 1, 0)$ or $(1, 1, 2)$ modulo 3. Because the point *i* is also on the line *bd*, the correct choice is $(1, 1, 0)$, and the fourth points on those two lines, namely *l* and *m*, must be $(1, 1, 2)$ and $(1, 2, 0)$, respectively. Continuing in this way, we can easily compute a complete set of coordinates for the geometry and thus verify that it is isomorphic to the stated projective plane. In this case, $k = (0, 1, 1)$, $c = (0, 2, 1)$, $f = (2, 1, 1)$, $j = (1, 0, 1)$, $g = (1, 0, 2)$, and $h = (1, 2, 1)$.

If we select one hyperplane in a projective geometry to be the "hyperplane at infinity", the "finite" points not on that hyperplane form an *affine geometry*, a subgeometry of the projective space. Figure 1.3c shows such an affine geometry, obtained by removing the line *jklm* from the projective geometry in Figure 1.3b.

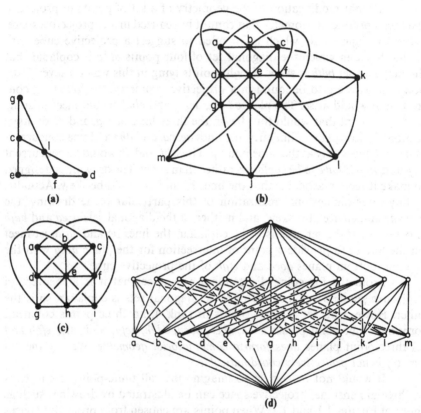

FIGURE 1.3. (a) A six-point geometry. (b) The projective plane over GF(3). (c) The affine plane over GF(3). (d) The lattice of flats of the projective plane over GF(3).

The geometry in Figure 1.1a (= Figure 1.3a) is also a subgeometry of the projective plane over GF(3). The choice of lettters as labels for its points was made in such a way as to show this embedding: Observe that the rank of any subset of the set *cdefgl* of points is the same when measured in the subgeometry or in the whole projective geometry *ab* . . . *m*.

The lattice of flats of the projective plane over GF(3) is drawn in Figure 1.3d. It has a property not shared by all geometric lattices. The reverse-order, or *opposite*, lattice is also geometric. Thus, if two flats of rank *k* have a join of rank *k* + 1, their intersection will have rank *k* − 1. In particular, any two distinct hyperplanes *cover* their intersection. Such geometric lattices are *modular*. Subgeometries (formed by selecting a subset of the set of points, and taking all joins of those elements) of modular geometries do not need to be modular. Note how, for instance, the lines *le* and *cf* meet at the point i in Figure 1.3d (or 1.3b), but have no common point in the subgeometry, Figure 1.3a. The "parallel" lines ae and dh of the affine configuration Figure 1.3c meet at the point l on the "line at infinity" selected in Figure 1.3b.

The operation of *projection*, central to the theory of projective geometry, involves two lattice operations. If a geometry G is to be projected from a center C onto a screen S, we use the lattice operation *join* to create from each flat A of G the flat $A \vee C$ (the bundle of *rays* of the projection) and then use the lattice operation *meet* to find the image $(A \vee C) \wedge S$ where these rays meet the screen. The first of these two operations is a natural operation on geometric lattices, and creates a *quotient* structure whose flats are certain of the flats of G. For instance, if the subgeometry in Figure 1.3a is projected from the point j in the projective plane (Figure 1.3b) as center, the distinct joins of flats of G with j are j itself, the four lines through j, and the entire plane. The sets of points of G that lie in these various flats are the empty set, l, cf, e, dg, and the entire set. It is this set of flats that forms the *quotient* of G associated with this projection. Geometrically, it is a four-point line in which two of the points are represented by pairs of elements.

This is where the distinction between matroids and combinatorial geometries becomes significant. The quotient structure arising from a projection is not, strictly speaking, a geometry. We must either say that the quotient geometry is in this case based on a four-element set of points or else introduce a broader concept that will admit multiple points and even "zero" points that are in the geometric closure of the empty set. This broader concept is that of a *matroid*. The quotient matroid of the projection under discussion has no zero elements, but if we were to project the same subgeometry from one of its own points as center, that point would be a zero element (or *loop*) of the quotient matroid.

1.1.B. Vector Geometries

Many of the basic concepts of combinatorial geometry are straightforward generalizations of concepts in linear algebra. Given a set E of vectors in a finite-dimensional vector space V, every subset A of E will have a well-defined rank $r(A)$ equal to the dimension of the vector space spanned by the vectors in A. Those subsets $A \subseteq E$ that are intersections with E of subspaces of V are the flats of a matroid. We call such a structure a *vector matroid* (or *vector geometry*).

The passage from a vector matroid to its associated combinatorial geometry is the familiar passage from vectors to points in projective space: The zero vector remains in the closure of the empty set, and vectors that are scalar multiples of one another yield the same projective point. (Think of projection of Euclidean n-dimensional space from the point with position vector zero as center; the quotient of this projection is a vector matroid.)

$$A = \begin{bmatrix} 1 & 1 & 1 & 1 & 1 & 0 \\ 0 & 1 & -1 & -1 & -1 & 0 \\ 1 & 1 & 1 & 1 & 1 & 1 \end{bmatrix} \qquad B = \begin{bmatrix} -2 & 1 & 1 & 0 & 0 & 0 \\ -2 & 1 & 0 & 1 & 0 & 0 \\ -2 & 1 & 0 & 0 & 1 & 0 \\ 0 & 0 & 1 & -1 & 0 & 0 \\ 0 & 0 & 1 & 0 & -1 & 0 \\ 0 & 0 & 0 & 1 & -1 & 0 \end{bmatrix}$$

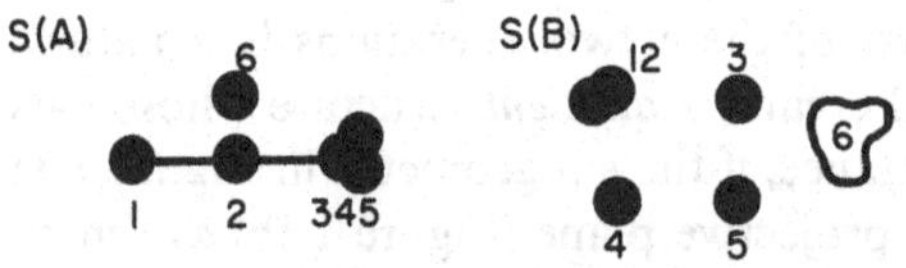

$$A' = \begin{bmatrix} 0 & 1 & -1 & -1 & -1 & 0 \\ 1 & 0 & 2 & 2 & 2 & 0 \\ 1 & 2 & 0 & 0 & 0 & 0 \\ 0 & 0 & 0 & 0 & 0 & 1 \end{bmatrix}$$

FIGURE 1.4. A vector matroid and its orthogonal matroid.

The notions of *bases* and *dependent* and *independent sets* arise from the example of vector geometries. A set of vectors is linearly dependent if and only if some nontrivial linear combination of them is equal to the zero vector. Although the values of the scalars used in linear combinations have no combinatorial meaning, the set of all dependent sets in a given vector geometry carries enough information to characterize the geometry up to combinatorial isomorphism. The minimal dependent sets, which we call *circuits*, are also a starting point for an axiomatization of combinatorial geometries.

Matrices provide particularly interesting examples of vector geometries. The rows of an *m*-by-*n* matrix (rank *k*) yield a vector matroid on *m* elements, rank *k*, and the columns yield a vector matroid on *n* elements, also of rank *k*. The matrix *A* in Figure 1.4, for instance, has rank 3. The three rows being independent, the corresponding geometry consists of three points not collinear, forming a triangle. Now consider the geometry of columns of this matrix *A*. Columns 3, 4, and 5 are multiples of one another, being equal; so they represent the same projective point. The first three columns form a singular 3-by-3 matrix, because there is a repeated row. So columns 1, 2, and 3 are dependent, and represent collinear points. The sixth column is independent of the other five; so the vector geometry is that labeled *S(A)* in Figure 1.4.

It is easy to recognize the *bases* (maximal independent sets) of such a vector geometry. A basis must *span* the geometry, and must contain no circuits, that is, no zero element, no pair of equal points, no three collinear points, no four coplanar points, and so forth. In the geometry of columns of the matrix *A*, 126, 136, 146, 156, 236, 246, and 256 are the only bases.

The six columns of the matrix *A* have many dependences. The coefficients of all minimal dependences (circuits) form the rows of the matrix *B*. There are only six circuits, three consisting of pairs 34, 35, and 45 of equal points, the other three consisting of triples 123, 124, and 125 of collinear points. No circuit contains the element 6. Note that every row of the matrix B is orthogonal to (has inner product zero with) every row of A, and the rows of A and B together span the entire six-dimensional vector space. Thus, the row spaces of the matrices A and B are orthogonal complements of one another.

The columns of the matrix B form a geometry $S(B)$ that we call the *orthogonal* $S^*(A)$ of the geometry of columns of A. The combinatorial relations between these orthogonal geometries $S(A)$ and $S(B)$ can be summarized briefly as follows:

(1) The complement of a basis for $S(A)$ is a basis for $S(B)$, and conversely,

(2) For any element p and any bipartition of the remaining elements into two sets X and Y, the element p depends on the set X in the geometry $S(A)$ if and only if p does not depend on the set Y in $S(B)$.

Let us take a close look at the geometry of columns of the matrix B. The sixth column being the zero vector, it is an element in the closure of the empty set and does not represent a point of the geometry. Columns 1 and 2 represent the same point. Columns 2, 3, 4, and 5 form a circuit, a dependence being given by the second row of the matrix A. Thus, the geometry of columns of B consists of four coplanar points 12, 3, 4, and 5 and a "phantom" element 6. Once the geometries $S(A)$ and $S(B)$ are drawn, we can check the foregoing orthogonality relations (1) and (2). Bases for the geometry $S(B)$ consist of three noncollinear points. There are seven such three-element sets, 345, 245, ..., and these are precisely the complements of the seven bases listed earlier for the geometry $S(A)$. Checking an example of property (2) of orthogonal geometries, notice that point 2 is in the closure of the set 345 in the geometry $S(B)$; so the point 2 is *not* in the closure of the complementary set 16 in the geometry $S(A)$.

Finally, look at the rows of the matrix B. These correspond to the circuits of the column geometry $S(A)$ and can be regarded as forming a geometry in their own right, the *geometry of circuits* of $S(A)$. This is not a combinatorial construction: Changes in the coordinatization of a geometry can change the linear relations between its circuits, without changing the circuits themselves (as subsets). Its combinatorial interest arises from the following fact: If we draw the lattice for the row geometry of the matrix B, we find an embedded copy of the lattice for $S(B)$, the column geometry of B, within it in inverted order (Figure 1.5). We say that the row geometry $R(B)$ is an *adjoint* of the column geometry $S(B)$.

Now let us take a second look at the two matrices in Figure 1.4. There is a certain lack of symmetry in this example. Although the rows of A are orthogonal to the rows of B, the rows of B have the special property that they come from the circuits of the column geometry $S(A)$; they are (up to a scalar factor in each vector) all the *minimal support vectors* in the space

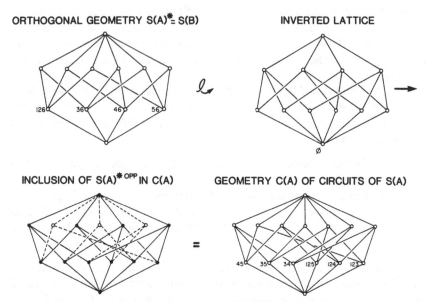

FIGURE 1.5. A lattice inclusion.

$R(A)^\perp$, the kernel of the linear transformation "left-multiplication by A." A similar construction would replace the matrix A by a matrix A' (Figure 1.4) whose rows are the minimal support vectors in the space $R(B)^\perp$, that is, by all minimal support vectors in the row space $R(A)$. The support sets of these minimal support vectors are combinatorially significant. The complements of these sets are the hyperplanes (copoints) of the column geometry $S(A)$. Thus, in the given example, the four lines 16, 26, 3456, and 12345 have as complements the minimal support sets 2345, 1345, 12, and 6, respectively.

These minimal support sets are called the *bonds* of the matroid $S(A)$. (The term *bond* had its origin in another context, namely in the theory of graphs, where the term refers to a link between two parts of a graph, as will be explained later.) *Bond* and *circuit* are dual notions, in the sense that the bonds of any matroid are the circuits of its orthogonal matroid.

The basic fact to remember is that the zero set of any vector in the row space of any matrix A is a flat in the column geometry $S(A)$, and the zero set of any vector orthogonal to all the rows is a flat of the orthogonal geometry. Furthermore, passage from a matrix A to the matrix A', whose rows are the minimal support vectors of the row space $R(A)$, does not change the column geometry $S(A) = S(A')$.

1.1.C. Function-Space Geometries

A *function-space geometry* (or *chain-group geometry*) $G(X, V)$ is a geometry on a set X of points, whose flats are determined as follows by a finite-dimensional vector space V of functions from a set X into a field F. A subset

$A \subseteq X$ is a flat of the geometry $G(X, V)$ if and only if the set A is an intersection

$$A = \bigcap_\alpha \ker f_\alpha$$

of subsets $\ker f_\alpha = f_\alpha^{-1}(0)$ for some family $\{f_\alpha\}$ of functions in V.

Function-space geometries are always representable as vector geometries, because *evaluation* at any point $p \in X$ is a linear functional defined on the vector space V and is thus an element (a vector) in the dual space V^* of V. The converse is also true: Every vector geometry is representable as a function-space geometry. If G is a vector geometry whose points are a set X of vectors in a vector space V, we can assume without loss of generality that these vectors span the vector space V. The vector geometry G is isomorphic to the function-space geometry $G(X, V^*)$ relative to the dual space V^* of linear functionals defined on V, restricted to X.

Function-space geometries provide a fresh look at vector geometries, from a new perspective. For instance, let V be the vector space $F_{(k)}[x_1, \ldots, x_n]$ of polynomials in n variables over a field F, having degree $\leq k$. This vector space of functions will induce a function-space geometry on any given subset of n-dimensional affine space. Taking that subset to be the entire affine space, we obtain a geometry A_k^n, the geometry of k-varieties in affine n-space over F. Taking F to be the field of real numbers, we find that the rank of the geometry A_k^n is equal to the binomial coefficient $N = \binom{n+k}{n}$. Each subset $Y \subseteq A^n$ has rank $r_k(Y)$ equal to N minus the dimension of the subspace consisting of those polynomials of degree $\leq k$ that are zero at all points of Y.

For instance, the function-space geometry A_2^2 defined on the real plane by quadratic polynomials in two variables has rank $6 = \binom{4}{2}$. Its hyperplanes, rank 5, are the plane conics (ellipses, hyperbolas, etc., and degenerate conics consisting of pairs of distinct lines). A single straight line has rank 3 because a pair of points is closed, but any conic containing three points of a straight line must contain the entire line. The geometry has two types of flats of rank 3: three noncollinear points, or one straight line. All types of flats for this geometry are shown in Figure 1.6.

By restricting our attention to a proper subspace of the vector space of quadratic polynomials in two variables, we can obtain some interesting *quotients* of the function-space geometry A_2^2. For instance, consider only those polynomials having coefficient 0 for the mixed term involving both variables. Each such conic has a symmetry of reflection with respect to each of the lines parallel to the coordinate axes, passing through the center of the conic. The resulting function-space geometry has some nonclosed sets of three noncollinear points, the closures of which have one additional point that completes a tetrahedral drawing with opposite sides having slopes that are negatives of each other (Figure 1.7).

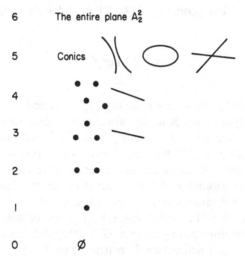

FIGURE 1.6. The function-space geometry A_2^2.

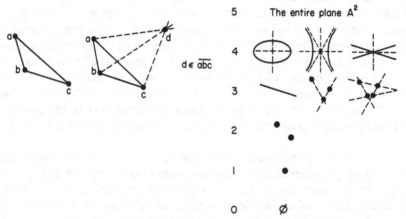

FIGURE 1.7. A function-space geometry from 0 mixed terms.

It is through the notion of function-space geometries that most of the interesting facts concerning many well-known geometries (hyperbolic geometries, Möbius geometries of circles, etc.) can be systematically studied.

1.2. FURTHER ALGEBRAIC EXAMPLES

1.2.A. Examples from Exterior Algebra

During the nineteenth century, Plücker and Grassmann showed how to coordinatize the flats of a projective space and began the study of linear families of such flats. The idea was to represent each flat of rank k lying in a

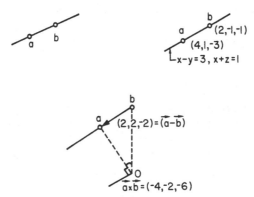

FIGURE 1.8. Plücker coordinates.

space of rank n by a $k \times n$ matrix whose rows are vectors representing an independent set of points on the flat in question, and then to observe that the set of $\binom{n}{k}$ determinants of $k \times k$ submatrices of this matrix depend, up to a common scalar multiple, only on the flat, not on the independent set of points chosen to represent it. In this way, each k-flat of a projective geometry of rank n is represented as a point in a projective geometry of rank $\binom{n}{k}$.

The method of deriving coordinates of lines in real projective 3-space is shown in Figure 1.8. For instance, the line with equations $x - y = 3$, $x + z = 1$ in 3-space contains finite points $a = (4, 1, -3)$, $b = (2, -1, -1)$, with projective coordinates $(4, 1, -3, 1)$ and $(2, -1, -1, 1)$. The 2×4 matrix of those coordinates yields six Plücker coordinates $(d_{14}, d_{24}, \ldots) = (2, 2, -2, -4, -2, -6) \cong (1, 1, -1, -2, -1, -3)$. The first three Plücker coordinates give a vector $\mathbf{a} - \mathbf{b}$ along the line; the last three form a vector $\mathbf{a} \times \mathbf{b}$ perpendicular to the plane containing both the line and the origin. The six Plücker coordinates of any line satisfy a quadratic relation, the "p-relation,"

$$d_{14}d_{23} + d_{24}d_{31} + d_{34}d_{12} = 0$$

[because $(\mathbf{a} - \mathbf{b}) \cdot (\mathbf{a} \times \mathbf{b}) = 0$]; so the set of lines in 3-space, thus coordinatized, forms a *subgeometry* of projective 5-space (rank 6) consisting of all points on a quadric surface in that space.

The geometry of lines in 3-space has some quite remarkable properties. A set of three distinct lines is dependent if and only if the lines are both coplanar and have a common point (i.e., they lie in a *flat pencil*). Four skew lines are dependent if and only if they lie in one family of rulings of a doubly-ruled quadric surface (i.e., in a *regulus*). Any two concurrent lines span their flat pencil; any three skew lines span a regulus; a set of lines is closed if and only if it is closed with respect to completion of flat pencils and reguli.

The combinatorial geometry of lines in 3-space is extremely useful in analyzing mechanical and statical configurations, because for a set of forces to be in equilibrium, the lines along which they act must be dependent.

1.2.B. Algebraic Dependence

The transcendence degree of an extension of a field is defined to be the cardinal number of a maximal set of algebraically independent transcendentals in the extension. The transcendence degree of a field K over a subfield F can be ascertained from the collection of fields M that are extensions of F and that are relatively algebraically closed in K (i.e., every element of K algebraic over M is in M).

An element b of a field K depends algebraically on a subfield M if and only if b is a solution of some polynomial equation $p(x) = 0$ with coefficients in the subfield M.

1.2.1. Proposition. Let F be an algebraically closed field, let $x_1, \ldots, x_n$ be independent transcendentals, and let $F(x_1, \ldots, x_n)$ be the associated extension field (the field of rational functions in n indeterminates, over F). Then the relatively algebraically closed subfields of $F(x_1, \ldots, x_n)$ containing F form a geometric lattice L of rank n.

Proof. The points of the lattice L can be represented as those relatively algebraically closed subfields $F(y)$, where y is a transcendental, that is, a single rational function in the indeterminates $x_1, \ldots, x_n$. For any set T of such transcendentals, a transcendental x depends algebraically on T when x is a solution of a polynomial equation in one variable, with coefficients in $F(T)$. Then the enlargement of each set T of transcendentals to the set of all transcendentals dependent algebraically on T is a closure operator. A set of transcendentals is closed if and only if it is the set of transcendentals in some relatively algebraically closed field, contained between F and $F(x_1, \ldots, x_n)$.

If a transcendental x depends algebraically on a set T of transcendentals, it depends also on a finite set T_f of transcendentals occurring in the coefficients of some polynomial equation. Thus, the closure operator has the finite-basis property (see Proposition 2.4.2, part cl5).

According to Chapter 2, Section 2.3, we must prove that the closure operator has the increase, monotonicity, idempotence, and exchange properties. The first two are trivial, and the third is a standard theorem in algebra. Now we shall prove exchange.

If a transcendental x depends algebraically on a set $T \cup y$ of transcendentals, but not on the set T, then x is the solution of some polynomial equation, in one variable ξ, with coefficients in the field $F(T \cup y)$. Multiply this polynomial by an appropriate element of $F(T \cup y)$ (to clear all occurrences of y in denominators of coefficients), and substitute a variable η for

y, so that the polynomial equation becomes a polynomial equation in two variables ξ and η over the field $F(T)$, with solution $\xi = x, \eta = y$. The variable η must occur nontrivially in this polynomial, for otherwise x would depend algebraically on T. Substituting $\xi = x$, we see that y depends algebraically on the set $T \cup x$, and the closure operator has the exchange property. $\square$

Consider the set of real symmetric functions in three variables x, y, and z. We wish to show that the three symmetric functions

$$[x] = x + y + z,$$
$$[x^2] = x^2 + y^2 + z^2,$$
$$[xy] = xy + xz + yz,$$

form a three-point line in the geometry of algebraic dependence of symmetric functions. It is easy to see that these functions are pairwise independent. That they are dependent follows from the fact that the polynomial $A^2 - B - 2C$, evaluated at $A = [x], B = [x^2]$, and $C = [xy]$, becomes the zero polynomial in the variables x, y, z; that is,

$$[x]^2 = (x + y + z)^2 = x^2 + y^2 + z^2 + 2xy + 2xz + 2yz$$
$$= [x^2] + 2[xy].$$

Algebraic dependence of such polynomials can also be seen as *generic* linear dependence. To see how, form the Jacobian matrix

$$\begin{bmatrix} \dfrac{\partial A}{\partial x} & \dfrac{\partial A}{\partial y} & \dfrac{\partial A}{\partial z} \\ \dfrac{\partial B}{\partial x} & \dfrac{\partial B}{\partial y} & \dfrac{\partial B}{\partial z} \\ \dfrac{\partial C}{\partial x} & \dfrac{\partial C}{\partial y} & \dfrac{\partial C}{\partial z} \end{bmatrix} = \begin{bmatrix} 1 & 1 & 1 \\ 2x & 2y & 2z \\ y+z & x+z & x+y \end{bmatrix} = J\begin{bmatrix} A & B & C \\ x & y & z \end{bmatrix}.$$

Taking some arbitrary values, say $x = 1$, $y = -1$, $z = 2$, we find that the Jacobian matrix becomes singular, with determinant zero. Is this because we have mistakenly chosen *special* values for the variables, values at which the Jacobian matrix has lower rank than usual, or will this happen for virtually any choice of values for x, y, and z? To find out, we evaluate the determinant of the Jacobian matrix itself, computing in the polynomial ring $R[x, y, z]$. Here we find

$$|J| = 2y(x + y) - 2z(x + z) + 2z(y + z)$$
$$- 2x(y + x) + 2x(x + z) - 2y(y + z) = 0.$$

The rank of a set P of such polynomials in the geometry of algebraic dependence is equal to the highest rank attained by their Jacobian matrix, evaluated at various points. Points (x, y, z) for which this maximum value is attained are called *regular points* for the *algebraic set* of solutions of the set P of polynomial equations. Linear dependence of rows of a Jacobian matrix, evaluated at a regular point, can also be viewed as linear dependence of the rows of the Jacobian matrix itself, using "scalars" from the integral domain $R[x, y, z]$. Thus, in the foregoing example, writing $\partial[x]$, $\partial[x^2]$, and $\partial[xy]$ for the rows of the Jacobian matrix, we derive from the polynomial equation $[x]^2 = [x^2] + 2[xy]$ the fact that

$$2[x] \, \partial[x] = \partial[x^2] + 2\partial[xy].$$

So the polynomials $2[x]$, -1, -2 are scalars for a linear dependence of the rows of the Jacobian matrix.

1.3. COMBINATORIAL EXAMPLES

1.3.A. Partitions

We turn now to a number of examples of combinatorial geometries that are definable directly in combinatorial terms. (We do not rule out the possibility that they may *also* be coordinatizable as sets of points in some projective space.) We begin with a glance at the geometry of *partitions*.

A *partition* π of a set X is a set of disjoint subsets of X (called its *blocks* or *parts*) whose union is all of X. The set P of partitions π of a finite set X has a natural partial order: We say a partition σ is a *refinement* of a partition τ if every part of σ is contained in some single part of τ. For example, $(aeg)(bf)(c)(dh)(i)$ is a refinement of $(adegh)(bf)(ci)$. In this order of refinement, the set of partitions of a finite set forms a geometric lattice. The points of this geometry are the elementary partitions formed by joining just one pair of elements. The hyperplanes of P are those partitions having just two parts, the *bipartitions* of X. The rank of a partition π is equal to the number of elements of X minus the number of parts of π. Figure 1.9 shows the lattice of partitions of a four-element set and its associated geometry of partitions. We shall show in the next section that the geometry of partitions is a vector geometry and that this example in particular consists of the six points of intersection of four general lines in the plane.

But first let us think about the geometry of partitions from a geometrical point of view. For any partition π of a set X, the set of pairs (a, b) of elements of X that are together in some part of π forms an *equivalence relation* on X. The surprising fact about partition geometries is that in constructing them we manage to place the pairs (a, b) of elements of X in a geometric configuration P in such a way that the *flat sections* of G are precisely the equivalence relations on X. The notion of an equivalence relation has become

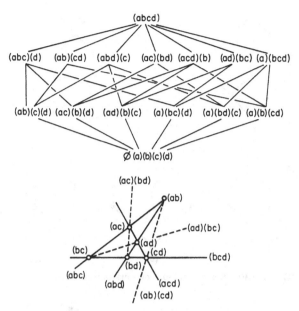

FIGURE 1.9. A partition geometry.

linearity in an appropriate geometric setting. This point of view will prove helpful in many combinatorial problems.

1.3.B. Graphic Geometries

Graphic combinatorial geometries are also a very natural class of geometries arising from a combinatorial setting. They are discussed in detail in Chapter 6. A linear *graph G* is a set X of *vertices* together with a subset $E \subseteq \binom{X}{2}$ of the set of unordered pairs of elements of X, the *edges* of the graph. A *cycle* in the graph G is a simple closed path of edges and vertices in G. The cycles of any graph are the circuits of a combinatorial geometry on the set E of edges of G. A basis for the geometry of a connected graph is simply any spanning tree of edges in the graph. A more general form of graph, admitting loops and multiple edges, yields a matroid: Its loops are in the closure of the empty set, and its multiple edges (edges between the same pair of vertices) represent the same geometric point. In Figure 1.10, subgraphs a and b are independent, b is a basis, c and d have nullity 1, d being a circuit, and e has nullity 2, rank 10.

There is a simple Galois coconnection between the lattice $\mathscr{B}(E)$ of subsets of E and the lattice $P(X)$ of partitions of the set X of vertices. Given any subset $A \subseteq E$ of edges of G, we associate a partition $\pi(A)$ of the vertices of G: Two vertices are in the same part of $\pi(A)$ if and only if they are connected by some path along edges in the subset A. This Galois coconnection represents the geometry of the graph G as a subgeometry of the geometry

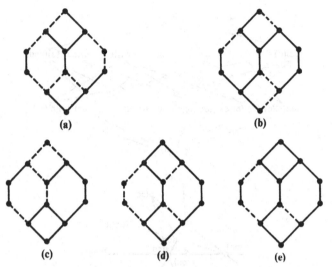

FIGURE 1.10.   Some subgraphs of a graph (shown by solid lines).

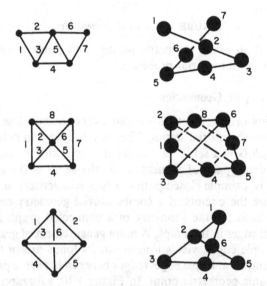

FIGURE 1.11.   Some graphs and their associated geometries.

P of partitions of X. Conversely, every subgeometry of a partition geometry is the geometry of a graph. In Figure 1.11 we give examples of some graphs and their associated combinatorial geometries.

In the geometry of a graph, the *closure* of a set X of edges in the graph is formed by completing cycles, that is, by adding those edges e such that $X \cup e$ contains a cycle containing the edge e.

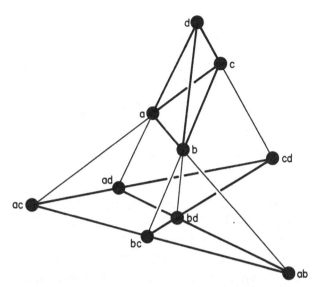

FIGURE 1.12. A free geometry on four points together with its hyperplane section.

Geometries of graphs are easily coordinatized. If the graph has n vertices $a, b, \ldots$, each edge (say ab) is represented by a vector in dimension n, having the a component $+1$, the b component -1, and all other components 0. The reader can easily check that minimal dependences among such vectors come from cycles in the graph.

This representation also reveals another striking fact about geometries of graphs. Let points $a, b, \ldots$ be coordinatized as projective points in a *free geometry* (each point spanning a new dimension), as in Figure 1.12.

	a	b	c	d	...	h	□
a	1	0	0	0	...	0	1
b	0	1	0	0	...	0	1
...							

If we intersect the lines formed by pairs of these points with the hyperplane H at infinity (last coordinate $= 0$), we obtain a new set of points

	a	b	c	d	...	h	□
ab	1	-1	0	0	...	0	0
ac	1	0	-1	0	...	0	0
...							

that are linearly isomorphic to the vectors coordinatizing the geometry of the graph. Thus, we have shown that graphic geometries are hyperplane sections of free geometries.

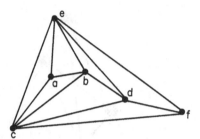

FIGURE 1.13. A triangulated sphere.

1.3.C. Simplicial Geometries

Let $\binom{X}{k}$ be the set of k-element subsets of an n-element set X. We call these k-element subsets $A \in \binom{X}{k}$ *simplices*. The *boundary* ∂A of a simplex $A = x_1 \ldots x_k$ is the formal linear combination

$$\partial A = \hat{x}_1 x_2 \ldots x_k - x_1 \hat{x}_2 \ldots x_k + \cdots + (-1)^{k-1} x_1 \ldots x_{k-1} \hat{x}_k,$$

where the circumflex symbol indicates those points that are deleted from the k-simplex A to form the various bounding $(k-1)$-simplices. Using this notion of boundary, we can form an $\binom{n}{k}$-by-$\binom{n}{k-1}$ matrix of coefficients $(0, \pm 1)$ occurring in these boundary formulas. This is the coordinatizing matrix of the *simplicial geometry* S_k^n of k-element subsets of an n-element set.

To each subset $T \subseteq \binom{X}{k}$ we can associate a simplicial complex, consisting of all the k-simplices in T, together with all simplices of rank $k-1$ and lower. If A is a single k-simplex not in the set T, then the rank increase $r(T \cup A) - r(T)$ is equal to 1 exactly when the Betti number β_{k-2} decreases on inclusion of the simplex A and is equal to 0 exactly when the Betti number β_{k-1} increases. For example, with $T_0 = \{abc, bcd, bde, cdf, def\}$, $T_1 = T_0 \cup abe$, $T_2 = T_1 \cup ace$, $T_3 = T_2 \cup cef$, then $r(T_i) = 5 + i$ for $i = 0, 1, 2$, because some nonbounding 1-cycle has been "filled in" with a disc, but $r(T_3) = r(T_2)$, because an 8-faced triangulated sphere has just been formed (Figure 1.13).

1.3.D. Transversal Geometries

The process of selecting distinct representatives of a family of subsets of a set gives rise to a matroid. If A_i, $i \in I$, is a family of subsets of a set X, then a subset $Y \subseteq X$ is a system of distinct representatives (SDR) if and only if there is a 1-1 function $f: Y \to I$ such that for all $y \in Y$, $y \in A_{f(y)}$. The SDRs of any family of subsets form the independent sets of a matroid, called the *transversal matroid* of the family A_i, $i \in I$. A chapter in the next volume will be devoted to these matroids.

These matroids can also be thought of as arising from binary relations. Given a binary relation ρ from a set Y into a set I (i.e., a subset of $Y \times I$),

FIGURE 1.14. Some bases of a matroid defined by the incidence relation of a graph.

then for each element $i \in I$, let $A_i = \{y \in Y; y\rho i\}$. Then $\{A_i; i \in I\}$ is a family of sets, and the independent sets of its transversal matroid are those subsets of Y that can be matched one-to-one into the set I, along the relation ρ.

Consider for a moment the relation of *incidence* between the edges and vertices of a linear graph. The transversal matroid of this binary relation provides an amusing example of a geometry on the edge-set of a graph, a geometry different from the graphic geometry described earlier. If no component of the graph is simply a tree, then every vertex can be matched to an incident edge, and the bases of the transversal geometry consist of those maximal subgraphs each component of which is a cycle together with some number of incident trees (Figure 1.14). What are the circuits of such a matroid? (See Exercise 1.7.)

1.3.E. A Geometry of Triples

We have already introduced some geometries based on the set of edges of a graph: the graphic geometry and a geometry arising from a matching between vertices and edges. We have also mentioned one geometry on the set of triples drawn from a given finite set: the simplicial geometry. Let us consider a second geometry on the set of triples of elements in a set.

This geometry comes from the geometric completion of a certain lattice. Begin with the Boolean algebra $\mathscr{B}_n$ of subsets of an n-element set X; let $T_2(\mathscr{B}_n)$ be the lattice obtained by identifying the zero element with all the one- and two-element subsets of X. This *lower-truncated lattice* $T_2(\mathscr{B}_n)$ is not geometric, because the join of two points (two triples) can have rank as high as 4. The process of *Dilworth completion* (see Chapter 7, Section 7.7) produces a new geometric lattice $D_2(\mathscr{B}_n)$, also of rank $n - 2$, whose flats include all the elements of the truncated lattice as a meet-closed subset.

A combinatorial description of this Dilworth completion is most easily given in terms of independent sets. A set T of triples taken from the set X is independent if and only if for any integer $k > 0$, the union of any k triples in T contains at least $k + 2$ elements of X. The sets of triples in Figure 1.15a–c are independent; in each case their closure is the entire set of triples on the given set of vertices. The triples shown in Figure 1.15d are not only independent but also closed. Figure 1.15e–g illustrates circuits in this geometry of

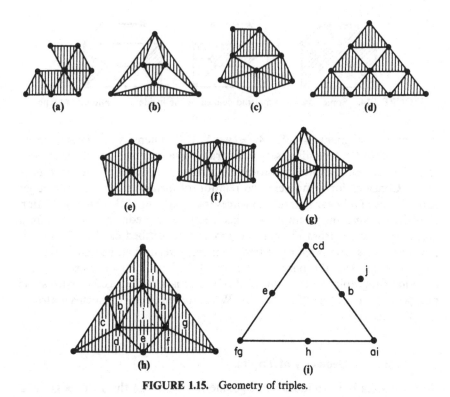

FIGURE 1.15. Geometry of triples.

triples. The 10 triples in Figure 1.15h form a geometry of rank 7. The set *abcdefg* is a basis.

Figure 1.15i shows the rank-3 geometry orthogonal to this geometry of 10 triples: The complements of its lines are the circuits of the geometry of triples. For instance, *fgj* and *abcdi* being lines in the orthogonal geometry, *abcdehi* and *efghj* are circuits in the geometry of triples.

1.3.F. Small Geometries

Let us look at a few small, but otherwise unrestricted, geometries. Given a five-element set, how many essentially different (nonisomorphic) geometries can be constructed on it? It will turn out that all such geometries can be represented in an affine space of dimension no greater than 4; so we draw them as such (Figure 1.16).

First, there is a unique geometry of rank 2 consisting of five points on a line. A rank-3 geometry on five points can have at most one line of four points, and at most two lines of three points. There are four possibilities, shown in the second row of Figure 1.16. A rank-4 geometry on five points has at most one nontrivial flat; this can be a three-point line or a four-point

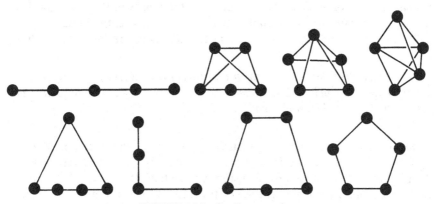

FIGURE 1.16. Small geometries.

TABLE 1.1 *Tabulation of* $g_{n,k}$

	Number of points							
Rank	8	7	6	5	4	3	2	1
1	0	0	0	0	0	0	0	1
2	1	1	1	1	1	1	1	
3	68	23	9	4	2	1		
4	617	49	11	3	1			
5	217	22	4	1				
6	40	5	1					
7	6	1						
8	1							
Total	950	101	26	9	4	2	1	1

plane or five points in general position on a flat of rank 4. If the geometry on five points has rank 5, the points must be in general position in a space of rank 5.

A similar count of the number $g_{n,k}$ of essentially different geometries of rank k on an n-element set, for $n = 1, \ldots, 8$, yields the tabulation in Table 1.1. Let g_n be the total $g_{n,1} + \cdots + g_{n,n}$. Then the recursion $g_{n+1} = K(g_n)^{3/2}$ seems approximately correct, on the basis of these very limited data, with a value of K of about 0.95. This would suggest that there are some twenty-eight thousand essentially different geometries on a nine-element set.

1.3.G. Coverings

A *covering* π of a set X is a family of subsets of X, with union X. A covering is a *partition of type n* (or an *n-partition*) of the set X if every member of π has at least n elements, and every n-element subset of X is contained in a *unique* member of π. The members of π are the *blocks* of the covering. For $n = 1$, an *n*-partition is an ordinary partition. For $n = 2$, the blocks can be called lines, and our requirement is that two points determine a unique line.

1.3.1. Proposition. *For any n-partition of a set X, the following subsets are the flats of a combinatorial geometry: the entire set X, the blocks, and all subsets with fewer than n elements.*

Proof. If a flat A of an *n*-partition has fewer than $n - 1$ elements, it is covered by flats of the form $A \cup x$, for any element $x \notin A$. So the complementary set $X \backslash A$ is trivially partitioned by inclusion in these flats. If the flat A has exactly $n - 1$ elements, each *n*-element set of the form $A \cup x$ is in a unique block containing A; so, again, the set $X \backslash A$ is partitioned by inclusion in such blocks. If the flat A has rank n, it is covered only by the flat X. So the flats of any *n*-partition form a geometric lattice. $\square$

From each geometric lattice L of rank $n + 1$, a geometric lattice $T(L)$ of rank n can be formed by identifying all the copoints of the lattice L with the element 1. If this *truncation operation* is performed on the lattice of flats of a partition of type n, the resulting lattice is composed of all subsets of a set X that have fewer than n elements, together with the entire set X. Such geometries have a particularly simple structure; they are also obtainable by truncating the Boolean algebra $\mathscr{B}(X)$ to rank n. The converse of this observation also holds:

1.3.2. Proposition. *A geometric lattice L of rank n + 1 is isomorphic to the lattice of flats of a partition of type n if and only if its truncation T(L) is isomorphic to the rank-n truncation of a Boolean algebra, if and only if every lattice interval $[0, x]$ to a coline x is distributive (and is therefore a Boolean algebra).*

Another way of saying this is that any general projection of the geometry of a partition of type n into a space of rank n consists of a set of points in *general position* in that space.

Despite the fact that partitions of type n account for all geometries of rank $n + 1$ with no nontrivial colines (colines with at least n points), these examples of combinatorial geometries may still seem rather special. However, partitions of type n actually *predominate* in any enumeration of small geometries of rank $n + 1$. For instance, there are 322 essentially different such geometries of rank 4 on eight points, out of a total of 617 different geometries of that rank and cardinality. It appears probable that the partitions of type n will also predominate in any asymptotic enumeration of geometries.

1.4. STRUCTURE AND RELATED GEOMETRIES

1.4.A. Structure Geometries

If a linear graph G is realized with its vertices in particular positions in k-dimensional Euclidean space R^k, it can be taken to represent a bar-and-joint structure, composed of rigid bars (the edges) connecting universal joints (the vertices). This mechanical model gives rise to a combinatorial geometry that depends not only on the dimension k of the space but also on the positions of the vertices of the graph G within that space.

These geometries are easily coordinatized. For a graph with e edges and n vertices in k-space R^k, we form a matrix M of scalars with e rows and nk columns, or, equivalently, an e-by-n matrix with k-vector entries. In the row corresponding to an edge between vertices a and b we place the vector $\mathbf{a} - \mathbf{b}$ in the (multiple) column a, the vector $\mathbf{b} - \mathbf{a}$ in the column b, and the zero vector in all other columns. The *structure geometry* of the graph G in the given position is the geometry of rows of this matrix M.

An n-vector of k-vectors, one k-vector for each vertex of the graph, forms a row orthogonal to all the rows of the matrix M if and only if these vectors are the velocities of the vertices in an *infinitesimal motion* of the structure. If $v_\mathbf{a}$ and $v_\mathbf{b}$ are the velocities of the vertices at the ends of an edge ab, the combined velocities form a vector orthogonal to the row for edge ab if and only if $(v_\mathbf{a} - v_\mathbf{b}) \cdot (\mathbf{a} - \mathbf{b}) = 0$, that is, if and only if the relative velocity of the ends is perpendicular to the bar, if and only if these velocities cause no first-order change in the length of the bar.

The independent sets of a structure geometry are (for each nonnegative integer i) those i-element sets of edges that remove exactly i infinitesimal degrees of freedom from the vertices, regarded as originally free to move with k degrees of freedom each in R^k. Thus, the complete graph K_4 on four vertices has rank 3 on the line R^1, has rank 5 in the plane R^2, and has rank 6 in 3-space R^3. The rank of a set of edges also has a statical interpretation: The row space they generate is the space of equilibrium loads supportable by the structure having only those bars.

In the special case of structures along a line, the structure geometry is just the geometry of the graph G. The coordinatizing row $\mathbf{a} - \mathbf{b}$, $\mathbf{b} - \mathbf{a}$, $0, \dots$ for the edge ab is a multiple of the row $1, -1, 0, \dots$ used to coordinatize that edge in the graphic geometry. Also, the number of degrees of freedom removed by bars between vertices free to move along a line depends only on how the vertices are partitioned by connectivity along edges in the given graph. Thus, we have the usual connection between edge-sets and vertex partitions in graphic geometries.

Dependent sets for structures in the plane R^2 are intimately associated with projections of spatial polyhedra, according to theorems of James Clerk Maxwell and Luigi Cremona. Four circuits in the structure geometries of

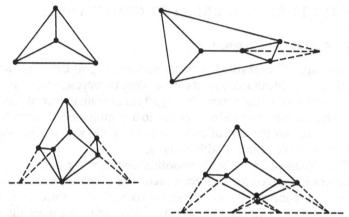

FIGURE 1.17.   Edge-sets that are circuits in the planar-structure geometry.

graphs in the plane are illustrated in Figure 1.17. For three of these circuits we have indicated certain *projective geometric properties* that are required in order that the plane structure be the projection of a spatial polyhedron, and are thus required in order that the graph in that given position be a dependent structure.

1.4.B.   Concurrence Geometries

The geometry of partitions on a finite set is an example of a geometry whose flats are all possible constructions of a certain type that can be made with a given set of elements. We consider another such example. Let $H_1, \ldots, H_k$ be a set of variable hyperplanes in R^n, each having a fixed intersection A_i with a fixed hyperplane H. As the hyperplanes vary, each with one degree of freedom about its fixed intersection A_i, a number of combinatorially distinct configurations can be formed in R^n. These configurations have a natural partial order with respect to specialization and form a geometric lattice of rank $k - n$, called the *concurrence geometry* for the fixed pattern $\{A_i\}$ of intersections on the hyperplane H.

For instance, given six variable lines meeting a fixed line in the plane at six general points, the concurrence geometry has rank 4 and consists of the $\binom{6}{3} = 20$ points of intersection of six general planes in R^3 (rank 4). The 20 points correspond to configurations in which three of the variable lines meet in a point (are *concurrent*) (Figure 1.18).

We have simply to replace the fixed pattern $\{A_i\}$ of hyperplane sections on H by a *polar geometry* G of points, and the concurrence geometry becomes the geometry $C(G)$ of circuits of G. In this geometry, a set of circuits is a flat if and only if it is the set of circuits that remain dependent in some lifting G' of G into a space of one higher dimension, in such a way that G

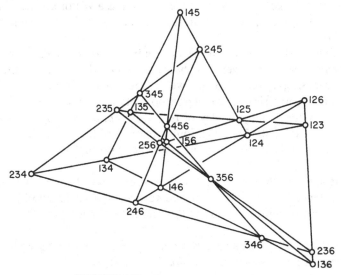

FIGURE 1.18. A concurrence geometry.

is a point (or parallel) projection of G'. For instance, letting G be six points in general position on a line, the points of $C(G)$ are the individual circuits of G (triples of collinear points). The corresponding plane configurations in which these circuits are preserved have those three points collinear, and the rest of the points in general position over their images on the line. The lines of $C(G)$ consist either of pairs of triples of points having no point or one point in common, or of all four triples in a quadruple of points. The corresponding plane configurations have either two three-point lines, nonintersecting or meeting in a point, or one four-point line.

EXERCISES

1.1. Give projective coordinates for the vertices of a projective cube in 3-space (rank 4), and find the coefficients of the linear dependence among the points of each of its six faces, and on each of its diagonal planes if such planes exist in the projective cube you have selected. Considered as functions from the set of vertices into the real numbers, and thus as vectors in a real vector space of rank 8, the six "face" dependences have what rank?

1.2. Consider a plane figure consisting of three parallel lines meeting two parallel lines to form a grid of six points. Find all flats in this six-point geometry, and draw the geometric lattice they form. List all the circuits and bases of this geometry.

1.3. Represent the geometry of Exercise 1.2 both as a vector geometry and as a function-space geometry. Can this be done over the two-element field GF(2)?

1.4. Relative to the function space of quadratic polynomials in variables x, y having coefficient zero for the xy term, find the closure of the three-point set $(3, 3)$, $(2, 0)$, $(-4, 6)$. Answer: It contains a fourth point $(-\frac{3}{2}, -\frac{3}{2})$.

1.5. Draw the geometric lattice L of partitions of the 5-element set $\{a, b, c, d, e\}$, and show how the geometry of the graph ab, ac, bc, bd, ce, de has its lattice of flats embedded as a join-sublattice of the lattice L.

1.6. Find all circuits and bases in the 3-simplicial geometry on a five-element set.

1.7. What are the circuits in the transversal matroid of the binary relation of incidence between the edges and vertices of a linear graph?

1.8. Find a vector representation of the geometry of triples in a five-element set.

1.9. Find all circuits of the structure geometry of the complete graph K_5 with vertices in positions $(0, 0)$, $(0, 1)$, $(0, 2)$, $(1, 1)$, $(1, 2)$.

CHAPTER 2

Axiom Systems

Giorgio Nicoletti and Neil White

In Whitney's 1935 article that laid the groundwork for the field of combinatorial geometries and matroids he gave four equivalent (cryptomorphic) definitions of the term *matroid*. Today we have a much larger number of definitions of matroid, all equivalent to the original. We shall state nine of them and prove their equivalence in this chapter. We shall then consider the specializations of these definitions in order to give the slightly more restrictive concept of combinatorial geometry.

The purpose of considering such a large number of equivalent definitions is twofold: First, we lay a useful foundation for the remainder of the book, because different applications can proceed most easily from different definitions. Second, we come to understand many of the interrelations among the basic concepts of the theory.

We shall denote by $\mathbb{N}$ the set of nonnegative integers, and by E a given finite set. The set of all subsets of E will be denoted by 2^E. For every subset $X \in 2^E$ we shall denote the cardinality of X by $|X|$. For the sake of readability, the subsets of 2^E (i.e., collections of subsets of E) will be called *families*, and they will be denoted by script capital letters, whereas the subsets of $2^{(2^E)}$ (i.e., collections of families) will be called *collections*, and they will be denoted by boldface capital letters.

We recall that an *antichain* (or *incomparable family*) in E is a family $\mathscr{A}$ of subsets of E such that, if $X, Y \in \mathscr{A}$ and $X \subseteq Y$, then $X = Y$.

2.1. BASIS AXIOMS

A family $\mathscr{B}$ of subsets of E will be called a *family of bases* (or *basic family*) for E, and its sets will be called *bases*, if the following axioms hold:

(b1) $\mathscr{B} \neq \varnothing$ (*nontriviality*).

(b2) $\mathscr{B}$ is an antichain in E (*incomparability*).

(b3) For every $X, Y \subseteq E$, $X \subseteq Y$, if there exist $B_1, B_2 \in \mathscr{B}$ such that $X \subseteq B_1$ and $B_2 \subseteq Y$, then there exists $B_3 \in \mathscr{B}$ such that $X \subseteq B_3 \subseteq Y$ (*middle basis axiom*).

We shall denote by $\mathbf{B}(E)$ the collection of all basic families for E. Then, a (*finite*) *matroid* $M(E)$ on E is a pair $(E, \mathscr{B})$, where $\mathscr{B} \in \mathbf{B}(E)$.

A principal example of a finite matroid is obtained from a vector space V by letting E be an arbitrary finite subset of V. Then if $\mathscr{B}$ is taken as the family of all maximal subsets of E that are linearly independent in V, axioms (b1), (b2), and (b3) can be easily verified for $\mathscr{B}$; hence we have a matroid $M_V(E)$. Such a matroid is a vector matroid, as discussed in Chapter 1.

An abstract matroid $M(E)$ will be called *coordinatizable* if it is isomorphic to some vector matroid. Which abstract matroids are coordinatizable is a question that will be examined in a chapter in the next volume.

A second example of a matroid, also discussed in Chapter 1, is obtained by letting E be the set of edges of a finite connected graph Γ. We say that a subset $B \subseteq E$ is a basis if B is a spanning tree of Γ. We can then see that axioms (b1), (b2), and (b3) are satisfied; hence we have a matroid, denoted $M_\Gamma(E)$. A matroid isomorphic to $M_\Gamma(E)$ for some connected graph Γ is called *graphic*. This example will be examined in much greater detail in Chapter 6. We have seen in Chapter 1 that a graphic matroid is actually a special type of vector matroid.

Henceforth we shall frequently use the notation $(A - b) \cup c$ instead of the more cumbersome $(A - \{b\}) \cup \{c\}$, and likewise for similar constructions.

Axiom (b3) can be replaced by many others (see Appendix of Matroid Cryptomorphisms). In particular, we have the following:

2.1.1. Proposition. *Let $\mathscr{B}$ be a nonempty antichain of subsets of E. The following statements are equivalent:*

(b3)   *For every $X, Y \subseteq E$, $X \subseteq Y$, if there exist $B_1, B_2 \in \mathscr{B}$ such that $X \subseteq B_1$ and $B_2 \subseteq Y$, then there exists $B_3 \in \mathscr{B}$ such that $X \subseteq B_3 \subseteq Y$ (middle basis axiom).*

(b3.1) *For every $B_1, B_2 \in \mathscr{B}$ and for every $b_1 \in B_1$ there exists $b_2 \in B_2$
        such that $(B_1 - b_1) \cup b_2 \in \mathscr{B}$*   *(weak basis-exchange axiom)*

Proof. Suppose (b3) holds. First of all we show that if $B_1, B_2 \in \mathscr{B}$ and
$|B_1 - B_2| = 1$, then $|B_2 - B_1| = 1$: Let $x = B_1 - B_2$ and $A = B_1 \cap B_2 =
B_1 - x$. Then there exists $y \in B_2 - A$. We have

$$B_2 \supseteq A \cup y \subseteq B_1 \cup y \supseteq B_1,$$

which implies that there exists $B_3 \in \mathbf{B}$ such that

$$(B_1 - x) \cup y = A \cup y \subseteq B_3 \subseteq B_1 \cup y.$$

By the antichain condition, $B_3 \neq B_1 \cup y$, and it follows that $B_3 = (B_1 - x) \cup
y = B_2$ and $|B_2 - B_1| = 1$.

Now let $B_1, B_2 \in \mathscr{B}$, $B_1 \neq B_2$, $x \in B_1 - B_2$, and set $X = B_1 - x$, $Y =
B_2 \cup X$. By hypothesis, there exists $B_3 \in \mathscr{B}$ such that $X \subseteq B_3 \subseteq Y$; obviously
we have $X = B_1 \cap B_3$ and $|B_1 - B_3| = 1$, which implies that $|B_3 - B_1| = 1$;
that is, there exists $y \in Y = X \cup B_2$ such that $B_3 = X \cup y$, and clearly $y \in B_2$.
Therefore, (b3.1) holds.

Conversely, suppose that (b3.1) holds: Now let $B_1, B_2 \in \mathscr{B}$, $B_1 \neq B_2$,
$X \subseteq B_1$, $B_2 \subseteq Y$, $X \subseteq Y$. In the family of all sets in $\mathscr{B}$ that contain X we
can choose a B_3 such that $B_2 \cap B_3$ is maximal. Suppose now that there
exists $x \in B_3 - Y$; then, by hypothesis, there exists $y \in B_2$ such that $(B_3 - x) \cup
y \in \mathscr{B}$ and $X \subseteq (B_3 - x) \cup y$, $B_2 \cap B_3 \subsetneqq [(B_3 - x) \cup y)] \cap B_2$, which con-
tradicts the assumption that $B_2 \cap B_3$ is maximal. Thus, $X \subseteq B_3 \subseteq Y$, and
(b3) holds. $\square$

2.1.2. Corollary. *Any two bases of a matroid $M(E)$ have the same car-
dinality. That is, if $\mathscr{B} \in \mathbf{B}(E)$ and $B_1, B_2 \in \mathscr{B}$, then $|B_1| = |B_2|$.*

Proof. This is an easy exercise, using (b3.1). $\square$

In the definition of matroid, the axiom (b3) has been chosen over the
others because of its self-duality. Indeed, if we reverse the order in the Boolean
algebra 2^E, axiom (b3) remains unchanged. This fact allows us to state a fundamental metatheorem in the theory of matroids. First, for every notion defined in the theory of matroids, its *dual* is the notion defined by reversing the
partial order in the Boolean algebra 2^E. Again, if p is a statement in the
theory of matroids, its dual is the statement p^* obtained from p by reversing
the partial order in 2^E and replacing each notion with its dual. Then we have
the following proposition:

2.1.3. Proposition (duality principle). *If p is a statement in the theory
of matroids that has been proved true, then also its dual p^* is true.*

Now, the foregoing description of duality applies to axiom systems and their consequent theorems. However, we also wish to define the dual of an individual matroid $M(E) = (E, \mathscr{B})$. This requires us to fix a particular anti-automorphism (order-reversing bijection) of 2^E, and the obvious choice is complementation. Thus, we have the following proposition.

2.1.4. Proposition. *Let $\mathscr{B}$ be a family of subsets of E, and let*

$$\mathscr{B}^* := \{E - B: B \in \mathscr{B}\}.$$

Then $\mathscr{B} \in \mathbf{B}(E)$ if and only if $\mathscr{B}^* \in \mathbf{B}(E)$.

Consequently, we are led to give the following definition:

Given a matroid $M(E) := (E, \mathscr{B})$, its *dual*, or *orthogonal matroid*, is the matroid

$$M^*(E) := (E, \mathscr{B}^*).$$

Most works on matroid theory use the antiautomorphism of complementation not only to define the dual of an individual matroid but also to define the idea of duality of statements. Thus, we shall shortly define the dual concept of an independent set to be a spanning set, whereas most authors define it to be the complement of a spanning set.

2.2. OTHER FAMILIES OF SUBSETS

Recall that a *descending family* (or *simplicial complex*) in E is a family $\mathscr{A}$ such that for every $X, Y \subseteq E$, if $X \subseteq Y$ and $Y \in \mathscr{A}$, then $X \in \mathscr{A}$. Dually, an *ascending family* in E will be a family $\mathscr{A}$ such that for every $X, Y \subseteq E$, if $X \supseteq Y$ and $Y \in \mathscr{A}$, then $X \in \mathscr{A}$. For any given family $\mathscr{A}$ of subsets of E we define the *upper cone*, the *lower cone*, the *max*, the *min*, and the *opposite* of $\mathscr{A}$, respectively, as follows:

$$\text{upp}(\mathscr{A}) := \{X \subseteq E: \text{there exists } A \in \mathscr{A}, X \supseteq A\},$$

$$\text{low}(\mathscr{A}) := \{X \subseteq E: \text{there exists } A \in \mathscr{A}, X \subseteq A\},$$

$$\text{max}(\mathscr{A}) := \{A \in \mathscr{A}: A \text{ is maximal in } \mathscr{A}\},$$

$$\text{min}(\mathscr{A}) := \{A \in A: A \text{ is minimal in } \mathscr{A}\},$$

$$\text{opp}(\mathscr{A}) := \{X \subseteq E: X \notin \mathscr{A}\}.$$

Clearly, the upper cone and the lower cone of $\mathscr{A}$ are always an ascending family and a descending family, respectively. The max and the min of $\mathscr{A}$ are both antichains in E. Lastly, the opposite of $\mathscr{A}$ is a descending family whenever $\mathscr{A}$ is an ascending family.

Let $M(E) := (E, \mathscr{B})$ be a finite matroid. A subset $X \subseteq E$ will be called an *independent set* in $M(E)$, or, dually, a *spanning set* in $M(E)$, if $X \in \text{low}(\mathscr{B})$, or $X \in \text{upp}(\mathscr{B})$, respectively. We shall set

$$\mathscr{I} := \text{low}(\mathscr{B}) \quad \text{and} \quad \mathscr{S} := \text{upp}(\mathscr{B}).$$

A subset $X \subseteq E$ will be called a *dependent set* in $M(E)$, or, dually, a *non-spanning set* in $M(E)$, if $X \in \text{opp}(\mathscr{I})$, or $X \in \text{opp}(\mathscr{S})$, respectively. We shall set

$$\mathscr{D} : \text{opp}(\mathscr{I}) \quad \text{and} \quad \mathscr{N} := \text{opp}(\mathscr{S}).$$

Finally, a subset $X \subseteq E$ will be called a *circuit* in $M(E)$, or, dually, a *hyperplane* in $M(E)$, if $X \in \text{min}(\mathscr{D})$, or $X \in \text{max}(\mathscr{N})$, respectively. We shall write

$$\mathscr{C} := \text{min}(\mathscr{D}) \quad \text{and} \quad \mathscr{H} := \text{max}(\mathscr{N}).$$

These concepts are familiar in the examples of coordinatized matroids and graphic matroids. In a coordinatized matroid $M_V(E)$, independent and dependent sets are linearly independent and linearly dependent subsets of E, spanning sets are subsets of E that span V, and hyperplanes are hyperplanes of V intersected with E, provided such an intersection spans the hyperplane in question. In a graphic matroid $M_\Gamma(E)$, where Γ is connected, independent sets are forests that are subgraphs of Γ, spanning sets are subgraphs containing spanning trees, and circuits are circuits of Γ.

The collections of all families defined earlier, when $\mathscr{B}$ lies in $\mathbf{B}(E)$, shall be denoted respectively as follows:

$$\mathbf{I}(E), \ \mathbf{S}(E), \ \mathbf{D}(E), \ \mathbf{N}(E), \ \mathbf{C}(E), \ \mathbf{H}(E).$$

Obviously, for every matroid $M(E)$ the families of circuits and hyperplanes are antichains, the families of dependent sets and spanning sets are ascending families, and the families of independent sets and nonspanning sets are descending families.

We explicitly note that, by the definitions, the independent sets of the dual matroid $M^*(E)$ are precisely the complements of the spanning sets of $M(E)$, and dually. Similarly, the complements of hyperplanes of $M(E)$, often called *bonds* of $M(E)$, are the circuits of $M^*(E)$.

A glance at the previous definitions will confirm that different matroids over E [i.e., different families in $\mathbf{B}(E)$] are related to different families of spanning sets, independent sets, etc. Thus, there exists a "natural" bijection from $\mathbf{B}(E)$ into each one of the collections defined earlier. Now, if we succeed in characterizing the families that belong to such collections, we are able to define matroids in different, but cryptomorphic, ways.

Indeed, let us suppose that, for example, $i1$, $i2$, and $i3$ are properties characterizing the families of independent sets of matroids over E. Then we can say that a matroid is (cryptomorphically) a pair $M(E) := (E, \mathscr{I})$, where $\mathscr{I}$ is a family satisfying $i1$, $i2$, and $i3$, now assumed as axioms. In this case we shall define a basis in $M(E)$ as a maximal set in $\mathscr{I}$.

Let us make a more precise definition of cryptomorphism. Let $u1$, $u2, \ldots, uh$ and $v1, v2, \ldots, vk$ be axiom systems concerning families $\mathscr{U}$ and

$\mathscr{V}$, respectively, of subsets of E (or, as we shall see later, families of operators $f: 2^E \to 2^E$, or, again, families of functions $f: 2^E \to \mathbb{N}$). Let $U(E)$ and $V(E)$ be the collections of all families $\mathscr{U}$ and $\mathscr{V}$ that satisfy the axiom systems $u1$, $u2, \ldots, uh$ and $v1, v2, \ldots, vk$, respectively. An *interpretation* is a function $\alpha: U(E) \to V(E)$, specifically given as a rule for constructing $\mathscr{V} := \alpha\mathscr{U}$ in terms of a given $\mathscr{U} \in U(E)$. By this definition α could be quite arbitrary, but we shall be interested only in interpretations that correspond to a "natural" connection between the concepts underlying U and V. We must, of course, check that $\alpha\mathscr{U} \in V(E)$ when we construct an interpretation, and this amounts to proving that the axioms $u1, u2, \ldots, uh$ for $\mathscr{U}$ imply axioms $v1, v2, \ldots, vk$ for $\alpha\mathscr{U}$. Now, the systems $u1, u2, \ldots, uh$ and $v1, v2, \ldots, vk$ will be called *crypto-morphic* if there exist interpretations

$$\alpha: U(E) \to V(E) \quad \text{and} \quad \beta: V(E) \to U(E).$$

that are two-sided inverses of each other.

In this case, the functions α and β are called *cryptomorphisms*. Thus, a cryptomorphism is a bijective interpretation whose inverse is also an interpretation. We remark that if $u1, u2, \ldots, uh$ and $v1, v2, \ldots, vk$ are cryptomorphic, then every theorem derived from the first system of axioms can be translated (sometimes awkwardly) via the cryptomorphism into a theorem derivable from the second system of axioms. We also note that cryptomorphism is an equivalence relation. In particular, cryptomorphic systems of axioms involving the same concept (as we saw for bases in Proposition 2.1.1) are cryptomorphic in a degenerate way; that is, they are equivalent axiom systems in the usual sense.

Cryptomorphisms also appear in other branches of mathematics. For example, open sets and the Kuratowski closure operator provide cryptomorphic axiomatizations of topology. However, the concept of basis of a topology does not provide another cryptomorphism, because a basis is not uniquely determined by a topology. Another example of cryptomorphism is provided by equivalence relations and partitions.

Part of the fascination of matroids and their slightly more specialized cousins, combinatorial geometries, lies in the large number of crypto-morphisms of the subject. After proving that the functions defined at the beginning of this section are cryptomorphisms relating circuits, dependent sets, independent sets, their duals (hyperplanes, nonspanning sets, spanning sets), and bases (which are self-dual), we shall then also look at other cryptomorphisms relating families of functions $f: 2^E \to \mathbb{N}$, families of operators $f: 2^E \to 2^E$, and other families of subsets of E. A complete listing of known cryptomorphisms is given in the Appendix.

Coming back to independent and spanning sets, we have the following:

2.2.1. Proposition. *A family $\mathscr{I}$ belongs to $\mathbf{I}(E)$, that is, it is the family of all independent sets for a matroid $M(E)$, if and only if the following conditions* *hold:*

    (i1) $\mathscr{I} \neq \varnothing$ (*nontriviality*).

    (i2) $\mathscr{I}$ *is a descending family.*

    (i3) *For every* $I_1, I_2 \in \mathscr{I}$, *if* $|I_1| < |I_2|$, *then there exists* $x \in I_2 - I_1$ *such that* $I_1 \cup x \in \mathscr{I}$ (*augmentation*).

Dually, a family $\mathscr{S}$ belongs to $\mathbf{S}(E)$, *that is, it is the family of all spanning sets for a matroid $M(E)$, if and only if the following conditions hold:*

    (s1) $\mathscr{S} \neq \varnothing$ (*nontriviality*).

    (s2) $\mathscr{S}$ *is an ascending family.*

    (s3) *For every* $S_1, S_2 \in \mathscr{S}$, *if* $|S_1| > |S_2|$, *then there exists* $x \in S_1 - S_2$ *such that* $S_1 - x \in \mathscr{S}$ (*reduction*).

Proof. We first show that conditions (i1), (i2), and (i3) characterize the families in $\mathbf{I}(E)$. Suppose that $\mathscr{B} \in \mathbf{B}(E)$, and set $\mathscr{I} := \mathrm{low}(\mathscr{B})$. Then $\mathscr{I}$ obviously satisfies (i1) and (i2). By Corollary 2.1.2, all the sets in $\mathscr{B}$ have the same cardinality. Now let $I_1, I_2 \in \mathscr{I}$, with $|I_1| < |I_2|$. Then there exist $B_1, B_2 \in \mathscr{B}$ such that $I_1 \subseteq B_1$ and $I_2 \subseteq B_2$. Setting $Y := B_2 \cup I_1$, we have $B_1 \supseteq I_1 \subseteq Y \supseteq B_2$. By axiom ($b3$) there exists $B_3 \in \mathscr{B}$ such that $I_1 \subseteq B_3 \subseteq Y$. It follows that $B_3 - I_1 \subseteq B_2$ and $|B_3 - I_1| > |B_2 - I_2|$. Hence, there exists $x \in I_2 \cap (B_3 - I_1)$. This implies that $\mathscr{I}$ satisfies condition (i3).

    Conversely, suppose that the family $\mathscr{I}$ satisfies conditions (i1), (i2), and (i3), and set $\mathscr{B} := \mathrm{max}(\mathscr{I})$. Obviously $\mathscr{B}$ satisfies conditions (b1) and (b2). Moreover, the sets in $\mathscr{B}$, because of (i3), have the same cardinality. Consequently, if $I \in \mathscr{I}$, $B \in \mathscr{B}$, and $|I| = |B|$, then $I \in \mathscr{B}$. Now let $X \subseteq B_1 \in \mathscr{B}$, $B_2 \subseteq Y$, $B_2 \in \mathscr{B}$, and $X \subseteq Y$. If $|X| < |B_2|$, then (i3) ensures the existence of $Z \subseteq B_2$ such that $X \cup Z \in \mathscr{I}$ and $|X \cup Z| = |B_2|$. Thus, $X \cup Z \in \mathscr{B}$, $X \subseteq X \cup Z \subseteq Y$, and (b3) holds. This shows that $\mathscr{B} \in \mathbf{B}(E)$, and $\mathscr{I}$ is the family of independent sets of the matroid $M(E) := (E, \mathscr{B})$.

    By the duality principle, conditions (s1), (s2), and (s3) characterize the families in $\mathbf{S}(E)$. $\square$

    We explicitly note that duality between (i3) and (s3) becomes more evident when these statements are translated in the language of the Boolean algebra 2^E. We shall refer to (i1), (i2), and (i3) as the *independence axioms*, and to (s1), (s2), and (s3) as the *span axioms*.

    2.2.2. Corollary. *The basis axioms, independence axioms, and span axioms are all cryptomorphic.*

Proof. In Proposition 2.2.1 we proved that $\mathrm{low}: \mathbf{B}(E) \to \mathbf{I}(E)$ and $\mathrm{max}: \mathbf{I}(E) \to \mathbf{B}(E)$ are interpretations, and clearly they are two-sided inverses of each other. Dually, $\mathbf{B}$ and $\mathbf{S}$ are cryptomorphic. $\square$

Let us now examine dependent sets and nonspanning sets.

2.2.3. Proposition. *A family $\mathscr{D}$ belongs to $\mathbf{D}(E)$, that is, it is the family of all dependent sets for a matroid over E, if and only if the following conditions hold:*

(d1) $\varnothing \notin \mathscr{D}$ *(nontriviality).*

(d2) *$\mathscr{D}$ is an ascending family.*

(d3) *For every $D_1, D_2 \in \mathscr{D}$, if $D_1 \cap D_2 \notin \mathscr{D}$, then for every $x \in E$, $(D_1 \cup D_2) - x \in \mathscr{D}$.*

Dually, a family $\mathscr{N}$ belongs to $\mathbf{N}(E)$, that is, it is the family of all nonspanning sets for a matroid over E, if and only if the following conditions hold:

(n1) $E \notin \mathscr{N}$ *(nontriviality).*

(n2) *$\mathscr{N}$ is a descending family.*

(n3) *For every $N_1, N_2 \in \mathscr{N}$, if $N_1 \cup N_2 \notin \mathscr{N}$, then for every $x \in E$, $(N_1 \cap N_2) \cup x \in \mathscr{N}$.*

Proof. Suppose that $\mathscr{D} \in \mathbf{D}(E)$; then $\mathscr{I} := \mathrm{opp}(\mathscr{D})$ satisfies axioms (i1), (i2), and (i3), which trivially imply conditions (d1) and (d2) for $\mathscr{D}$. Now let $D_1, D_2 \in \mathscr{D}$ such that $D_1 \cap D_2 \in \mathscr{I}$. Then $D_1 - D_2 \neq \varnothing \neq D_2 - D_1$; it follows that $(D_1 \cup D_2) - x \in \mathscr{D}$ for every $x \notin D_1 \cap D_2$. Assume now that $x \in D_1 \cap D_2$ and $(D_1 \cup D_2) - x \in \mathscr{I}$. By axiom (i3), because $D_1 \cap D_2 \in \mathscr{I}$, $|D_1 \cap D_2| < |(D_1 \cup D_2) - x|$, there exists $I \in \mathscr{I}$ such that $D_1 \cap D_2 \subseteq I$, $I \subseteq [(D_1 \cup D_2) - x] \cup (D_1 \cap D_2) = D_1 \cup D_2$, and $|I| = |(D_1 \cup D_2) - x|$. This implies that $D_1 \subseteq I$ or $D_2 \subseteq I$, which is in contradiction with the hypothesis. Then (d3) holds.

Conversely, let us suppose that conditions (d1), (d2), and (d3) hold for a family $\mathscr{D}$ and set $\mathscr{I} := \mathrm{opp}(\mathscr{D})$. Then trivially $\mathscr{I}$ satisfies (i1) and (i2). Now let $I_1, I_2 \in \mathscr{I}$, with $|I_1| < |I_2|$; if $|I_1 - I_2| = 0$, then (i3) obviously holds. Suppose now that (i3) holds whenever $|I_1 - I_2| = n$, and let $I_1, I_2 \in \mathscr{I}$ such that $|I_1 - I_2| = n + 1$. Let $y \in I_1 - I_2$, and set $I_1' := I_1 - y$; then $I_1' \in \mathscr{I}$, $|I_1'| < |I_2|$, and $|I_1' - I_2| = n$. By hypothesis we can find $I_2' \in \mathscr{I}$ such that $I_1' \subseteq I_2' \subseteq I_1' \cup I_2$ and $|I_2'| = |I_1'| + 1$; by iteration we can assume $|I_2'| = |I_2|$. This implies the existence of $x_1, x_2 \in I_2' - I_1$, $x_1 \neq x_2$. By (d3), either $I_1 \cup x_1 \in \mathscr{I}$ or $I_1 \cup x_2 \in \mathscr{I}$, because $(I_1 \cup x_1 \cup x_2) - y \subseteq I_2' \in \mathscr{I}$. Then, by induction, (i3) always holds. This shows that $\mathscr{I} \in \mathbf{I}(E)$ and $\mathscr{D} \in \mathbf{D}(E)$.

By the duality principle, conditions (n1), (n2), and (n3) characterize the families in $\mathbf{N}(E)$. $\square$

We shall refer to (d1), (d2), and (d3) as the *dependence axioms* and to (n1), (n2), and (n3) as the *nonspan axioms*. The nonspan axioms are the least frequently used of the cryptomorphisms that we are considering, but they are included because they have a natural place in the scheme of axiomatizations in Theorem 2.2.6.

Looking at circuits and hyperplanes, we have the following:

2.2.4. Proposition. *A family $\mathscr{C}$ belongs to $\mathbf{C}(E)$, that is, it is the family of all circuits for a matroid over E, if and only if the following conditions are satisfied:*

(c1) $\varnothing \notin \mathscr{C}$ *(nontriviality).*

(c2) $\mathscr{C}$ *is an antichain*   *(incomparability).*

(c3) *For every $C_1, C_2 \in \mathscr{C}$, such that $C_1 \neq C_2$, and for every $x \in E$, there exists $C_3 \in \mathscr{C}$ such that $C_3 \subseteq (C_1 \cup C_2) - x$ (elimination).*

Dually, a family $\mathscr{H}$ belongs to $\mathbf{H}(E)$, that is, it is the family of all hyperplanes for a matroid over E, if and only if the following conditions hold:

(h1) $E \notin \mathscr{H}$ *(nontriviality).*

(h2) $\mathscr{H}$ *is an antichain*   *(incomparability).*

(h3) *For every $H_1, H_2 \in \mathscr{H}$, such that $H_1 \neq H_2$, and for every $x \in E$, there exists $H_3 \in \mathscr{H}$ such that $(H_1 \cap H_2) \cup x \subseteq H_3$ (covering).*

Proof. Suppose $\mathscr{C} \in \mathbf{C}(E)$, and set $\mathscr{D} := \mathrm{upp}(\mathscr{C})$. Then $\mathscr{D}$ satisfies properties (d1), (d2), and (d3), which trivially imply (c1) and (c2). Let $C_1, C_2 \in \mathscr{C}, C_1 \neq C_2$; then $C_1, C_2 \in \mathscr{D}$ and $C_1 \cap C_2 \notin \mathscr{D}$. By property (d3) we deduce that for every $x \in E$, the set $(C_1 \cup C_2) - x$ belongs to $\mathscr{D}$; thus, there exists $C_3 \in \mathscr{C}$ such that $C_3 \subseteq (C_1 \cup C_2) - x$, and property (c3) holds for $\mathscr{C}$.

Conversely, let $\mathscr{C}$ be a family satisfying properties (c1), (c2), and (c3), and set $\mathscr{D} := \mathrm{upp}(\mathscr{C})$. Then $\mathscr{D}$ obviously satisfies properties (d1) and (d2). Now let $D_1, D_2 \in \mathscr{D}, D_1 \cap D_2 \notin \mathscr{D}$, and let $C_1, C_2 \in \mathscr{C}$ such that $C_1 \subseteq D_1, C_2 \subseteq D_2$; then $C_1 \neq C_2$. For every $x \in E$, if $x \notin C_1 \cap C_2$, the set $(D_1 \cup D_2) - x$ is obviously in $\mathscr{D}$, because $C_i \subseteq (D_1 \cup D_2) - x$ with $i = 1$ or $i = 2$. Now let $x \in C_1 \cap C_2$. By (c3) there exists $C_3 \in \mathscr{C}$ such that $C_3 \subseteq (C_1 \cup C_2) - x \subseteq (D_1 \cup D_2) - x$, which implies that $(D_1 \cup D_2) - x \in \mathscr{D}$, and (d3) always holds. This shows that $\mathscr{D} \in \mathbf{D}(E)$ and $\mathscr{C} \in \mathbf{C}(E)$.

By the duality principle, properties (h1), (h2), and (h3) characterize the families in $\mathbf{H}(E)$. $\square$

We shall refer to (c1), (c2), and (c3) as the *circuit axioms* and to (h1), (h2), and (h3) as the *hyperplane axioms*. Axioms (c3) and (h3) can be replaced by seemingly stronger axioms. We have, indeed, the following:

2.2.5. Proposition. *Let $\mathscr{A}$ be an antichain. Then (c3) holds for $\mathscr{A}$ if and only if the following property holds:*

(c3.1) *For every $C_1, C_2 \in \mathscr{A}$ such that $C_1 \neq C_2$, and for every $x \in C_1 \cap C_2, y \in C_1 - C_2$, there exists $C_3 \in \mathscr{A}$ such that $y \in C_3 \subseteq (C_1 \cup C_2) - x$ (strong elimination).*

Dually, $\mathscr{A}$ satisfies (h3) *if and only if the following property holds:*

(h3.1) *For every $H_1, H_2 \in \mathscr{A}$ such that $H_1 \neq H_2$, and for every*
$x \notin H_1 \cup H_2, y \in H_2 - H_1$, *there exists $H_3 \in \mathscr{A}$ such that*
$y \notin H_3 \supseteq (H_1 \cap H_2) \cup x$ (*sharp covering*).

Proof. Obviously (c3.1) implies (c3). To prove that (c3) implies (c3.1), we proceed by induction on $|C_1 \cup C_2|$, noting that (c3.1) is easily proved from (c3) if $|C_1 \cup C_2| \leq 3$. Assume that (c3.1) is true for all pairs of circuits whose union has cardinality less than $|C_1 \cup C_2|$, and let $x \in C_1 \cap C_2$, $y \in C_1 - C_2$. By (c3), there exists $C_3 \in \mathscr{A}$, $C_3 \subseteq C_1 \cup C_2 - x$. We can assume $y \notin C_3$, for otherwise we are done. Now $|C_2 \cup C_3| < |C_1 \cup C_2|$, because $y \notin C_3 \cup C_2$.

Because $C_3 \not\subseteq C_1$, there exists $z \in C_3 \cap (C_2 - C_1) = (C_2 \cap C_3) - C_1$. By the induction hypothesis, there exists $C_4 \in \mathscr{A}$, $C_4 \subseteq C_3 \cup C_2 - z$, $x \in C_4$. Now $x \in C_1 \cap C_4$, $y \in C_1 - C_4$, and $|C_1 \cup C_4| < |C_1 \cup C_2|$, because $z \notin C_1 \cup C_4$. Again, using the induction hypothesis, there exists $C_5 \in \mathscr{A}$, $y \in C_5$, $C_5 \subseteq C_1 \cup C_4 - x \subseteq C_1 \cup C_2 - x$, proving strong elimination.

By the duality principle, (h3) and (h3.1) are equivalent. $\square$

The preceding results can be restated as follows:

2.2.6. Theorem. *Circuit, dependence, independence, basis, span, non-span, and hyperplane axioms are cryptomorphic by means of the following cryptomorphisms:*

$$C(E) \overset{\text{upp}}{\underset{\text{min}}{\rightleftarrows}} D(E) \overset{\text{opp}}{\underset{\text{opp}}{\rightleftarrows}} I(E) \overset{\text{max}}{\underset{\text{low}}{\rightleftarrows}} B(E) \overset{\text{upp}}{\underset{\text{min}}{\rightleftarrows}} S(E) \overset{\text{opp}}{\underset{\text{opp}}{\rightleftarrows}} N(E) \overset{\text{max}}{\underset{\text{low}}{\rightleftarrows}} H(E).$$

The foregoing diagram suggests that one might give some new axiomatizations of matroids by characterizing the collections defined through repeated applications of operators upp, opp, max, and their duals. Whether or not such characterizations will be useful remains to be seen.

2.3. CLOSURE AND RANK

It is now worthwhile to address our attention to two more concepts, namely, closure and rank.

A (matroid) *closure operator* over the finite set E is an operator cl: $2^E \rightarrow 2^E$ satisfying the following:

(cl1) For every $X \subseteq E$, $X \subseteq \text{cl}(X)$ (*increase*).
(cl2) For every $X, Y \subseteq E$, if $X \subseteq Y$ then $\text{cl}(X) \subseteq \text{cl}(Y)$
    (*monotonicity*).

(cl3) For every $X \subseteq E$, $\text{cl}[\text{cl}(X)] = \text{cl}(X)$ (*idempotence*).

(cl4) For every $X \subseteq E$ and for every $y, z \in E$, if $y \in \text{cl}(X \cup z) - \text{cl}(X)$, then $z \in \text{cl}(X \cup y) - \text{cl}(X)$ (*exchange*).

The collection of all closure operators will be denoted by $\mathbf{cl}(E)$.

It would be more consistent with previous notation to use $\mathscr{cl}$ for the closure operator of a particular matroid rather than cl, just as we used $\mathscr{B}$ for the family of bases of a particular matroid, but cl is commonly accepted notation because it does not refer to a collection of subsets of E. There should be no confusion caused by this, and similar remarks apply to r in the following axiom systems.

In later chapters, an alternative notation for $\text{cl}(X)$ is $\bar{X}$.

A *rank function* over E is a function $r: 2^E \to \mathbb{N}$ satisfying the following:

(r1) For every $X \subseteq E$, $0 \le r(X) \le |X|$ (*cardinality bound*).

(r2) For every $X, Y \subseteq E$, if $X \subseteq Y$ then $r(X) \le r(Y)$ (*monotonicity*).

(r3) For every $X, Y \subseteq E$, $r(X \cup Y) + r(X \cap Y) \le r(X) + r(Y)$ (*semimodularity*).

The collection of all rank functions over E will be denoted by $\mathbf{r}(E)$.

We shall refer to (cl1), (cl2), (cl3), and (cl4) and (r1), (r2), and (r3) as the *closure axioms* and *rank axioms*, respectively.

Note that $\mathbf{cl}(E)$ and $\mathbf{r}(E)$ are nonempty collections, because the identity operator on E is an element of $\mathbf{cl}(E)$, and cardinality is an element of $\mathbf{r}(E)$.

As an alternative to the "global" rank axioms just given, we also have the following *local rank axioms*, which are often more suitable for inductive proofs.

A *local rank function* over E is a function $r: 2^E \to \mathbb{N}$ satisfying the following:

(r'1) $r(\varnothing) = 0$ (*normalization*).

(r'2) For every $X \subseteq E$ and for every $y \in E$, $r(X) \le r(X \cup y) \le r(X) + 1$ (*unit increase*).

(r'3) For every $X \subseteq E$ and for every $y, z \in E$, if $r(X) = r(X \cup y) = r(X \cup z)$, then $r(X) = r(X \cup y \cup z)$ (*local semimodularity*).

The collection of all local rank functions over E will be denoted by $\mathbf{r}'(E)$.

Several other variations of rank-function axioms are possible, such as (r'1), (r'2), and (r3). See the Appendix.

Now we shall show that closure, rank, and local rank axioms are cryptomorphic to each of the previous axiom systems. In order to do this we define some maps that will turn out to be interpretations.

For any given family $\mathscr{H} \in \mathbf{H}(E)$ of hyperplanes, the operator $\alpha(\mathscr{H})$: $2^E \to 2^E$ is defined as follows: For every $X \subseteq E$,

$$[\alpha(\mathscr{H})](X) := \cap\{H \in \mathscr{H} : X \subseteq H\},$$

with the convention that the intersection of the empty family of subsets of E is E itself.

Again, for any closure operator $\text{cl} \in \mathbf{cl}(E)$, the family $\beta(\text{cl})$ is defined as follows:

$$\beta(\text{cl}) := \max\{X \subseteq E : X = \text{cl}(X) \neq E\}.$$

Then, we have the following:

2.3.1. Proposition. *Closure and hyperplane axioms are cryptomorphic by means of the following cryptomorphisms:*

$$\mathbf{H}(E) \underset{\beta}{\overset{\alpha}{\rightleftarrows}} \mathbf{cl}(E).$$

Proof. First of all, we show that α is a map from $\mathbf{H}(E)$ into $\mathbf{cl}(E)$. Let $\mathcal{H} \in \mathbf{H}(E)$ be a family of hyperplanes, and set $\text{cl} := \alpha(\mathcal{H})$; then cl obviously satisfies (cl1), (cl2), and (cl3). Let $X \subseteq E$, $y, z \in E$ such that $y \in \text{cl}(X \cup z) - \text{cl}(X)$. Then for every hyperplane $H \in \mathcal{H}$, if $X \cup z \subseteq H$, then $y \in H$, but there exists $H_1 \in \mathcal{H}$ such that $X \subseteq H_1$, $y \notin H_1$, $z \notin H_1$. Suppose now that there exists $H_2 \in \mathcal{H}$ such that $X \cup y \subseteq H_2$ and $z \notin H_2$. Then $X \subseteq H_1 \cap H_2$, $z \notin H_1 \cup H_2$, $y \in H_2 - H_1$, and by (h3.1) we have that there exists $H_3 \in \mathcal{H}$ such that $X \subseteq H_3$, $X \cup z \subseteq (H_1 \cap H_2) \cup z \subseteq H_3$, $y \notin H_3$, which contradicts $y \in \text{cl}(X \cup z) - \text{cl}(X)$. It follows that for every $K \in \mathcal{H}$, if $X \cup y \subseteq K$, then $z \in K$, which implies $z \in \text{cl}(X \cup y) - \text{cl}(X)$. This shows that cl is a closure operator, and α maps $\mathbf{H}(E)$ into $\mathbf{cl}(E)$.

Conversely, let $\text{cl} \in \mathbf{cl}(E)$ be a closure operator over E, and set $\mathcal{H} := \beta(\text{cl})$; then $\mathcal{H}$ obviously satisfies (h1) and (h2). Now let $H_1, H_2 \in \mathcal{H}$, $H_1 \neq H_2$, and let $x \in E$, $x \notin H_1 \cup H_2$. Set $K := \text{cl}[(H_1 \cap H_2) \cup x]$, and suppose $K = E$. Let $h_1 \in H_1 - H_2$. Then $h_1 \in \text{cl}[(H_1 \cap H_2) \cup x] - \text{cl}(H_1 \cap H_2)$, and, by (cl4), $x \in \text{cl}[(H_1 \cap H_2) \cup h_1] - \text{cl}(H_1 \cap H_2) \subseteq \text{cl}(H_1) - \text{cl}(H_1 \cap H_2)$, which contradicts $x \notin H_1 \cup H_2$. Then $K \neq E$.

In the finite family of all $X \subseteq E$ such that $X = \text{cl}(X) \neq E$ it is now possible to find H_3 such that $K \subseteq H_3$, H_3 maximal. Hence, $x \in H_3 \supseteq H_1 \cap H_2$, proving that $\mathcal{H}$ is a family of hyperplanes of a matroid, and β maps $\mathbf{cl}(E)$ into $\mathbf{H}(E)$.

Finally, we leave as an exercise the proof that α and β are two-sided inverses of each other, and hence that they are cryptomorphisms. $\square$

For any given family of independent sets $\mathcal{I} \in \mathbf{I}(E)$, we define a function $\gamma(\mathcal{I}) : 2^E \to \mathbb{N}$ as follows: For every $X \subseteq E$, we set

$$[\gamma(\mathcal{I})](X) := \max\{|I| : I \subseteq X, I \in \mathcal{I}\}.$$

The identity map over the collection $\mathbf{r}(E)$ of all ranks over E will be denoted by δ. Finally, for any given local rank function $r \in \mathbf{r}'(E)$, we define a family

$\varepsilon(r)$ as follows:

$$\varepsilon(r) := \{X \subseteq E : r(X) = |X|\}.$$

Then, we have the following:

2.3.2. Proposition. *Rank, local rank, and independence axioms are cryptomorphic by means of these cryptomorphisms:*

$$\mathbf{I}(E) \xrightarrow{\gamma} \mathbf{r}(E) \xrightarrow{\delta} \mathbf{r}'(E) \xrightarrow{\varepsilon} \mathbf{I}(E).$$

Proof. Let $\mathscr{I} \in \mathbf{I}(E)$ be a family of independent sets, and let $r := \gamma(\mathscr{I})$; then the cardinality bound (r1) and monotonicity (r2) easily follow from (i1), (i2), and the definition of γ. To prove semimodularity (r3), let $X, Y \subseteq E$. Let $r(X \cap Y) = u, r(X \cup Y) = v$, and let $U \in \mathscr{I}, U \subseteq X \cap Y, |U| = u; V \subseteq X \cup Y$, $V \in \mathscr{I}, |V| = v$. By monotonicity (r2), $u \le v$, and if $u < v$, applying augmentation (i3), we can adjoin $v - u$ distinct elements of $V - U$ to U to obtain $V' \in \mathscr{I}, |V'| = v, U \subseteq V' \subseteq X \cup Y$. Now $V' \cap X \in \mathscr{I}, V' \cap Y \in \mathscr{I}$, and $V' = U \cup (V' - X) \cup (V' - Y)$, because U is a maximal independent subset of $X \cap Y$; hence, $r(X) + r(Y) \ge |V' \cap X| + |V' \cap Y| = |U| + |V' - Y| + |U| + |V' - X| = |U| + |V'| = r(X \cap Y) + r(X \cup Y)$. This shows that γ maps $\mathbf{I}(E)$ into $\mathbf{r}(E)$.

To prove that δ maps $\mathbf{r}(E)$ into $\mathbf{r}'(E)$, let r be a rank function over E. From the cardinality bound we immediately get normalization (r'1). If $Z \subseteq E$ and $x \in E$, then $r(Z) \le r(Z \cup x)$ follows from monotonicity (r2), and because the cardinality bound implies $0 \le r(x) \le 1$, using semimodularity (r3), we have $r(Z \cup x) \le r(Z \cup x) + r(Z \cap x) \le r(Z) + r(x) \le r(Z) + 1$, proving unit increase (r'2). If $Z \subseteq E$, $x \in E$, $y \in E$, and $r(Z \cup x) = r(Z \cup y) = r(Z)$, then $r(Z \cup x \cup y) = r[(Z \cup x) \cup (Z \cup y)] \le r(Z \cup x) + r(Z \cup y) - r[(Z \cup x) \cap (Z \cup y)] = r(Z)$, which proves local semimodularity (r'3).

Now we show that ε maps $\mathbf{r}'(E)$ into $\mathbf{I}(E)$. Let $r \in \mathbf{r}'(E)$ be a local rank function, and set $\mathscr{I} := \varepsilon(r)$. Then $\mathscr{I}$ satisfies (i1) as an immediate consequence of normalization. By induction, from the unit-increase axiom we easily see that $0 \le r(X) \le |X|$ for every $X \subseteq E$; hence, $X \notin \mathscr{I}$ if and only if $r(X) < |X|$. If $X \subseteq Y$ and $X \notin \mathscr{I}$, using induction with unit increase again, we see that $r(Y) \le r(X) + |Y - X| < |X| + |Y - X| = |Y|$; hence, $Y \notin \mathscr{I}$, proving (i2). Finally, let $U, V \in \mathscr{I}$, with $|V| > |U|$. If $r(U \cup x) = r(U)$ for all $x \in V$, then we can argue inductively from local semimodularity that $r(U \cup V) = r(V) = r(U)$. Because this contradicts $|U| < |V|$, there exists $x \in V$ for which $r(U \cup x) = r(U) + 1 = |U \cup x|$, whence $U \cup x \in \mathscr{I}$, proving augmentation (i3), and showing that ε maps $\mathbf{r}'(E)$ into $\mathbf{I}(E)$.

It remains to be shown that each of γ, δ, and ε has a two-sided inverse. It is easy to see that $\varepsilon \delta \gamma$ is the identity map on $\mathbf{I}(E)$. For $r \in \mathbf{r}(E)$, consider $\gamma \varepsilon \delta(r) \in \mathbf{r}(E)$. For $X \subseteq E$,

$$[\gamma \varepsilon \delta(r)](X) = \max\{|Y| : Y \subseteq X, |Y| = r(Y)\}.$$

Thus, $\gamma\varepsilon\delta(r) = r$ if for every $X \subseteq E$ there exists $Y \subseteq X$ such that $|Y| = r(Y) = r(X)$, where $r \in \mathbf{r}(E)$, and hence $r \in \mathbf{r}'(E)$. But if Y is a maximal subset of X satisfying $|Y| = r(Y)$, then $r(Y \cup x) = r(Y)$ for all $x \in X$, and inductively by $(r'3)$, $r(X) = r(Y) = |Y|$, proving that $\gamma\varepsilon\delta$ is the identity map on $\mathbf{r}(E)$. The same argument shows that $\delta\gamma\varepsilon$ is the identity on $\mathbf{r}'(E)$; hence, $\gamma^{-1} = \varepsilon\delta$, $\delta^{-1} = \gamma\varepsilon$, and $\varepsilon^{-1} = \delta\gamma$ are all two-sided inverses.   □

As mentioned earlier, there are more axiom systems for matroids that are cryptomorphic to those we have listed. These systems include bonds, flats (or closed sets), open sets, and so forth. These systems and some of their cryptomorphisms are covered in the exercise section at the end of this chapter, and a fairly complete listing is to be found in the Appendix of Matroid Cryptomorphisms.

2.4. COMBINATORIAL GEOMETRIES AND INFINITE MATROIDS

We now state the conditions that must be added to each of the axiom systems given earlier in order to get the corresponding axiom systems for combinatorial geometries. Proving that these nine systems are again cryptomorphic to each other (using restrictions of the cryptomorphisms already considered) is an easy exercise.

Combinatorial geometries are a very natural restriction of the collection of matroids to consider. Among vector matroids, they constitute the class in which no two vectors are scalar multiples of each other, and in which the zero vector is excluded. If we then pass from the vector space to the corresponding projective space by the usual process of identifying scalar multiples, the elements of a vector combinatorial geometry remain distinct. Thus, every subset of a projective geometry is a combinatorial geometry (but not conversely!), and, in general, combinatorial geometries are often visualized as geometric configurations (see Exercise 2.3). Combinatorial geometries are also a very natural class among graphic matroids. They correspond to graphs with no loops or multiple edges.

A matroid $M(E) := (E, \mathscr{B})$, with $\mathscr{B} \in \mathbf{B}(E)$, will be called a *combinatorial geometry* (or *simple matroid*) if it satisfies any of the equivalent conditions of the following proposition. In this case, $M(E)$ is often denoted $G(E)$.

2.4.1. Proposition. *Let $M(E) := (E, \mathscr{B})$, with $\mathscr{B} \in \mathbf{B}(E)$, be a matroid over E, and let $\mathscr{C}$, $\mathscr{D}$, $\mathscr{I}$, $\mathscr{S}$, $\mathscr{N}$, and $\mathscr{H}$ be the families of its circuits, dependent sets, independent sets, spanning sets, nonspanning sets, and hyperplanes, respectively, and let cl and $\mathbf{r}$ be the closure operator and the rank function of $M(E)$. Then the following statements are equivalent:*

(c4)   *If $C \in \mathscr{C}$ then $|C| \geq 3$.*
(d4)   *If $D \in \mathscr{D}$ then $|D| \geq 3$.*

(i4)   For every $x, y \in E$, $\{x, y\} \in \mathcal{I}$.

(b4)   For every $x, y \in E$ there exists $B \in \mathcal{B}$ such that $\{x, y\} \subseteq B$.

(s4)   For every $x, y \in E$ there exists $S \in min(\mathcal{S})$ such that $\{x, y\} \subseteq S$.

(n4)   For every $x, y \in E$, $x \neq y$, there exists $N \in max(\mathcal{N})$ such that $x \in N$, $y \notin N$.

(h4)   For every $x, y \in E$, $x \neq y$, there exists $H \in \mathcal{H}$ such that $x \in H$, $y \notin H$.

(c15) $cl(\varnothing) = \varnothing$ and $cl(x) = x$ for every $x \in E$.

(r4 = r'4) For every $X \subseteq E$, if $|X| \leq 2$, then $r(X) = |X|$.

Thus far we have considered only matroids in which the set E is finite. If we now allow E to be infinite, our theory remains largely unchanged provided we insist that every basis is finite. However, the dual of such a matroid does not have finite bases. To allow duality, the finite-basis assumption must be relaxed; such axiom systems will be considered in a chapter on infinite matroids in a later volume. For now, we content ourselves with describing the various cryptomorphic versions of the finite-basis assumption, again leaving the proof as an exercise.

The finite-basis assumption is used in Crapo and Rota (1970) and is stronger than the finitary assumption of Klee (1971) and Welsh (1976), which amounts to assuming that all circuits are finite, though bases may be infinite.

2.4.2. Proposition. Let E be an arbitrary set. Then the basis, independence, span, dependence, nonspan, circuit, hyperplane, closure, and rank axioms on E, each with one additional axiom as follows, are cryptomorphic.

(c5)   If $X \subseteq E$, $|X| = \infty$, then there exists $C \in \mathcal{C}(E)$ such that $C \subseteq X$.

(d5)   If $X \subseteq E$, $|X| = \infty$, then $X \in \mathcal{D}$.

(i5)   For every $I \in \mathcal{I}(E)$, $|I| < \infty$.

(b5)   For every $B \in \mathcal{B}(E)$, $|B| < \infty$.

(s5)   If $S \in \mathcal{S}(E)$, then there exists $S' \in \mathcal{S}(E)$, $S' \subseteq S$, $|S'| < \infty$.

(n5)   If $X \subseteq E$, $X \notin \mathcal{N}(E)$, then there exists $X' \notin \mathcal{N}(E)$, $X' \subseteq X$, $|X'| < \infty$.

(h5)   If $X \subseteq E$, and if for every $H \in \mathcal{H}(E)$, $X \nsubseteq H$, then there exists $X' \subseteq X$, $|X'| < \infty$, such that for every $H \in \mathcal{H}(E)$, $X' \nsubseteq H$.

(cl6)   If $X \subseteq E$, there exists $X' \subseteq X$, $|X'| < \infty$, $cl(X') = cl(X)$.

(r5)   If $X \subseteq E$, there exists $X' \subseteq X$, $|X'| < \infty$, $r(X') = r(X)$.

EXERCISES

Many of these exercises are covered in later chapters or in the Appendix of Matroid Cryptomorphisms.

2.1. If $\mathcal{B}$ is a nonempty antichain of subsets of E, prove that (b3) and (b3.1) are equivalent to the following apparently stronger axiom:

(b3.2) For every $B_1, B_2 \in \mathcal{B}$, and for every $b_1 \in B_1$, there exists $b_2 \in B_2$ such that $(B_1 - b_1) \cup b_2 \in \mathcal{B}$ and $(B_2 - b_2) \cup b_1 \in \mathcal{B}$.

2.2. Prove Corollary 2.1.2.

2.3. Let E be a set and $\mathcal{L}$ a collection of subsets of E (called lines) such that the following hold:

(l1) There exist three elements of E that are not contained in a common line.

(l2) For every $e_1, e_2 \in E$, there exists a unique line $l \in \mathcal{L}$ such that $e_1 \cup e_2 \subseteq l$.

Prove that $(E, \mathcal{B})$ forms a matroid of rank 3, where $\mathcal{B} = \{B \subseteq E: |B| = 3,$ and $B \nsubseteq l$ for every $l \in \mathcal{L}\}$.

2.4. Prove that if $cl \in \mathbf{cl}(E)$, and $X \subseteq E$ is such that $cl(X) = X$, then $X = \cap\{H: X \subseteq H, H \in \beta(cl)\}$. Thus complete the proof of Proposition 2.3.1.

2.5. Prove Proposition 2.4.1.

2.6. Prove Proposition 2.4.2.

2.7. A *flat* in a matroid $M(E)$ is a set $X \subseteq E$ such that $cl(X) = X$. Prove that the following axioms for the family $\mathcal{F}$ of flats are cryptomorphic to the other axiom systems for matroids. Specifically, find cryptomorphisms in both directions between **F** and each of **H**, **cl**, **r**, and **C**.

(F1) $E \in \mathcal{F}$.

(F2) If $F_1, F_2 \in \mathcal{F}$, then $F_1 \cap F_2 \in \mathcal{F}$.

(F3) If $F \in \mathcal{F}$, let $\mathcal{K}_F$ be the family of flats that cover F; that is, $\mathcal{K}_F = \{F': F' \in \mathcal{F}, F' \supsetneq F, \text{and } F' \supsetneq G \supsetneq F \text{ implies } G \notin \mathcal{F}\}$. Then $\{F' - F: F' \in \mathcal{K}_F\}$ is a partition of $E - F$.

2.8. Find an axiom system for open sets that is cryptomorphic to all the axiom systems for matroids, where a set is *open* if it is the complement of a flat.

REFERENCES

Brylawski, T., and Kelly, D. (1980). *Matroids and Combinatorial Geometries.* University of North Carolina, Chapel Hill.

Crapo, H., and Rota, G.-C. (1970). *On the Foundations of Combinatorial Theory: Combinatorial Geometries,* preliminary edition. M.I.T. Press, Cambridge, Mass.

Klee, V. (1971). The greedy algorithm for finitary and cofinitary matroids. *Combinatorics, Proc. Symposia in Pure Math., Amer. Math. Soc.* **19**: 137–52.

Nicoletti, G. (1979). Generating cryptomorphic axiomatizations of matroids, in *Lecture Notes in Mathematics, Vol. 792, Geometry and Differential Geometry,* edited by R. Artzy and I. Vaisman, pp. 110–13. Springer-Verlag, Berlin.

Welsh, D. (1976). *Matroid Theory.* Academic Press, London.

Whitney, H. (1935). On the abstract properties of linear dependence. *Amer. J. Math.* **57**: 509–33.

CHAPTER 3

Lattices

Ulrich Faigle

One of the ways to describe a matroid $M(E)$ in Chapter 2 was by means of a closure operator cl on the set of subsets of E. This closure operator distinguishes a *closed set* or *flat* of the matroid $M(E)$ as a set $T \subset E$ with the property $T = \mathrm{cl}(T)$. In this chapter we want to study the collection $L(M)$ of flats of $M(E)$ and find out how much of the structure of $M(E)$ is reflected in the structure of $L(M)$.

$L(M)$ is (partially) ordered by set-theoretic inclusion. Furthermore, there are two natural binary operations on $L(M)$: For every pair T_1, T_2 of flats of $M(E)$, define $T_1 \vee T_2$ as the smallest flat containing both T_1 and T_2, and $T_1 \wedge T_2$ as the largest flat contained in both T_1 and T_2. [It is not difficult to see that these operations are well defined. In fact, $T_1 \vee T_2 = \mathrm{cl}(T_1 \cup T_2)$, and $T_1 \wedge T_2 = T_1 \cap T_2$.]

The appropriate setting for study of this algebraic structure on $L(M)$ is *lattice theory*. So we shall briefly introduce the concept of a lattice and then proceed to investigate special classes of lattices. The guiding example thereby is the lattice of subspaces of a finite-dimensional vector space. Such a lattice is, in particular, *modular*. Hence, we shall look at the concept of modularity in a lattice, with special emphasis on the notion of a *modular pair* of elements. This leads to *semimodular* and *M-symmetric* lattices. With the additional property that every element of a lattice be generated by *atoms*, we arrive at the class of *geometric* lattices. This class now provides all the tools we need to study matroids from a lattice point of view: The theories of geometric lattices and of combinatorial geometries are equivalent (Theorem 3.4.1). Indeed, this equivalence is another cryptomorphism, as discussed in Chapter 2. Therefore, the decomposition theory for geometric lattices can directly be interpreted to yield a decomposition theory for matroids.

Chapter 1 introduced combinatorial geometries as structures sharing the "same" properties as point sets in projective space. We have obtained geometric lattices by retaining basic properties of atomistic modular lattices. The final section of this chapter is devoted to showing that in both cases we really start with the same object: An indecomposable atomistic modular lattice can be thought of as a projective geometry, and vice versa.

A note about our terminology: Because the atoms of a modular geometric lattice can be viewed as points of a projective geometry, atomistic lattices are sometimes simply called *point lattices*. In our lattice-theoretic approach to matroid theory, however, we shall adhere to the terms more commonly used in lattice theory and thus give preference to *atom* and *atomistic*.

3.1. POSETS AND LATTICES

A *(partially) ordered set* or *poset* $P = (P, \leq)$ is a set P endowed with a binary relation $\leq$ ("less than or equal to") satisfying the following:

(PO1) $x \leq x$ for all $x \in P$ *(reflexivity)*.
(PO2) $x \leq y$ and $y \leq x$ imply $x = y$ *(antisymmetry)*.
(PO3) $x \leq y$ and $y \leq z$ imply $x \leq z$ *(transitivity)*.

Every subset of a poset P is, of course, again a poset with respect to the induced order. In particular, for every two elements $x, y \in P$, $x \leq y$, the *interval* $[x, y] := \{z \in P : x \leq z \leq y\}$ is a poset. We say that y *covers* x if $[x, y] = \{x, y\}$. We say that x and y are *comparable* if either $x \leq y$ or $y \leq x$.

For visual conception it is convenient to picture a poset by its *Hasse diagram*, whose points represent the elements of the poset and whose lines indicate the covering relation. In Figure 3.1 we display three posets by their Hasse diagrams.

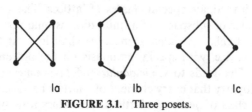

la   lb   lc
FIGURE 3.1. Three posets.

A nonempty subset C of the poset P is a *chain* if any two elements of C are comparable with respect to the order induced by P; $l(C) := |C| - 1$ is the *length* of the chain C. C is *maximal* if for every $x \in P - C$, $C \cup x$ is not a chain.

A familiar feature of the collection of subspaces of a vector space is that the dimension of a subspace S is the length of a maximal chain of subspaces of S. Viewed this way, the dimension of a subspace finds its analogue in the height of an element in an arbitrary poset.

If a is an element of the poset P, we define the *height* $h(a)$ by

$$h(a) := \sup\{l(C): C \text{ chain in } P(a)\},$$

where $P(a) = \{x \in P: x \leq a\}$; $h(P) = \sup\{h(a): a \in P\}$ is the *length* of P.

The height function of a poset P is of particular importance when $h(P)$ is finite. As we shall see in Section 3.3, many properties of lattices of finite length can be stated as properties of the associated height functions.

The element x of the poset P is *maximal* (*minimal*) if $x \leq y$ ($x \geq y$) implies $x = y$ for all $y \in P$; x is the *greatest* (*least*) element of P if $y \leq x$ ($y \geq x$) for all $y \in P$. Thus, the greatest element, if it exists, is maximal, but the converse is not necessarily true (Figure 3.1a).

We now come to the central notion of this chapter.

A *lattice* $L = (L, \vee, \wedge)$ is a poset L such that for every two elements $x, y \in L$, the poset $\{z \in L: z \geq x, z \geq y\}$ has a least element $x \vee y \in L$, the *join* of x and y, and the poset $\{z \in L: z \leq x, z \leq y\}$ has a greatest element $x \wedge y \in L$, the *meet* of x and y.

Join and meet are associative and commutative binary operations on the lattice L and thus turn the poset L into an algebraic structure. Moreover, the order on L can be recaptured from the join and meet operations:

$$x \leq y \quad \text{if and only if } x \vee y = y$$

$$\text{if and only if } x \wedge y = x.$$

The collection of subsets of a set, ordered by containment, forms a lattice, and so does the collection of subspaces of a vector space. We remarked at the beginning of this chapter that the collection of flats of a matroid is a lattice. However, not every poset is also a lattice. The Hasse diagram of Figure 3.1a does not represent a lattice.

The notions of join and meet readily extend to arbitrary nonempty subsets of a poset P. So if A is a nonempty subset of P and the poset $\{z \in P: z \geq a$ for every $a \in A\}$ has a least element $\vee A$, we call $\vee A$ the *join* of A. Similarly, we can talk about the *meet* $\wedge A$ of A if it exists. It is not difficult to see that in a finite lattice, join and meet exist for every nonempty subset.

The example of the rational numbers Q with the usual order shows that the join $\vee A$ or meet $\wedge A$ of an infinite subset A of a lattice need not always be defined. That is, a lattice need not be *complete*. This situation cannot arise when there are no infinite chains in the lattice.

3.1.1. Proposition. *If L is a lattice with no infinite chains, then L is complete. In particular, L has the finite-basis property; that is, for every subset $A \subset L$, there is a finite subset $A_f \subset A$ such that $\vee A = \vee A_f$.*

Proof: Let A be any nonempty subset of L, and choose $a_0 \in A$. If $a \leq a_0$ for all $a \in A$, we are done. If not, choose $a_1 \in A$, with $a_1 \not\leq a_0$, and let $\bar{a}_1 = a_0 \vee a_1$. Then $a_0 < \bar{a}_1$. If possible, choose $a_2 \in A$, with $a_2 \not\leq \bar{a}_1$, and let $\bar{a}_2 = \bar{a}_1 \vee a_2$. Then $a_0 < \bar{a}_1 < \bar{a}_2$, and so forth. Because L contains no infinite chains, this procedure must terminate after a finite number of steps with an element $\bar{a}_n$, necessarily the join of A.

The existence of the meet of A is demonstrated the same way. □

For a matroid $M(E)$ with E infinite, $L(M)$ has the finite-basis property precisely when M has finite bases (see Proposition 2.4.2).

Note that a complete lattice L possesses a greatest element $1 = \vee L$ and a least element $0 = \wedge L$.

A subset L' of a lattice L is a *join-subsemilattice* of L if $x \vee y \in L'$ for every $x, y \in L'$. Similarly, we define a *meet-subsemilattice*. A *sublattice* of the lattice L is a subset $L' \subset L$ that is both a join- and a meet-subsemilattice. Thus, a subset L' of L need not be a sublattice of L even if L' is a lattice in its own right with respect to the induced order. However, every interval of a lattice apparently is a sublattice.

The concept of a sublattice is sometimes useful if one wishes to characterize certain lattices in terms of excluded substructures. We illustrate this with the two important classes of distributive and modular lattices.

A lattice L is said to be *distributive* if L satisfies

(D) $x \wedge (y \vee z) = (x \wedge y) \vee (x \wedge z)$ for all $x, y, z \in L$.

More generally, a lattice L is *modular* if L satisfies

(M) $x \geq z$ implies $x \wedge (y \vee z) = (x \wedge y) \vee z$ for all $x, y, z \in L$.

For example, the lattice of subsets of a set is distributive, and we invite the reader to show that the lattice of subspaces of a finite-dimensional vector space, although in general not distributive, is modular.

Every sublattice of a modular (distributive) lattice is modular (distributive). So the "only if" part of the next characterization theorem is obvious. The "if" part is more surprising.

3.1.2. Theorem. (a) *A lattice L is modular if and only if L does not contain the lattice N_5 of Figure 3.1b as a sublattice.* (b) *A lattice L is distributive if and only if L is modular and does not contain the lattice M_3 of Figure 3.1c as a sublattice.*

Proof. We just sketch the main ideas for the proof of statement (a). The proof of statement (b) can be carried out similarly. The computations, however, are a little more involved; see, for example, Crawley and Dilworth (1973).

First observe that if $x \geq z$, the relation $x \wedge (y \vee z) \geq (x \wedge y) \vee z$ holds in every lattice. Suppose now that $x \wedge (y \vee z) > (x \wedge y) \vee z$. Then one can check that the elements $y, y \vee z, x \wedge y, x \wedge (y \vee z)$, and $(x \wedge y) \vee z$ form a sublattice N_5 of L. $\square$

3.2. MODULARITY

In the last section we gave the lattice of subspaces of a finite-dimensional vector space as an example of a modular lattice. We shall now investigate the concept of modularity in a lattice in more detail. Our goal is to show in the subsequent sections that modularity is a suitable concept to study combinatorial geometry from a purely lattice-theoretic point of view.

An element a of an arbitrary lattice L is a *modular element* of L if

(ME) $x \geq z$ implies $x \wedge (a \vee z) = (x \wedge a) \vee z$ for every $x, z \in L$.

More generally, we say that two elements a and b form a *modular pair*, denoted $(a, b)M$, if

(MP) $b \geq z$ implies $b \wedge (a \vee z) = (b \wedge a) \vee z$ for every $z \in L$.

Thus, a lattice is modular exactly when every element is a modular element or, equivalently, when every pair of elements is a modular pair.

Modular pairs can be characterized in terms of associated order-preserving maps.

3.2.1. Proposition. *Let a and b be elements of the lattice L. Then the following conditions are equivalent:*

(i) $(a, b)M$.
(ii) $\phi_b \colon [a, a \vee b] \to [a \wedge b, b] \colon x \to x \wedge b$ *is surjective.*
(iii) $\psi_a \colon [a \wedge b, b] \to [a, a \vee b] \colon x \to x \vee a$ *is injective.*

Proof. (i) implies (ii): For every $x \in [a, a \vee b]$ we have $\phi_b(a \vee x) = (a \vee x) \wedge b = x \vee (a \wedge b) = x$, because $b \geq x$.

(ii) implies (iii): Suppose there are elements $x, y \in [a \wedge b, b]$ such that $x \neq y$ but $\psi_a(x) = \psi_a(y)$. Then we must have $x < x \vee y$ or $y < x \vee y$. Let us assume the first, and choose $z \in [a, a \vee b]$ such that $\phi_b(z) = z \wedge b = x$. Thus, $z \geq a$ and $z \geq x$; hence, $z \geq a \vee x = a \vee y \geq y$. Consequently, $z \geq x \vee y$, and therefore $\phi_b(z) \geq (x \vee y) \wedge b = x \vee y > x$, a contradiction to the choice of z.

(iii) implies (i): If $(a, b)M$ does not hold, there exists $x \leq b$ such that $x \vee (a \wedge b) < (x \vee a) \wedge b$. But then $x \vee a = (a \vee x) \vee (a \wedge b) = \psi_a[x \vee (a \wedge b)] \leq$

$\psi_a[(x \lor a) \land b] = [(x \lor a) \land b] \lor a \le x \lor a$. Hence, $\psi_a[x \lor (a \land b)] = \psi_a[(x \lor a) \land b]$, and ψ_a is seen to be not injective. $\quad\square$

If L is a modular lattice, the order-preserving maps ϕ_b and ψ_a are inverses of each other (see Exercise 3.5). In particular, the intervals $[a \land b, b]$ and $[a, a \lor b]$ are *isomorphic*. Therefore, we obtain the following as an immediate consequence of Proposition 3.2.1:

3.2.2. Theorem (transposition principle). *A lattice L is modular if and only if for every two elements a and b of L, the intervals $[a \land b, b]$ and $[a, a \lor b]$ are isomorphic via the isomorphisms ϕ_b and ψ_a.*

If we require the transposition principle to hold only for elements a and b such that b covers $a \land b$, we arrive at the class of (upper) semimodular lattices. That is, a lattice L is *semimodular* if for all elements a and b of L,

(SM) b covers $a \land b$ implies $a \lor b$ covers a.

The notion of a modular pair suggests another generalization of the idea of a modular lattice. A lattice L is called *M-symmetric* if for all elements a and b of L,

(MS) $(a, b)M$ implies $(b, a)M$.

Although at first sight M-symmetry seems to point in a different direction than semimodularity, our next proposition will reveal the class of M-symmetric lattices to be contained in the class of semimodular lattices. This containment is proper (see Exercise 3.7). However, in the next section we shall see that within the class of lattices of finite length, every semimodular lattice is also M-symmetric.

3.2.3. Proposition. *If the lattice L is M-symmetric, then L is semimodular.*

Proof. Let $x, y \in L$ be elements such that x covers $x \land y$. Suppose $x \lor y$ does not cover y; that is, suppose there exists $z \in L$ with $y < z < x \lor y$. Because $x \land y \le x \land z \le x \land (x \lor y) = x$, we have $x \land z = x \land y$ ($x \land z = x$ would imply $x \le z$ and hence $x \lor y \le z$). Similarly, $x \lor z = x \lor y$. But then the map ϕ_x: $[z, x \lor y] \to [x \land y, x]$ is surjective. So Proposition 3.2.1 allows us to conclude $(z, x)M$. From the M-symmetry of L, we therefore obtain $(x, z)M$. Now, $y \le z$ implies $z = z \land (x \lor y) = (z \land x) \lor y = y$, a contradiction to the choice of z. $\quad\square$

Theorem 3.1.2 raises the question whether or not a characterization of semimodular lattices by excluded finite sublattices is possible. Dilworth has given a negative answer: Every finite lattice can occur as a sublattice of a semimodular lattice; see Crawley and Dilworth (1973, Chapter 14). For a characterization of semimodular lattices in terms of "saturated" sublattices, see Vilhelm (1955).

3.3. SEMIMODULAR LATTICES OF FINITE LENGTH

Thus far we have no finiteness condition imposed on the lattices under consideration. From now on, however, we shall restrict our attention to lattices of finite length. This loss of generality offers its reward: The property of a lattice of finite length to be semimodular can be understood as a property of the associated height function.

Recalling Proposition 3.1.1, we observe that a lattice L of finite length is complete with greatest element 1 and least element 0. Then $h(L)$ is the length of a longest chain between 0 and 1. But we point out that $h(L)$ is not necessarily equal to the length of an arbitrary maximal chain (see the lattice N_5 of Figure 3.1b). Fortunately, such complications cannot occur in semimodular lattices of finite length.

3.3.1. Theorem (Jordan-Dedekind chain condition). *Let L be a semimodular lattice of finite length and C and C' any two maximal chains in L. Then $l(C) = l(C')$.*

Proof. We proceed by induction on $h(L)$. The theorem obviously holds if $h(L) \leq 1$. For the two maximal chains $C = \{0 < c_1 < \ldots < 1\}$ and $C' = \{0 < c_1' < \ldots < 1\}$, we can assume $c_1 \neq c_1'$. If not, we obtain the theorem from the induction hypothesis on the semimodular lattice $[c_1, 1]$, restricting C and C' to $[c_1, 1]$. Because C and C' are maximal chains, c_1 and c_1' both cover $c_1 \wedge c_1' = 0$, and therefore $c_1 \vee c_1'$ covers both c_1 and c_1'. Let C'' be any maximal chain from $c_1 \vee c_1'$ to 1 in $[c_1 \vee c_1', 1]$. Then $C - 0$ and $C'' \cup c_1$ are maximal chains in $[c_1, 1]$. Hence, by the induction hypothesis, $l(C) - 1 = l(C'') - 1$, and, similarly, $l(C') - 1 = l(C'') - 1$ (Figure 3.2). $\square$

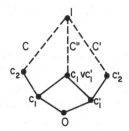

FIGURE 3.2. Maximal chains in the proof of Theorem 3.3.1.

Theorem 3.3.1 already hints at the importance of the height function for semimodular lattices of finite length. In fact, those lattices can be characterized by their height function.

3.3.2. Theorem. *Let L be a lattice of finite length. Then the following three conditions on L are equivalent:*

(i)   *L is semimodular.*
(ii) *If C is a maximal chain in the interval $[a, b]$, then for every $c \in L$, $\{x \vee c : x \in C\}$ is a maximal chain in $[a \vee c, b \vee c]$.*
(iii) *$h(a \vee b) + h(a \wedge b) \leq h(a) + h(b)$ for all $a, b \in L$.*

Proof. (i) implies (ii): Apparently, it suffices to show that $y \vee c$ covers or equals $x \vee c$ if y covers x in C. Clearly, $y \vee c = x \vee c$ if $y \leq x \vee c$. If $y \not\leq x \vee c$, then we must have $x = y \wedge (x \wedge c)$, and therefore $y \vee (x \vee c) = y \vee c$ covers $x \vee c$ by semimodularity.

(ii) implies (i): If x covers $x \wedge y$, then $y = (x \wedge y) \vee y < x \vee y$ is, by assumption, a maximal chain between y and $x \vee y$; that is, $x \vee y$ covers y.

(i) and (ii) imply (iii): By the Jordan-Dedekind chain condition, any maximal chain C in $[a \wedge b, b]$ satisfies $l(C) = h(b) - h(a \wedge b)$. The same is true for the interval $[(a \wedge b) \vee a, b \vee a]$. Thus, by (ii), $h(b) - h(a \wedge b) \geq h(a \vee b) - h(a)$, because $(a \wedge b) \vee a = a$.

(iii) implies (i): Suppose $x \in L$ is of smallest height so that the semimodularity condition (SM) fails for x. Then $[0, x]$ must be a semimodular lattice. Let y be any element such that x covers $x \wedge y$. Now, $h(x) - h(x \wedge y) = 1$, because all maximal chains in $[0, x]$ are of equal length. But then $h(x \vee y) - h(y) \leq 1$. Consequently, because $x \vee y \neq y$, $x \vee y$ must cover y.   □

By the definition of closure in a matroid, the height of a flat equals its matroid rank. So Theorem 3.3.2 implies that the lattice of flats of a matroid is semimodular. And we remark that for this reason the "height" in a lattice is often also called "rank."

Our next observation yields a characterization of modular pairs in semimodular lattices in terms of the height function.

3.3.3. Theorem. *Let a and b be two arbitrary elements of the semimodular lattice L of finite length. Then $(a, b)M$ if and only if $h(a \vee b) + h(a \wedge b) = h(a) + h(b)$.*

Proof. Assume $(a, b)M$, and let C be a maximal chain in $[a \wedge b, b]$. By Theorem 3.3.2, $\psi_a(C)$ is a maximal chain in $[a, a \vee b]$. Moreover, by Proposition 3.2.1, ψ_a is injective. Hence, $l(C) = l[\psi_a(C)]$, implying $h(a \vee b) - h(a) = h(b) - h(a \wedge b)$, because the Jordan-Dedekind chain condition holds in L.

Conversely, if $(a, b)M$ does not hold, as in the proof "(iii) implies (i)" of Proposition 3.2.1, there are elements $z, z' \in [a \wedge b, b]$, $z < z'$, with $\psi_a(z) =$

$\psi_a(z')$. Let C be a maximal chain in $[a \wedge b, b]$ containing z and z'. Then $l(C) > l[\psi_a(C)]$. Hence, $h(a \vee b) - h(a) < h(b) - h(a \wedge b)$. $\qquad\square$

Consequently, using Proposition 3.2.3, we obtain the following:

3.3.4. Corollary. *Let L be a lattice of finite length. Then L is semi-modular if and only if L is M-symmetric.*

Another immediate consequence of Theorem 3.3.3 is the following:

3.3.5. Corollary. *Let L be a lattice of finite length. Then L is modular if and only if for every $a, b \in L$, $h(a \vee b) + h(a \wedge b) = h(a) + h(b)$.*

3.4. GEOMETRIC LATTICES

It is an almost trivial observation about vector spaces that every subspace can be expressed as a sum of one-dimensional subspaces. However, as the example of a finite chain of length $l > 1$ shows, the analogous statement is generally false even for distributive lattices.

If we now further restrict our attention to semimodular lattices of finite length for which the analogous statement is true, we are in a position to deal with combinatorial geometry as a branch of lattice theory.

Let L be a lattice of finite length. An *atom* of L is an element covering 0. In other words, the atoms of L are just the elements a of L with $h(a) = 1$. L is *atomistic* if every element can be expressed as a join of atoms. Note that by Proposition 3.1.1, every element of the atomistic lattice L is a join of a finite number of atoms ($0 \in L$ is obtained as the empty join).

We call an atomistic semimodular lattice (of finite length) *geometric*.

Because every flat of a matroid is the join of the flats of rank 1 that it contains, the lattice of flats of a matroid is geometric. In fact, every geometric lattice can be thought of as the lattice of flats of some matroid.

3.4.1. Theorem. *Let A be the set of atoms of the geometric lattice L. For any subset $X \subset A$, let $r(X) := h(\vee X)$. Then r satisfies the local rank axioms (see Chapter 2, Section 2.3) of a combinatorial geometry $G_L(A)$. Moreover, $X \subset A$ is a flat of $G_L(A)$ if and only if $X = \{a \in A: a \leq x\}$ for some $x \in L$.*

Proof. To show axiom (r'2), let $z := \vee Z$ for $Z \subset A$. Then $h(z) \leq h(z \vee x)$ for every $x \in A$, because $z \leq z \vee x$. By the semimodular property of the height function, we have $h(z \vee x) + h(z \wedge x) \leq h(z) + h(x)$. Thus, $h(z \vee x) \leq h(z) + 1$.

For axiom (r'3), observe that, by assumption, $z \vee x = z$ and $z \vee y = z$. Hence, $z \vee x \vee y = z$, and (r'3) follows.

If $X \subset A$ is an arbitrary set of atoms, let $x := \vee X$. Then for every $a \in A$, $r(X \cup a) = r(X)$ if and only if $h(x \vee a) = h(x)$ if and only if $a \leq x$. $\qquad\square$

Theorem 3.4.1 suggests that we interpret the concepts introduced for matroids in Chapter 2 in arbitrary geometric lattices. So we say that a set $\{x_1, \ldots, x_n\}$ of atoms of a geometric lattice is *independent* if $h(x_1 \vee \cdots \vee x_n) = n$. A *basis* is a maximal independent set and a *circuit* a minimal dependent ($=$ not independent) set of atoms. Thus, in view of Theorem 3.4.1 and the propositions of Chapter 2, the rank, circuit, independence, and basis axioms for matroids are satisfied in every geometric lattice.

If $L = L(M)$ is the lattice of flats of a matroid M of finite rank (and possibly infinite cardinality), the atoms of L correspond to the flats of rank 1 of M; that is, the atoms of L can be thought of as the points of the combinatorial geometry $\bar{M}$ associated with M, that is, the *simplification* of M, as described in Chapter 7, Section 7.4. Because a set of atoms is independent in $G_L(A)$ if and only if the corresponding set of points is independent in $\bar{M}$, we conclude that $G_L(A)$ and $\bar{M}$ are isomorphic. We also observe that, by the theorem, L is isomorphic to the lattice of flats of $G_L(A)$. Theorem 3.4.1 therefore establishes a one-to-one correspondence between the isomorphism classes of geometric lattices and the isomorphism classes of combinatorial geometries. In other words, the theories of geometric lattices and combinatorial geometries are equivalent.

So let us look at geometric lattices in more detail.

It follows immediately from the definition that every interval of a semimodular lattice again is a semimodular lattice. Thus, the next proposition is straightforward.

3.4.2. Proposition. *Every interval* $[a, b]$ *of a geometric lattice L is a geometric lattice.*

Proof. By the foregoing remark, we must show only that the interval $[a, b]$ of L is atomistic. So let $x \in [a, b]$, $x \neq a$, be an arbitrary element, and consider atoms $x_1, \ldots, x_n$ of L such that $x = x_1 \vee \cdots \vee x_n$. Then $x = x \vee a = (x_1 \vee a) \vee \cdots \vee (x_n \vee a)$. By the semimodular property of L, $x_i \vee a$ covers a if $x_i \wedge a = 0$. If $x_i \wedge a = x_i$, then $x_i \vee a = a$. Thus, we get $x = \vee\{(x_i \vee a): x_i \not\leq a\}$ as a join of atoms of $[a, b]$. □

Proposition 3.4.2 describes one way to derive a geometric lattice from a given one. Another important construction is the content of Proposition 3.4.3.

3.4.3. Proposition. *Let B be a subset of atoms of the geometric lattice L and $L(B)$ the set of elements of L that can be expressed as joins of elements of B. Then the join-subsemilattice $L(B)$ of L is a geometric lattice with respect to the order induced by L.*

Proof. $L(B)$ is an atomistic lattice of finite length. To verify the semimodular property (SM), choose arbitrary elements $x, y \in L(B)$ such that y covers $x \wedge y$

in $L(B)$. Then there must exist an element $b \in B$ such that $b \le y$ and $b \nleq x$; that is, $y = (x \wedge y) \vee b$ and $x \vee y = (x \wedge y) \vee x \vee b = x \vee b$. Now $x \vee b$ covers x in L because L is geometric. Thus, a fortiori, $x \vee b$ must cover x in $L(B)$. □

In view of the aforementioned correspondence between geometric lattices and combinatorial geometries, one may expect Proposition 3.4.2 and Proposition 3.4.3 to reflect natural constructions on matroids. Indeed, as we shall see in Chapter 9, passing to a "minor" of a matroid corresponds to passing to a join-subsemilattice, in the sense of Proposition 3.4.3, of an interval in the associated lattice of flats.

The preceding discussion of geometric lattices has presented geometric lattices as generalizations of atomistic modular lattices of finite length by dropping the requirement of global modularity in favor of the weaker assumption of M-symmetry. Nevertheless, as the properties of the height function show, many features of the lattice of subspaces of a finite-dimensional vector space are still retained. We give another illustration of this fact.

A lattice L is *complemented* if every element $x \in L$ has a *complement* $y \in L$, where $x \vee y = 1$ and $x \wedge y = 0$. L is *relatively complemented* if every interval of L is complemented.

3.4.4. Proposition. *A geometric lattice L is relatively complemented. In fact, a complement y of the element $x \in L$ can always be chosen so as to satisfy $h(x \vee y) + h(x \wedge y) = h(x) + h(y)$.*

Proof. In view of Proposition 3.4.2 it suffices to show that a geometric lattice L is complemented.

Let $x \in L$ be arbitrary, and let $x_1, \ldots, x_n$ be independent atoms of L such that $x = x_1 \vee \cdots \vee x_n$. Choose atoms y_i of L such that $\{x_1, \ldots, x_n, y_1, \ldots, y_k\}$ is a basis, and let $y = y_1 \vee \cdots \vee y_k$. Then $x \vee y = 1$, and $h(x \wedge y) \le h(x) + h(y) - h(x \vee y) = 0$ implies $x \wedge y = 0$. □

We remark that the converse of Proposition 3.4.4 is also true in the following sense: A relatively complemented semimodular lattice of finite length is geometric (see Exercise 3.10).

The concept of complementation in a geometric lattice L gives rise to an equivalence relation on the set of atoms of L that will prove useful for the decomposition theory in the next section.

Let us call two atoms a and b of a geometric lattice L *perspective*, in notation $a \sim b$, if they share a common complement $x \in L$: $a \vee x = b \vee x = 1$ and $a \wedge x = b \wedge x = 0$.

We leave it as an easy exercise (Exercise 3.12) to show that $a \sim b$ if and only if there is an element $x \in L$ such that $a \vee x = b \vee x$ and $a \wedge x = b \wedge x = 0$.

There is another way of looking at perspectivity.

3.4.5. Proposition. *Let a and b be atoms of the geometric lattice L. Then a ~ b if and only if there is a circuit C of L containing both a and b.*

Proof. Assume $a \sim b$, and choose $x \in L$ so as to minimize $h(x) = m$ under the condition $a \vee x = b \vee x$ and $a \wedge x = b \wedge x = 0$. Let $x_1, \ldots, x_m$ be the elements of a basis of $[0, x]$. Then $C = \{a, b, x_1, \ldots, x_m\}$ is dependent, because $h(a \vee b \vee x) = h(a \vee x) = m + 1 < m + 2$. We must show that every proper subset of C is independent.

Set $x^i := x_1 \vee \cdots \vee x_{i-1} \vee x_{i+1} \vee \cdots \vee x_m$, $1 \le i \le m$. Then $b \le a \vee x^i$ will imply $a \vee x^i = b \vee x^i$, because $h(a \vee x^i) = 1 + (m-1) = h(b \vee x^i)$, a contradiction to the choice of x. So $a \vee b \vee x^i = a \vee x = b \vee x$, $1 \le i \le m$. Because $h(a \vee b \vee x^i) = m + 1$, every proper subset of C must therefore be independent.

Conversely, if $C = \{a, b, x_1, \ldots, x_m\}$ is a circuit of L, set $x := x_1 \vee \cdots \vee x_m$. Now $a \vee x = a \vee b \vee x = b \vee x$, because $h(a \vee x) = h(b \vee x) = h(a \vee b \vee x)$, and $a \wedge x = 0 = b \wedge x$, because $h(a \wedge x) \le h(a) + h(x) - h(a \vee x) = 0$, and, similarly, $h(b \wedge x) = 0$.   □

We are now in a position to substantiate our claim that perspectivity defines an equivalence relation.

3.4.6. Proposition. *In a geometric lattice L, perspectivity of atoms is an equivalence relation.*

Proof. Perspectivity certainly is reflexive and symmetric. Suppose it is not transitive. Using the characterization of Proposition 3.4.5, we shall show that this assumption leads to a contradiction.

Choose circuits C_1 and C_2 of L such that $|C_1 \cup C_2|$ is minimized subject to the condition that there are atoms $a, b, c \in L$, $a, b \in C_1$, $b, c \in C_2$, but a and c lie in no common circuit. By the strong circuit-elimination axiom (see Proposition 2.2.5) there is a circuit $C_1' \subset (C_1 \cup C_2) - b$, with $a \in C_1'$. Because $C_1 - C_2$ is an independent set, $C_1' \subset C_1 - C_2$ is impossible. Hence, C_1' and C_2 intersect. By the choice of C_1 and C_2, we must have $|C_1' \cup C_2| = |C_1 \cup C_2|$, and, in particular, $C_1 - C_2 \subset C_1'$. Similarly, we can find a circuit $C_2' \subset (C_1 \cup C_2) - b$ such that $c \in C_2'$. Now C_2' must meet $C_1 - C_2$, and consequently C_1' and C_2' have a nonempty intersection. But then $|C_1' \cup C_2'| < |C_1 \cup C_2|$, in contradiction to the choice of C_1 and C_2.   □

3.5. DECOMPOSITION OF GEOMETRIC LATTICES

The *direct product* $P = P_1 \times P_2$ of two posets P_1 and P_2 is the Cartesian product $P = P_1 \times P_2$ ordered by

$$(x_1, x_2) \le (y_1, y_2) \quad \text{if} \quad x_1 \le y_1 \text{ and } x_2 \le y_2.$$

One immediately verifies that P is a lattice if and only if both P_1 and P_2 are lattices. Moreover, the join and meet operations are

$$(x_1, x_2) \vee (y_1, y_2) = (x_1 \vee y_1, x_2 \vee y_2)$$

and

$$(x_1, x_2) \wedge (y_1, y_2) = (x_1 \wedge y_1, x_2 \wedge y_2).$$

In particular, we observe the following:

3.5.1. Proposition. *Let $L = L_1 \times L_2$ be the direct product of two lattices L_1 and L_2. Then L is geometric (M-symmetric, semimodular, modular, distributive) if and only if both L_1 and L_2 are geometric (M-symmetric, semimodular, modular, distributive).*

Proof. Exercise 3.13.   □

Proposition 3.5.1 suggests that we look for geometric lattices that cannot be represented as direct products of smaller lattices. Those geometric lattices can then be viewed as building blocks for geometric lattices in general.

We say that a lattice L with 0 and 1 is *decomposable* if there are non-trivial intervals $L_1 = [0, z_1]$ and $L_2 = [0, z_2]$ such that for every element $x \in L$, there is a unique pair $(x_1, x_2) \in L_1 \times L_2$, with $x = x_1 \vee x_2$. Note that this is equivalent to saying that L is isomorphic to $L_1 \times L_2$.

3.5.2. Theorem. *A geometric lattice L is indecomposable if and only if every two atoms of L are perspective.*

Proof. We first note that if $A \cup p$ is a set of atoms of L with $p \leq \vee A$, then $p \sim a$ for some $a \in A$. Indeed, choose a minimal independent subset $A' \subset A$ such that $p \leq \vee A'$. Then $A' \cup p$ is a dependent set of atoms, but every proper subset of $A' \cup p$ is independent, as one easily verifies. So $A' \cup p$ is a circuit. Hence, $p \sim a$ for every $a \in A'$. This settles the case $p \notin A$. The case $p \in A$ is trivial.

Assume now that not every pair of atoms of L is perspective, and choose an atom $a \in L$. Let $[a] = \{b \in L : b \sim a\}$ and $z_1 = \vee[a]$. We have just shown that $[a]$ is the set of atoms of $[0, z_1]$. Letting $z_2 \in L$ be equal to the join of all atoms of L that are not in $[0, z_1]$, we claim that $L_1 = [0, z_1]$ and $L_2 = [0, z_2]$ yield a decomposition of L.

To see this, choose an arbitrary $x \in L$, and set $x_1 = x \wedge z_1$ and $x_2 = x \wedge z_2$. Because the sets of atoms of L_1 and L_2 are disjoint, the atoms of $[0, x_1]$ and $[0, x_2]$ are complementary subsets of the atoms of $[0, x]$. So $x = x_1 \vee x_2$. Let X_1 be a basis in $[0, x_1]$ and X_2 a basis in $[0, x_2]$. If $X_1 \cup X_2$ were dependent, $X_1 \cup X_2$ would contain a circuit necessarily consisting of atoms of L_1 *and* L_2, in contradiction to the fact that no atom of L_1 is perspective with an atom of L_2. So $X_1 \cup X_2$ must be a basis in $[0, x]$. Hence, $h(x) = h(x_1) + h(x_2)$. If $x_1' \leq z_1$ and $x_2' \leq z_2$ satisfy $x = x_1' \vee x_2'$, then $x_1' \leq x_1$ and $x_2' \leq x_2$. Hence, $h(x_1') + h(x_2') \geq h(x) = h(x_1) + h(x_2)$ implies $h(x_1') = h(x_1)$ and $h(x_2') = h(x_2)$, and therefore $x_1' = x_1$ and $x_2' = x_2$.

For the converse, assume L to be decomposable, with corresponding intervals L_1 and L_2. Choose arbitrary atoms $a \in L_1$ and $b \in L_2$. If a and b were perspective, we could find an element $x \in L$ such that $a \vee x = b \vee x$ and $a \wedge x = b \wedge x = 0$. But then the representation $x = x_1 \vee x_2$ with respect to L_1 and L_2 would imply $(a \vee x_1) \vee x_2 = x_1 \vee (x_2 \vee b)$. Hence, $a \vee x_1 = x_1$, and thus $a \leq x_1 \leq x$, a contradiction. $\qquad\square$

3.5.3. Corollary. *A geometric lattice L is isomorphic to the direct product of a finite number of indecomposable geometric lattices, and the factors occurring in such a product are uniquely determined by L up to order.*

Proof. Exercise 3.14. $\qquad\square$

3.6. PROJECTIVE GEOMETRY AND MODULAR GEOMETRIC LATTICES

In Chapter 1, combinatorial geometries were motivated by the example of the structure of an arbitrary set of points in a finite-dimensional projective space. In our lattice-theoretic approach to combinatorial geometry, on the other hand, we arrived at geometric lattices by abstracting some basic properties of atomistic modular lattices of finite length.

We shall now show that both approaches to combinatorial geometry really start out at the same point: A finite-dimensional projective space and an indecomposable atomistic modular lattice of finite length are essentially the same thing.

To this end, it is convenient to understand a projective space as arising from an incidence structure as follows.

We consider two disjoint sets of "points" and "lines" together with an incidence relation between those two sets. We say that two lines *intersect* if they are incident with a common point. A *subspace* of this incidence structure is a set of points that contains with every two points incident with a line all the points incident with this line. A subspace is *spanned* by a set of points if it is the smallest subspace containing this set. The incidence structure is a *projective geometry* of *projective dimension* $n - 1$ if the following four axioms are satisfied:

(PG1) Two distinct points are incident with one and only one line.
(PG2) If a line intersects two sides of a triangle (not at their intersection), then it also intersects the third line (Figure 3.3).

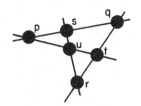

FIGURE 3.3. Axiom (PG2).

(PG3) Every line is incident with at least three points.
(PG4) The set of all points is spanned by n points, but not by fewer
than n points.

It is easily seen that the collection of subspaces of a projective geometry
forms an atomistic lattice. In fact, it is well known that for $n \geq 4$, the lattice
of subspaces is isomorphic to the lattice of subspaces of some n-dimensional
vector space (see, e.g., Baer 1952) and hence is modular geometric. It follows
from Exercise 3.15 that the lattice of subspaces is modular geometric for
$n \leq 3$ as well.

Now let M be a modular geometric lattice. We can call an element
x of M with $h(x) = 2$ a *line*, and the atoms of $[0, x]$ the *points* on the line x.
In this way we obtain an incidence structure $PG(M)$ that apparently satisfies
axiom (PG1). To convince ourselves that the other axioms also hold, we first
need a lemma.

3.6.1. Lemma. Let x and y be arbitrary elements and z an atom of the
modular geometric lattice M such that $z \leq x \vee y$ and $z \nleq y$. Then there is an
atom $x' \leq x$ such that $z \leq x' \vee y$.

Proof. Suppose that the lemma fails, and let x, y, z form a counterexample
such that $h(x \vee y)$ is minimal, and set $u = y \vee z$. So $h(u) = h(y) + 1$.

If $u = x \vee y$, then obviously there is an atom $x' \leq x$ such that $u = x' \vee y$,
that is, $z \leq x' \vee y$.

If $u < x \vee y$, let $\bar{x} = u \wedge x$. Then $\bar{x} \vee y = (u \wedge x) \vee y = u \wedge (x \vee y) = u$,
because M is modular.

Because $h(\bar{x} \vee y) < h(x \vee y)$, there is an atom $x' \leq \bar{x} \leq x$ such that $z \leq$
$x' \vee y$, a contradiction to the choice of $x \vee y$. □

Lemma 3.6.1 has important consequences.

3.6.2. Theorem. Let M be a modular geometric lattice. Then the subspaces of the incidence structure $PG(M)$ are exactly the flats of the combinatorial geometry $G_M(A)$.

Proof. Clearly, every flat of $G_M(A)$ is a subspace of PG(M). On the other hand, suppose there were atoms $x_1, \ldots, x_n$ generating a subspace that is properly contained in the flat of $G_M(A)$ generated by $x_1, \ldots, x_n$.

Consider $x = u_{n-1} \vee x_n$, where $u_{n-1} = x_1 \vee \cdots \vee x_{n-1}$. Then Lemma 3.6.1 says that for every atom z of M, $z \leq x$ if and only if $z \leq x' \vee x_n$ for some atom $x' \leq u_{n-1}$. Thus, by induction on n, we conclude that the subspace of PG(M) and the flat of $G_M(A)$ generated by $x_1, \ldots, x_n$ coincide, contradicting the choice of $x_1, \ldots, x_n$. $\square$

3.6.3. Corollary. *The incidence structure* PG(M) *satisfies axiom* (PG2).

Proof. Let the atoms p, q, s, t, u of M be as in Figure 3.3. We must show that there exists an atom $r \leq q \vee t$ such that $u \leq r \vee s$. But this follows immediately from Lemma 3.6.1 with $x = q \vee t$, $y = s$, and $z = u$. $\square$

In view of Theorem 3.6.2, the integer n in axiom (PG4) is the length $h(M)$ of the lattice M. It remains to deal with axiom (PG3).

3.6.4. Theorem. *Let M be a modular geometric lattice and p and q two atoms of M. Then $p \sim q$ if and only if the line $p \vee q$ contains at least three points.*

Proof. If $p \vee x = q \vee x$ and $p \wedge x = q \wedge x = 0$ for some $x \in L$, then $h[x \wedge (p \vee q)] = h(x) + h(p \vee q) - h(x \vee p \vee q) = 1$. So there must be a third point $s \leq p \vee q$, and $s \leq x$; that is, $p \neq s \neq q$.

Conversely, if the line $p \vee q$ contains a third point s, then s is a complement of both p and q in $[0, p \vee q]$; that is, $p \sim q$. $\square$

The decomposition theory of the preceding section thus yields the following:

3.6.5. Corollary. *The modular geometric lattice M is indecomposable if and only if the incidence structure* PG(M) *is a projective geometry.*

Stated loosely, Corollary 3.6.5 describes the structure of modular geometric lattices: Every modular geometric lattice is the direct product of a finite number of projective geometries.

EXERCISES

3.1. Show that every distributive lattice is modular.
3.2. Show that a distributive lattice of finite length is finite.
3.3. Complete the proof of Theorem 3.1.2.

3.4. Show that for a modular lattice M, the *dual* M^d (i.e., M with the inverted order) is also a modular lattice.

3.5. Show that for a modular lattice M, the maps ϕ_b and ψ_a of Proposition 3.2.1 are inverses of each other.

3.6. Recall that if U and V are subspaces of a finite-dimensional vector space W, then $(U + V)/V$ and $U/(U \cap V)$ are isomorphic. Give a lattice-theoretic interpretation of this "first isomorphism theorem."

3.7. Give an example of a semimodular lattice that is not M-symmetric.

3.8. Show that an atomistic lattice L of finite length is geometric if and only if every atom of L is a modular element of L.

3.9. Let L be a geometric lattice. Then L is distributive if and only if $G_L(A)$ has no circuits [i.e., $G_L(A)$ is the free matroid on the atoms of L].

3.10. Show that a relatively complemented semimodular lattice of finite length is geometric.

3.11. Let x and y be elements of the geometric lattice L. Then (x, y) M if and only if y is a minimal relative complement of x in $[x \wedge y, x \vee y]$.

3.12. Show that the atoms a and b of the geometric lattice L are perspective if and only if there is an element $x \in L$ such that $a \vee x = b \vee x$ and $a \wedge x = b \wedge x = 0$.

3.13. Prove Proposition 3.5.1.

3.14. Prove Corollary 3.5.3.

3.15. Show that any two distinct lines of a projective geometry of projective dimension $n - 1 \leq 2$ intersect in a point. Conclude that the lattice of subspaces is modular.

REFERENCES

Baer, R. (1952). *Linear Algebra and Projective Geometry*. Academic Press, New York.

Birkhoff, G. (1935). Abstract linear dependence in lattices. *Amer. J. Math.* **57**: 800–4.

Birkhoff, G. (1967). *Lattice Theory*, 3rd ed. American Mathematical Society Colloquium Publication 25, Providence, R.I.

Crawley, P., and Dilworth, R. P. (1973). *Algebraic Theory of Lattices*. Prentice-Hall, Englewood Cliffs, N.J.

Grätzer, G. (1978). *General Lattice Theory*. Birkhäuser, Basel.

Maeda, F., and Maeda, S. (1970). *Theory of Symmetric Lattices*. Grundlehren Band 173. Springer, Heidelberg.

Suzuki, M. (1967). *Structure of a Group and the Structure of Its Lattice of Subgroups*. Ergebnisse Band 10. Springer, Heidelberg.

Vilhelm, V. (1955). The selfdual kernel of Birkhoff's condition in lattices with finite chains (in Russian, with English summary). *Czech. Math. J.* **5**: 439–50.

CHAPTER 4

Basis-Exchange Properties

Joseph P. S. Kung

4.1. BRACKET IDENTITIES AND BASIS-EXCHANGE PROPERTIES

In this chapter we shall explore the connections between bracket or determinantal identities and basis-exchange properties in matroids. To see this
connection, we first recall the weak basis-exchange axiom from Chapter 2
[axiom (b3.1)].

> Let B_1 and B_2 be bases in a matroid. Then for every element $x \in B_1$,
> there exists an element $y \in B_2$ such that $(B_1 - x) \cup y$ is also a basis.

This axiom can be regarded as the combinatorial analogue of the Laplace
expansion for determinants. Let V be a vector space of dimension d. If
$x_1, \ldots, x_d$ are d vectors in V, their *bracket* (or *determinant*) is defined by

$$[x_1, \ldots, x_d] = \det(x_{ij})_{1 \le i, j \le d},$$

where (x_{ij}) are the coordinates of the vector x_i relative to a chosen basis of
V. Brackets satisfy the Laplace expansion

$$[x_1, \ldots, x_d][y_1, \ldots, y_d]$$
$$= \sum_{i=1}^{d} (-1)^{i-1} [y_i, x_2, \ldots, x_d][y_1, \ldots, y_{i-1}, x_1, y_{i+1}, \ldots, y_d].$$

The Laplace expansion can be interpreted in terms of basis exchange by
observing that a set $\{x_1, \ldots, x_d\}$ of vectors is a basis if and only if its bracket

$[x_1, \ldots, x_d]$ is nonzero. Hence, if $B_1 = \{x_1, \ldots, x_d\}$ and $B_2 = \{y_1, \ldots, y_d\}$ are bases, the left-hand side of the Laplace expansion is nonzero. Hence, there must be at least one nonzero term on the right-hand side. This implies that at least one of the sets $(B_1 - x_1) \cup y_i$ is a basis. We thus obtain the weak basis-exchange axiom.

However, the Laplace expansion translates combinatorially into a stronger exchange property.

4.1.1. Proposition (Symmetric basis-exchange axiom). *Let B_1 and B_2 be bases. Then for every element $x \in B_1$, there exists an element $y \in B_2$ such that $(B_1 - x) \cup y$ and $(B_2 - y) \cup x$ are both bases.*

The tool for proving this proposition is the notion of a fundamental circuit.

4.1.2. Lemma. *Let B be a basis and x an element not in B. Then there exists a unique circuit C contained in $B \cup x$.*

Proof. Because B is a maximal independent set, $B \cup x$ is dependent and contains at least one circuit. Because B is independent, every circuit contained in $B \cup x$ contains x. Now suppose that C_1 and C_2 are two distinct circuits contained in $B \cup x$. Because $x \in C_1 \cap C_2$, the circuit-elimination axiom [axiom (c3), Chapter 2] implies that there exists a circuit $C_3 \subseteq C_1 \cup C_2 - x \subseteq B$, contradicting the assumption that B is a basis. $\square$

The unique circuit C in Lemma 4.1.2 is called the *fundamental circuit of x relative to the basis B*.

Proof of Proposition 4.1.1. Let C be the fundamental circuit of x relative to B_2, and let D be the set of elements y in B_2 such that $(B_1 - x) \cup y$ is a basis. Note that y is in D if and only if y is not contained in the closure $\mathrm{cl}(B_1 - x)$, or $\overline{B_1 - x}$. Now, if $y \in D \cap (C - x)$, then $(B_2 - y) \cup x$ is a basis, because the element y is removed from the unique circuit C contained in $B_2 \cup x$. Hence, it suffices to show that the set $D \cap (C - x)$ is nonempty. Suppose it is empty. Then $(C - x) \subseteq \overline{B_1 - x}$, and x is dependent on $B_1 - x$, contradicting the assumption that B_1 is a basis. $\square$

Our interpretation of the Laplace expansion can be generalized to obtain a basis-exchange property (which may or may not hold in all matroids) from any bracket identity. Explicitly, we have, in order of complexity, the following interpretations:

(1) $[x_1, \ldots, x_d] \neq 0$ if and only if $\{x_1, \ldots, x_d\}$ is a basis.
(2) Let $M = [x_1, \ldots, x_d][y_1, \ldots, y_d] \ldots [z_1, \ldots, z_d]$ be a monomial in brackets. Then $M \neq 0$ if and only if every subset in the family $\hat{M}$ $= \{\{x_1, \ldots, x_d\}, \{y_1, \ldots, y_d\}, \ldots, \{z_1, \ldots, z_d\}\}$ is a basis. If this is the case, we say that the family M is *basic*.

(3) Let $M_1 + M_2 + \ldots + M_k = 0$ be a bracket identity. Then, either none of the families $\hat{M}_i$, $1 \leq i \leq k$, is basic or at least two of them are basic.

Note that whereas the first two interpretations retain all the information as far as being bases is concerned, the third interpretation does not.

Because the Laplace expansion is not the only bracket identity, we could, in principle, have many "abstract theories of linear dependence," depending on which bracket identity we chose to interpret as a basis-exchange axiom. For example, we can choose the *multiple Laplace expansion*.

$$\sum \pm [y_{\sigma(1)}, \ldots, y_{\sigma(k)}, x_{k+1}, \ldots, x_d][x_1, \ldots, x_k, y_{\sigma(k+1)}, \ldots, y_{\sigma(d)}] = 0,$$

the sum being over all permutations σ of $\{1, \ldots, d\}$ for which $\sigma(1) < \sigma(2) < \cdots < \sigma(k)$ and $\sigma(k + 1) < \sigma(k + 2) < \cdots < \sigma(d)$, and the exact value of the sign is irrelevant. This translates into the *multiple-exchange property* (or *subset-exchange property*):

Let B_1 and B_2 be bases and $A_1 \subseteq B_1$. Then there exists a subset $A_2 \subseteq B_2$ such that $(B_1 - A_1) \cup A_2$ and $(B_2 - A_2) \cup A_1$ are both bases.

However, we would not get a different abstract theory of linear dependence with the multiple-exchange property as an axiom, for the multiple-exchange property is deducible from the (single) basis-exchange axiom.

This is to be expected, because by the second fundamental theorem of projective invariant theory [see Weyl (1946) and Désarménien, Kung, and Rota (1978) for more details], every bracket identity can be deduced algebraically from the (single) Laplace expansion. However, because our interpretation does not retain all the information about basic collections of subsets, the question arises which bracket identities yield translations into basis-exchange properties holding in all matroids and which do not. This is perhaps the central question in the study of basis-exchange properties. A reasonable answer to this question could be in the form of a decision procedure. Such a decision procedure would yield much insight into the relation between matroid theory and linear algebra and should be useful in coordinatization problems. The main objective in this chapter is to discuss the partial results obtained in this direction.

4.2. THE EXCHANGE GRAPH

In this section we shall describe the basic technique for proving basis-exchange properties. This is the technique of the exchange graph (or the arrow notation). This technique is also basic to the matroid partition and intersection algorithms, which will be covered in a later volume.

Let $A \subseteq E$ and $B_1, B_2, \ldots, B_m$ be bases of the matroids $G_1, G_2, \ldots, G_m$

on E. (These matroids are not necessarily distinct.) The *exchange graph* for these given data is the directed graph with no loops on the vertex set A such that for vertices a and b there is an arc directed from a to b labeled by the basis B_i whenever b is in B_i, and $(B_i - b) \cup a$ is a basis of G_i, that is, if b can be *replaced* by a in B_i to yield another basis of G_i. We indicate this situation by

$$a \to_{B_i} b.$$

4.2.1. Lemma. *The following are equivalent:*

(i) $a \to_{B_i} b$.
(ii) *b is in the fundamental circuit of a relative to B_i in G_i.*
(iii) *a is in the bond $E - \overline{B_i - b}$.*

The proof is easy and is left as an exercise.

Note that the exchange graph may have several arcs labeled by different bases from a to b.

Let $a_1 \to_A a_2 \to_B \cdots \to_F a_k$ be a directed path in the exchange graph. It is said to *have no shortcuts* if (i) the vertices in it are all distinct and (ii) $a_i \to a_j$ is not an arc labeled by $\{A, B, \ldots, F\}$ (the bases occurring as labels in the path) unless $i = j - 1$ or $j < i$. Similarly, a directed cycle $a_1 \to_A a_2 \to_B \cdots \to a_k \to_F a_1$ *has no shortcuts* if (a) the vertices in it are distinct and (b) $a_i \to a_j$ is not an arc labeled by $A, B, \ldots,$ or F unless $i \equiv j - 1 \pmod{k}$.

The main result is the following lemma:

4.2.2. Lemma **(no-shortcut lemma).** *Let* $a \to_A b \to_B \cdots \to_F f$ $[a \to_A b \to_B \cdots e \to_F a]$ *be a directed path [cycle] with no shortcuts. Suppose that D is a basis occurring as a label and that $a_1 \to b_1, a_2 \to b_2, \ldots, a_r \to b_r$ are all the edges in the path [cycle] labeled by D. Then*

$$(D - \{b_1, \ldots, b_r\}) \cup \{a_1, \ldots, a_r\}$$

is a basis.

Proof. We may suppose that the arcs $a_i \to b_i$ are arranged in their order of appearance from left to right. Now, observe that the sets

$$D_p = (D - \{b_p, \ldots, b_r\}) \cup \{a_p, \ldots, a_r\}$$

obtained by replacing b_i with a_i in D, starting from the rightmost arc, are all bases. For $p = r$, this follows from the definition of $a_r \to_D b_r$. Suppose now that the assertion is true down to $q + 1$. The fundamental circuit of a_q relative to D is contained in $D - \{b_{q+1}, \ldots, b_r\}$, because if it were not, there would be an s, $q + 1 \leq s \leq r$, such that $a_q \to b_s$, contradicting the assumption that the path [cycle] has no shortcuts. Hence, we can still replace b_q by a_q in D_{q+1} to obtain a basis D_q. In particular, D_1 is a basis. $\square$

4.3. MULTIPLE AND ALTERNATING EXCHANGES

So far, the work done on translating determinantal identities has concentrated on discovering stronger basis-exchange properties valid in all matroids. These basis-exchange properties can be conveniently divided into two categories: multiple and alternating. The determinantal identity behind multiple exchanges is the multiple Laplace expansion (see Section 4.1).

4.3.1. Theorem (multiple basis exchange). *Let B_1 and B_2 be bases in a matroid G, and let $A_1 \subseteq B_1$. Then there exists a subset $A_2 \subseteq B_2$ such that $(B_1 - A_1) \cup A_2$ and $(B_2 - A_2) \cup A_1$ are both bases.*

Proof. We first augment the independent set A_1 to a basis $A_1 \cup C_1$ using elements C_1 from the basis B_2. [This is possible by axiom (i3) in Section 2.2 of Chapter 2, the independent-set augmentation axiom.] Similarly, we augment $B_1 - A_1$ to a basis $(B_1 - A_1) \cup C_2$ using elements C_2 from B_2. If $C_1 \cup C_2 = B_2$, then, on setting $A_2 = C_2$, we are done. Otherwise, there exists $x \in B_2$, $x \notin C_1 \cup C_2$. We show now how to "insert" x into $A_1 \cup C_1$ or $(B_1 - A_1) \cup C_2$.

Consider the exchange graph on the vertex set $C_1 \cup C_2 \cup x$ with bases $A_1 \cup C_1$ and $(B_1 - A_1) \cup C_2$. A vertex y is said to be *reachable (from x)* if there exists a directed path from x to y. Let U be the set of vertices reachable from x. We observe the following fact: there exists a vertex in U that occurs in both C_1 and C_2. To prove this fact, assume that no such vertex exists. Let $U_1 = U \cap C_1$ and $U_2 = U \cap C_2$. By assumption, $U_1 \cap U_2 = \varnothing$. Because $x \notin U_1$ or U_2, this implies

$$|U_1| + |U_2| = |U| - 1. \tag{4.1}$$

Consider now the closure of the independent set $A_1 \cup U_1$. This contains x, as $A_1 \cup U_1$ contains the fundamental circuit of x relative to $A_1 \cup C_1$. Moreover, $\overline{A_1 \cup U_1}$ contains all the other elements of U. To see this, let u be an element in U not in $A_1 \cup U_1$ (and hence not in the basis $A_1 \cup C_1$). The elements in the fundamental circuit of u relative to $A_1 \cup C_1$ are all joined to u and hence are reachable from x. In particular, they are all in $A_1 \cup U_1$, and $u \in \overline{A_1 \cup U_1}$ as claimed. We conclude that $U \subseteq \overline{A_1 \cup U_1}$ and

$$r(A_1 \cup U) = r(\overline{A_1 \cup U_1}) = |A_1 \cup U_1|.$$

Similarly,

$$r([B_1 - A_1] \cup U) = |(B_1 - A_1) \cup U_2|.$$

Adding these and using equation (4.1), we obtain

$$r(A_1 \cup U) + r([B_1 - A_1] \cup U) = |B_1| + |U| - 1.$$

On the other hand, by the semimodular inequality [axiom (r3) in Section 2.3],

$$r(A_1 \cup U) + r([B_1 - A_1] \cup U)$$
$$\geq r(A_1 \cup [B_1 - A_1] \cup U) + r([A_1 \cup U] \cap \{[B_1 - A_1] \cup U\})$$
$$\geq |B_1| + |U|.$$

This is a contradiction.

We can now insert x as follows. Let y be a vertex in U that occurs in both C_1 and C_2 and is at the minimum distance from x. Let

$$x = x_0 \to x_1 \to \cdots \to x_i = y$$

be a path of minimum length from x to y. This path can have no shortcuts, and we can perform the replacements as indicated in the no-shortcut lemma to obtain bases $A_1 \cup C_1'$ and $(B_1 - A_1) \cup C_2'$ such that $C_1' \cup C_2' = C_1 \cup C_2 \cup x \subseteq B_2$. As B_2 is finite, we must eventually obtain $C_1' \cup C_2' = B_2$.

This concludes the proof. $\square$

Our proof is essentially a rephrasing of the proof of the matroid partition theorem. The reader may find it of interest to guess what the matroid partition theorem is from this proof.

4.3.2. Corollary. *Let A and B be bases of the matroid G, and let $\{B_1, \ldots, B_k\}$ be a partition of B. Then there exists a partition $\{A_1, \ldots, A_k\}$ of A such that*

$$(B - B_i) \cup A_i, \qquad 1 \leq i \leq k.$$

are all bases.

There are two easy proofs. The first is by induction on k, the number of blocks in the partition of B. The second is a straightforward generalization of the proof of the theorem.

Alternating basis-exchange properties arise from the following fact in linear algebra: If $f(x_1, \ldots, x_p)$ is an alternating multilinear form, then f is zero on any p-tuple of linearly dependent vectors. Let f be the multilinear form

$$f(x_1, \ldots, x_p) = \sum_\sigma \text{sgn } \sigma \begin{bmatrix} x_{\sigma(1)}, x_{\sigma(2)}, & \cdots & , x_{\sigma(k)}, a_{k+1}, & \cdots & , a_d \\ x_{\sigma(k+1)}, & \cdots & , x_{\sigma(k+l)}, b_{l+1}, & \cdots & , b_d \\ \vdots & & & & \\ x_{\sigma(k+l+\cdots+m+1)}, & \cdots, & x_{\sigma(k+l+\cdots+m+n)}, c_{n+1}, & \cdots, & c_d \end{bmatrix}$$

where (a) $p = k + l + \cdots + m + n$, (b) the vectors $a_{k+1}, \ldots, a_d, b_{l+1}, \ldots, b_d, \ldots, c_{n+1}, \ldots, c_d$ are chosen constant vectors, (c) the summation is over all permutations σ of $\{1, \ldots, p\}$ such that

$$\sigma(1) < \sigma(2) < \cdots < \sigma(k),$$

$$\sigma(k + 1) < \sigma(k + 2) < \cdots < \sigma(k + l),$$

$$\vdots$$

$$\sigma(k + l + \cdots + m + 1) < \sigma(k + l + \cdots + m + 2) < \cdots < \sigma(p)$$

[these permutations are called (*algebraic*) *shuffles*], and (d) superposition of rows means multiplication of their brackets. The form **f** so defined is in fact alternating and is therefore zero on any p-tuple of dependent vectors. Hence, if one of the summands on the right is nonzero, there must be at least another nonzero summand to cancel it.

To translate this into the language of bases, we need a definition. Let S be a set and $\pi = \{S_1, \ldots, S_m\}$ a partition of S. A *nontrivial combinatorial shuffle* σ of S relative to π is a permutation of S for which there exists i and $x \in S_i$ such that $\sigma(i) \notin S_i$. Thus, a nontrivial combinatorial shuffle is a permutation that actually moves an element from one block to another. Note that an algebraic shuffle that is not the identity is a nontrivial combinatorial shuffle.

4.3.3. Theorem (alternating basis exchange). *Let $B_1, \ldots, B_m$ be bases of a matroid G on the finite set E with each basis partitioned into two blocks.*

$$B_i = X_i \cup Y_i, \qquad 1 \le i \le m,$$

such that the sets X_i are pairwise disjoint. Suppose that $X = X_1 \cup X_2 \cup \cdots \cup X_m$ is dependent in G. Then there exists a nontrivial combinatorial shuffle σ of X relative to $\{X_1, X_2, \ldots, X_m\}$ such that for all i, $\sigma X_i \cup Y_i$ is a basis.

Proof. Consider the exchange graph on the vertex set X with the bases $B_1, \ldots, B_m$. Because the subset X is dependent, X contains a circuit C. Observe that because a bond is the complement of a *closed* set, the intersection of the circuit C with any bond D cannot be a single-element set; that is, $|C \cap D| \ne 1$. For each $a \in C$ such that $a \in B_i$, consider the bond $D_a = E - B_i - a$. Because a is certainly contained in $C \cap D_a$, there must be, by our observation, another point $b \in C$, with $b \in B_j$ say, such that b is in the bond D_a; that is to say (Lemma 4.2.1), $b \to_{B_j} a$. Thus, from any vertex in C, we can always retreat to another vertex in C through an arc. Because C is finite, this implies that the exchange graph contains a directed cycle.

Choose a directed cycle of minimum length:

$$a \to_{B_i} b \to_{B_j} c \ldots d \to_{B_k} a.$$

The permutation σ on X defined by sending $a \to b$, $b \to c$, $\ldots$, $d \to a$, and keeping every vertex not in the cycle fixed, is a nontrivial combinatorial shuffle, because consecutive arcs in the cycle cannot be labeled by the same basis. Moreover, by the no-shortcut lemma, the sets $\sigma X_i \cup Y_i$ are all bases. This proves the theorem. □

Note that in our proof no use is made of the fact that the bases B_i are all from the same matroid. An analysis of the proof shows that we have proved the following stronger proposition:

4.3.4. Proposition. *Let $B_1, \ldots, B_m$ be bases of the matroids $G_1, \ldots, G_m$ on the finite set E, with each basis partitioned into two blocks.*

$$B_i = X_i \cup Y_i, \quad 1 \leq i \leq m,$$

such that the sets X_i are pairwise disjoint. Suppose that $X = \bigcup X_i$ contains a subset X' such that X' is a union of circuits in each of the matroids G_i. Then there exists a nontrivial combinatorial shuffle σ relative to the partition $\{X_1, X_2, \ldots, X_m\}$ such that for all i, $\sigma X_i \cup Y_i$ is a basis in G_i.

HISTORICAL NOTES

The point of view that matroid theory descends combinatorially from projective invariant theory was first put forward in Rota (1971) and developed further in Whiteley (1973, 1977, 1979). Our account differs slightly from these. The exchange graph is due independently to Knuth (1973), Greene and Magnanti (1975), and Lawler (1975). The multiple-exchange property was first proved by Brylawski (1973) and Greene (1973). Other proofs are in Woodall (1974) and Greene and Magnanti (1975). Our proof follows that of Greene and Magnanti (1975). A special case of the alternating exchange property was first proved in Greene (1974); the version presented here can be found in Kung (1978).

One topic we have not covered is the theory of bracket rings. A discussion of this will be presented in a chapter on coordinatizations in a later volume.

EXERCISES

4.1. Translate into the language of bases the *condensation identity*

$$[x_1, \ldots, x_d][y_1, \ldots, y_d]^{d-1}$$
$$= \det([y_1, \ldots, y_{j-1}, x_i, y_{j+1}, \ldots, y_d])_{1 \leq i, j \leq d}.$$

Is the translation valid for all matroids?

*4.2. (Kung 1978) Develop analogues of matroids for the other classical groups, namely, the general linear group acting on vectors and covectors, the orthogonal group, and the symplectic group.

4.3. (Krogdahl 1977) Let B be a basis of the matroid G on the set E. Consider the exchange graph Γ on the vertices E and the basis B of G.

(a) Observe that Γ is a bipartite graph with bipartition B and $E - B$. Observe further that if we reverse the edges of Γ, we obtain the exchange graph on E and the basis $E - B$ of the dual G^*.

(b) Show that the connected components of Γ as an undirected graph and the connected components of G as a matroid are the same. Hence, show that G and G^* have the same connected components as matroids.

*4.4. (Kung 1978) Show the orthogonally dual version of Theorem 4.3.3. Under the same initial hypotheses, suppose that $X = \bigcup X_i$ is a nonspanning set. Then there exists a nontrivial combinatorial shuffle σ of X relative to $\{X_1, \ldots, X_m\}$ such that for all i, $\sigma X_i \cup Y_i$ is a basis.

Base-orderable matroids (Brualdi and Scrimger 1968; Brualdi 1969; Bondy 1972; Ingleton 1975)

*4.5. (a) Let B_1 and B_2 be two bases of a matroid. Show that there exists a bijection $\varepsilon: B_1 \to B_2$ such that for all x in B_1, $(B_1 - x) \cup \varepsilon(x)$ is a basis. (*Hint:* Construct a relation and use the marriage theorem or use Corollary 4.3.2.)

A natural strengthening of (a) is to require that for all x in B_1, both $(B_1 - x) \cup \varepsilon(x)$ and $(B_2 - \varepsilon(x)) \cup x$ are bases. Such a bijection ε is called an *exchange ordering*. A matroid is called *base-orderable* if there exists an exchange ordering for any pair of bases. A further strengthening of (a) is to require that for all subsets $A \subseteq B_1$, $(B_1 - A) \cup \varepsilon(A)$ and $(B_2 - \varepsilon(A)) \cup A$ are both bases. Such a bijection ε is called a *multiple-exchange ordering*. Matroids for which there exists a multiple-exchange ordering for any pair of bases are called *strongly base-orderable*. The original motivation for base orderability was to generalize transversal matroids.

(b) Consider a relation R between the sets S and T. Let G be the transversal matroid on S (see Chapter 1). Suppose further that rank $G = |T|$. Let B_i, $i = 1, 2$, be bases of G. By definition of G, there exist bijections $\beta_i: B_i \to T$. Show that the bijection $\beta_1\beta_2^{-1}$ is a multiple-exchange ordering.

Not all matroids are base-orderable.

(c) Show that the circuit matroid $M(K_4)$ of the complete graph on four vertices is not base-orderable. Show that the Vámos cube (see Chapter 7, Exercise 7.6) is base-orderable but not strongly base-orderable. Conclude that base orderability does not arise from a determinantal identity.

(d) Show that the class of (strongly) base-orderable matroids is closed under duality, minors, direct sums, matroid joins, and matroid induction. In particular, gammoids (to be discussed in a later volume) are strongly base-orderable.

(e) Consider the class of binary base-orderable matroids. Show that this class is defined by the excluded minors: the four-point line and the circuit matroid $M(K_4)$ of the complete graph on four vertices. Hence, show that this class is precisely the class of series-parallel networks (see Chapter 6).

(f) Show that the class of base-orderable matroids cannot be characterized by a finite family of excluded minors.

Bases in oriented matroids (Gutierrez Novoa 1965; Las Vergnas 1978)

*4.6. Find a set of signed ordered basis-exchange axioms for oriented matroids.

*4.7. Show that there are exactly two assignments of signs to the ordered bases of an oriented matroid compatible with the orientation.

Unique basis-exchange properties (Suguira 1978; White 1980)

*4.8. Let B_1 and B_2 be bases of a matroid G, and $x \in B_1$. We say that $x \in B_1$ can be *uniquely exchanged* for $y \in B_2$ if y is the unique element in B_2 such that $(B_1 - x) \cup y$ and $(B_2 - y) \cup x$ are both bases (i.e., the exchange graph on the vertex set $B_2 \cup x$ and the bases B_1 and B_2 has exactly one unique directed cycle, namely, $x \rightleftarrows y$). Unique exchanges are important because in any coordinatization of G over a field, $[B_1][B_2] = \pm[(B_1 - x) \cup y][(B_2 - y) \cup x]$. Here, $[B]$ is the bracket $[x_1, \ldots, x_d]$, where $B = \{x_1, \ldots, x_d\}$.

Now consider the set of all n-tuples of bases $(B_1, \ldots, B_n)$ of G as the vertices of a graph. Let E_1 be the edges

$$(B_1, \ldots, B_{i-1}, (B_i - x) \cup y, B_{i+1}, \ldots, B_{j-1},$$
$$(B_j - y) \cup x, B_{j+1}, \ldots, B_n) \sim (B_1, \ldots, B_n) \qquad (4.2)$$

where $x \in B_i$ is uniquely exchanged for $y \in B_j$. Let E_2 be the edges in E_1 together with the edges

$$(B_{\pi(1)}, \ldots, B_{\pi(n)}) \sim (B_1, \ldots, B_n),$$

where π is a permutation of $\{1, \ldots, n\}$. Clearly, if $(B_1, \ldots, B_n) \sim (B'_1, \ldots, B'_n)$ is an edge in E_1 or E_2, then the two n-tuples of bases *have the same underlying multiset*; that is, the multiset union $B_1 \cup \cdots \cup B_n$ equals the multiset union $B'_1 \cup \cdots \cup B'_n$. Let $\Gamma_1(G)$ [respectively $\Gamma_2(G)$] be the graph with edge set E_1 [respectively E_2]. We say that a matroid is in UE(1) [respectively UE(2)] if all pairs of n-tuples of bases with the same underlying multiset are contained in the same connected component.

(a) Prove that UE(1) and UE(2) are closed under minors, direct sums, and orthogonal duality.

(b) Show that UE(1) is the class of series-parallel networks. [Compare with Exercise 4.5(e) describing all binary base-orderable matroids.]

**(c) Prove or disprove: All graphic matroids are in UE(2). All unimodular matroids are in UE(2).

> Now let E'_1 be the edges in equation (4.2), except that we only require $x \in B_i$ to be exchanged (not necessarily uniquely) for $y \in B_j$, and let E'_2 be $E'_1 \cup E_2$. The classes TE(1) and TE(2) are defined analogously in terms of E'_1 and E'_2, respectively.

(d) Prove that TE(1) and TE(2) are closed under minors, direct sums, and orthogonal duality.

**(e) Prove or disprove: TE(1) and TE(2) are the class of all matroids.

(f) Define and study the classes UE(3) and TE(3) by allowing multiple exchanges in the definition of the edges in equation (4.2).

> *Matroid basis graphs* (Cummins 1966; Dowling 1969; Holzmann and Harary 1972; Maurer 1973b; 1973c; Holzmann, Norton, and Tobey 1973; Donald, Holzmann, and Tobey 1977; Astié-Vidal 1980, 1980a)

*4.9. The *(matroid) basis graph* of a matroid G is the undirected graph with vertex set the collection of all bases of G such that a basis B_1 is connected to another basis B_2 whenever the symmetric difference $B_1 \triangle B_2$ has cardinality exactly 2.

(a) Show that given any edge in a basis graph there is a Hamiltonian cycle containing it and another Hamiltonian cycle not containing it. (*Hint:* Use a contraction-and-deletion argument.)

(b) Show that two matroids G and H have isomorphic basis graphs if and only if $G = G_1 \oplus G_2 \oplus \cdots \oplus G_n$ and $H = H_1 \oplus H_2 \oplus \cdots \oplus H_n$ and there is a permutation π of $\{1, \ldots, n\}$ such that $G_i \simeq H_{\pi(i)}$ or $G_i \simeq H^*_{\pi(i)}$. (Here, H^* is the dual of H.)

(c) Characterize matroid basis graphs. Show by an example that this should be a complicated affair.

(d) Relate constructions in matroids (such as direct sums, minors) to basis graphs.

(e) Characterize basis graphs of binary matroids. (Unsolved) Characterize basis graphs of other classes of matroids.

(f) A *complete* matroid basis graph is the basis graph of a truncated Boolean algebra. Characterize these graphs.

(g) Explore the relation between the automorphism group of a matroid and the automorphism group of its basis graph.

> *Basis monomial ring* (White 1977)

*4.10. Consider a matroid G on the set S. Let $\{x_s : s \in S\}$ be a set of indeterminates, one for each element of S. The *basis monomial ring* M_G of G

is the subring of the polynomial ring $k[x_s]$ (k a field) generated by the monomials $\{\prod_{s \in B} x_s: B$ a basis of $G\}$. Show that M_G is a Cohen-Macaulay ring.

Complementary bases (Magnanti 1974; Greene, Kleitman, and Magnanti 1974: Kleitman 1976)

*4.11. (a) Let G be a matroid on the set E, where $|E| = 2n$; the elements of E are labeled $x_1, x_1', x_2, x_2', \ldots, x_n, x_n'$, and $\{x_1, x_2, \ldots, x_n\}$ is a basis. Show that if G has no circuit of size k or smaller, then G has at least 2^k bases B' such that for each i, B' contains either x_i or x_i'. Such bases are called complementary bases.

(b) Generalize (a) to the following: Let G be a matroid on E. Suppose that $\{A_1, \ldots, A_n\}$ is a family of pairwise disjoint subsets of E that has a system of distinct representatives that is also an independent set of G. If $k \leq n$, G has no circuit of size k or smaller, and $|A_i| \leq t$ for every i, then $\{A_i\}$ has at least t^k independent systems of distinct representatives.

REFERENCES

Astié-Vidal, A. (1980). The automorphism group of a matroid. Combinatorics 79 (Proceedings of a colloquium, Université de Montréal, Montréal, Quebec, 1979), Part II, Ann. Discrete Math, **9**: 205–16.

(1980a). Factor group of the automorphism group of a matroid basis graph with respect to the automorphism group of the matroid. Discrete Math. **32**: 217–24.

Bondy, J. A. (1972). Transversal matroids, base-orderable matroids, and graphs. Quart. J. Math. Oxford Ser. (2) **23**: 81–9.

Brualdi, R. A. (1969). Comments on bases in dependence structures. Bull. Australian Math. Soc. **1**: 161–7.

Brualdi, R. A., and Scrimger, E. B. (1968). Exchange systems, matchings, and transversals. J. Combin. Theory **5**: 244–57.

Brylawski, T. H. (1973). Some properties of basic families of subsets. Discrete Math. **6**: 333–41.

Cummins, R. L. (1966). Hamiltonian circuits in tree graphs. IEEE Trans. Comput. Theory **13**: 82–90.

Davies, J. (1975–76). On bases of independence structures intersecting in a set of prescribed cardinality. J. London Math. Soc. **12**: 455–8.

Dawson, J. (1978). A remark on an exchange theorem for bases. J. Math. Anal. Appl. **62**: 354–5.

Désarménien, J., Kung, J. P. S., and Rota, G.-C. (1978). Invariant theory, Young bitableaux, and combinatorics. Adv. Math. **27**: 63–92.

Donald, J. D., Holzmann, C. A., and Tobey, M. D. (1977). A characterization of complete base graphs. J. Combin. Theory Ser. B **22**: 139–58.

Dowling, T. A. (1969). A characterization of the T_m graph. J. Combin. Theory **6**: 251–63.

Gabow, H. (1976). Decomposing symmetric exchanges in matroid bases. Math. Programming **10**: 271–6.

Greene, C. (1973). A multiple exchange property for bases. *Proc. Amer. Math. Soc.*
   39: 45–50.
   (1974). Another exchange property for bases. *Proc. Amer. Math. Soc.* **46**: 155–6.
Greene, C., Kleitman, D. J., and Magnanti, T. L. (1974). Complementary trees and
   independent matchings. *Stud. Appl. Math.* **53**: 57–64.
Greene, C., and Magnanti, T. L. (1975). Some abstract pivot algorithms. *SIAM J.
   Appl. Math.* **29**: 530–9.
Gutierrez Novoa, L. (1965). On *n*-ordered sets and order completeness. *Pacific J.
   Math.* **15**: 1337–45.
Holzmann, C. A., and Harary, F. (1972). On the tree graph of a matroid. *SIAM J.
   Appl. Math.* **22**: 187–93.
Holzmann, C. A., Norton, P. G., and Tobey, M. D. (1973). A graphical representation
   of matroids. *SIAM J. Appl. Math.* **25**: 618–27.
Ingleton, A. W. (1975). Non-base-orderable matroids, in *Proceedings of the Fifth
   British Combinatorial Conference*, pp. 355–9. Congressus Numerantium, No. 15,
   Utilitas Math., Winnipeg, Manitoba.
Kleitman, D. J. (1972). Finding uncomplemented trees. *Stud. Appl. Math.* **51**: 309–10.
   (1976). More on complementary trees. *Discrete Math.* **15**: 373–8.
Knuth, D. E. (1973). Matroid partitioning. Stanford Technical Report CS-73-342,
   Stanford University.
Krogdahl, S. (1977). The dependence graph for bases in matroids. *Discrete Math.*
   19: 47–59.
Kung, J. P. S. (1978). Bimatroids and invariants. *Adv. Math.* **30**: 238–49.
   (1978a). Alternating basis exchanges in matroids. *Proc. Amer. Math. Soc.* **71**: 355–8.
Kundu, S., and Lawler, E. L. (1973). A matroid generalization of a theorem of
   Mendelsohn and Dulmage. *Discrete Math.* **4**: 159–63.
Las Vergnas, M. (1978). Bases in oriented matroids. *J. Combin. Theory Ser. B* **25**: 283–9.
Lawler, E. L. (1975). Matroid intersection algorithms. *Math. Programming* **9**:31–56.
   (1976). *Combinatorial Optimization: Networks and Matroids.* Holt, Rinehart and
   Winston, New York.
Lawrence, J. (1982). Oriented matroids and multiply ordered sets. *Linear Algebra Appl.*
   48: 1–12.
McDiarmid, C. J. H. (1975). An exchange theorem for independence structures. *Proc.
   Amer. Math. Soc.* **47**: 513–14.
Magnanti, T. L. (1974). Complementary bases of a matroid. *Discrete Math.* **8**: 355–61.
Maurer, S. B. (1973a). Basis graphs of pregeometries. *Bull. Amer. Math. Soc.* **79**: 783–6.
   (1973b). Matroid basis graphs. I. *J. Combin. Theory Ser. B* **14**: 216–40.
   (1973c). Matroid basis graphs. II. *J. Combin. Theory Ser. B* **15**: 121–45.
   (1973d). Intervals in matroid basis graphs. *Discrete Math.* **11**: 147–59.
   (1975). A maximum-rank minimum-term-rank theorem for matroids. *Linear Algebra
   Appl.* **10**: 129–37.
Rota, G.-C. (1971). Combinatorial theory and invariant theory. Notes for N.S.F.
   Advanced Science Seminar in Combinatorial Theory, Bowdoin College, Brunswick Maine.
Suguira, H. (1978). Bracket rings of series-parallel networks. Preprint, Nagoya University.
Weyl, H. (1946). *The Classical Groups*, 2nd edition. Princeton University Press.
White, N. L. (1974). A basis extension property. *J. London Math. Soc.* **7**: 662–4.
   (1975a). The bracket ring of a combinatorial geometry. I. *Trans. Amer. Math. Soc.*
   202: 79–95.
   (1975b). The bracket ring of a combinatorial geometry. II. Unimodular geometries.
   Trans. Amer. Math. Soc. **214**: 233–48.
   (1977). The basis monomial ring of a matroid, *Adv. Math.* **24**: 292–7.

(1980). A unique exchange property for bases. *Linear Algebra Appl.* **31**: 81–91.

Whiteley, W. (1973). Logic and invariant theory. I: Invariant theory of projective properties. *Trans. Amer. Math. Soc.* **177**: 121–39.

(1977). Logic and invariant theory. III: Axioms systems and basic syzygies. *J. London Math. Soc.* **15**: 1–15.

(1979). Logic and invariant theory. IV. Invariants and syzygies in combinatorial geometry. *J. Combin. Theory Ser. B* **26**: 251–67.

Woodall, D. R. (1974). An exchange theorem for bases of matroids. *J. Combin. Theory Ser. B* **16**: 227–8.

CHAPTER 5

Orthogonality

Henry Crapo

5.1. INTRODUCTION

The notion of orthogonality of combinatorial geometries and of matroids is an abstraction of the usual notion of orthogonality in vector spaces and of perpendicularity in Euclidean geometry. We shall show, for instance, how, relative to any fixed basis X for an n-dimensional vector space T, every subspace $V \subseteq T$ defines a vector geometry on the set X, and orthogonal complementary subspaces V and $V^\perp$ define *orthogonal geometries*.

We begin this chapter with the abstract combinatorial definition of orthogonality. After some necessary preliminary work on vector geometries, we shall show how orthogonal complementary subspaces give rise to orthogonal geometries, and what this construction implies for matrices and for dual vector spaces. We shall bring the chapter to a close with two further examples of orthogonality of combinatorial geometries as they arise in the study of simplicial geometries and of structure geometries.

Two notes of caution are in order. First, the operation $G \to G^*$ taking each matroid G to its orthogonal matroid G^* is an operation of period 2. This is not so for geometries. If a geometry G has bonds only of cardinality 1 or 2, these become circuits of cardinality 1 or 2 in the orthogonal matroid. That is, they become loops or multiple points that disappear in the passage from the matroid G^* to its associated geometry. These elements will not reappear in the geometry G^{**}, if orthogonality is taken to be an operation on geometries, rather than on matroids.

Second, geometries of finite rank on infinite sets have orthogonal structures that, where they are definable at all, have infinite rank and have circuits of infinite cardinality. Such a theory will be set forth in a chapter on infinite matroids in a later volume, but for now we must restrict our attention to geometries orthogonal to one another on *finite sets*.

5.2. ORTHOGONAL GEOMETRIES

For each subset A of a geometry $G(E)$ on a finite set E, the *nullity* $n(A)$ is the difference

$$n(A) = |A| - r(A) \tag{5.1}$$

between the cardinality of A and its geometric rank. Then the *orthogonal rank* $r^*(A)$ of the set A is given by the difference

$$r^*(A) = n(E) - n(E \setminus A) \tag{5.2}$$

between the nullities of the entire set E and of the complement $E \setminus A$ of the set A. After proving that the orthogonal rank function r^* is the rank function of a matroid, we shall show how orthogonality of geometries manifests itself in terms of circuits, bonds, and bases.

5.2.1. Theorem. *If $G(E)$ is a geometry (or matroid) on a finite set E, with geometric rank $r(A)$ defined for each subset $A \subseteq E$, then the orthogonal rank function r^* defined by equations (5.1) and (5.2) is a geometric rank function, defining a matroid $G^*(E)$ on the same set E.*

Proof. The orthogonal rank of the empty set is zero, by equation (5.2). From equations (5.1) and (5.2) it follows that for any subset $A \subseteq E$,

$$r^*(A) = |A| + r(E \setminus A) - r(E). \tag{5.3}$$

If a set B covers a set A in the Boolean algebra $\mathscr{B}(E)$ of subsets of the set E, then

$$
\begin{aligned}
r^*(B) &- r^*(A) \\
&= |B| - |A| - [r(E \setminus A) - r(E \setminus B)] \\
&= 1 - [r(E \setminus A) - r(E \setminus B)],
\end{aligned}
$$

the term in brackets being equal to either 0 or 1. Thus, $r^*(B)$ is equal either to $r^*(A)$ or to $r^*(A) + 1$, and the function r^* is a unit-increase function.

It remains to prove that r^* is a semimodular function. For any subsets A, B of the set E, equation (5.3) yields

$$
\begin{aligned}
r^*(A \cap B) &+ r^*(A \cup B) \\
&= |A \cap B| + |A \cup B| + r[(E \setminus A) \cup (E \setminus B)] + r[(E \setminus A) \cap (E \setminus B)] \\
&\quad - 2r(E) \leq |A| + |B| + r(E \setminus A) + r(E \setminus B) - 2r(E) \\
&= r^*(A) + r^*(B).
\end{aligned}
$$

Note that the essential step in this proof is to observe that cardinality is a modular function, whereas r is semimodular, on the Boolean algebra $\mathcal{B}(E)$.

Because r^* is a unit-increase function and is semimodular on $\mathcal{B}(E)$, with value zero for the empty set and with integer values for all subsets $A \subseteq E$, r^* is itself the geometric rank function for a matroid on the set E.

□

Figure 5.1 illustrates some pairs of orthogonal geometries. Note that in the fourth example, the two geometries are isomorphic, but not with respect to the identity function. The permutation that exchanges the elements c and e provides an isomorphism between G and G^*.

5.2.2. Proposition. *If G and G^* are orthogonal matroids on a finite set E, then*

$$r(G) + r(G^*) = |E|.$$

Proof. $r(E) + r^*(E) = r(E) + [|E| - r(E) + r(\phi)] = |E|.$ □

5.2.3. Proposition. *The operation $G \to G^*$, taking each matroid G to its orthogonal matroid, has period 2. That is, $G^{**} = G$ for any matroid G.*

Proof. For any subset $A \subseteq E$ of elements of a matroid $G(E)$,

$$\begin{aligned}
r^{**}(A) &= |A| + r^*(E\backslash A) - r^*(E) \\
&= |A| + [|E\backslash A| + r(A) - r(E)] - [|E| - r(E)] \\
&= r(A),
\end{aligned}$$

because $|A| + |E\backslash A| - |E| = 0$.   □

5.2.4. Proposition. *Let G and G^* be orthogonal matroids on a set E, let p be any element of the set E, and let A, B be complementary subsets of the set $E\backslash p$. Thus, $A, \{p\}, B$ is a three-part partition of the set E. Then p is in the G-closure of the set A if and only if p is not in the G^*-closure of the set B.*

Proof. Let $\bar{B}^*$ denote the G^*-closure of the set B. Then, by equations (5.1) and (5.2),

$p \in \bar{A}$ if and only if $r(A \cup p) = r(A)$

if and only if $n(A \cup p) > n(A)$ if and only if $r^*(B) < r^*(B \cup p)$

if and only if $p \notin \bar{B}^*$.   □

The converse of Proposition 5.2.4 is also true, in a very strong sense. In order for this relation to hold between two closure operators, they must both be *geometric*, and thus must define a pair of orthogonal geometries. To

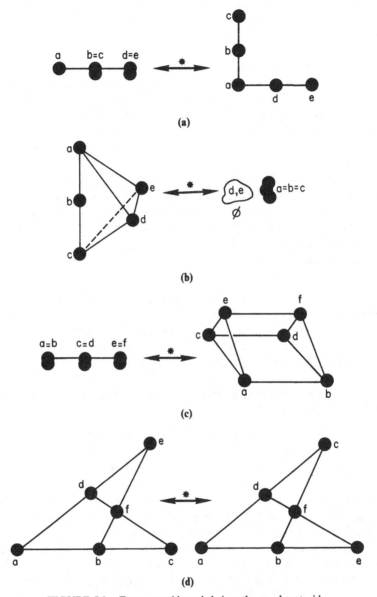

FIGURE 5.1. Four matroids and their orthogonal matroids.

set the stage for this converse theorem, we define two closure operators J, K on the Boolean algebra $\mathscr{B}(E)$ of a finite set E to be orthogonal if and only if for every three-part partition A, $\{p\}$, B of E we have

$$p \in J(A) \quad \text{if and only if} \quad p \notin K(B).$$

5.2.5. Proposition. *Two closure operators J and K on a finite Boolean algebra $\mathcal{B}(E)$ are orthogonal if and only if both closures are geometric and they define orthogonal geometries.*

Proof. Assuming J and K are orthogonal, we prove that the closure operator J has the exchange property. Assume $p \in J(A \cup q)\backslash J(A)$, and let $B = E\backslash(A \cup p \cup q)$. Then $p \notin K(B)$, but $p \in K(B \cup q)$. Thus, $K(B) \neq K(B \cup q)$, and so $q \notin K(B)$. So $q \in J(A \cup p)$, and J has the exchange property. By symmetry, so does the operator K.

For any subset $A \subseteq E$, let $x_1, \ldots, x_k$ be a linear order of the elements of the set A. Let $A_0 = \phi$ and $A_i = \{x_1, \ldots, x_i\}$ for $i = 1, \ldots, k$. Then the difference $n(E) - n(E\backslash A)$ that appears in formula (5.2) is equal to the number of indices $i \in \{1, \ldots, k\}$ for which $J(E\backslash A_i) = J(E\backslash A_{i-1})$, that is, the number of indices i for which $x_i \in J(E\backslash A_i)$, or, equivalently, for which $x_i \notin K(A_{i-1})$. This is the number of indices $i \in \{1, \ldots, k\}$ for which $r^*(A_{i-1}) < r^*(A_i)$; so it is equal to the rank $r^*(A)$. Consequently, $n(E) - n(E\backslash A) = r^*(A)$, and the geometries defined by the geometric closure operators J and K are orthogonal. $\qquad\square$

Propositions 5.2.4 and 5.2.5 provide a very convenient definition of the orthogonal of any given geometry. In the first part of Figure 5.1, observe that the element d is in the G-closure of the set ac, but it is not in the G-closure of the set bc. Now look at the orthogonal geometry. The element d is not in the G^*-closure of the set be, but it is in the G^*-closure of the set ae.

5.2.6. Theorem. *If G and G^* are orthogonal matroids on a finite set E, then for any subset $A \subseteq E$,*

>*A is a circuit of G if and only if A is a bond of G^*.*
>*A is a basis of G if and only if $E\backslash A$ is a basis of G^*.*
>*A is independent in G if and only if $E\backslash A$ spans G^*.*

Proof. The independent subsets A in G are those for which $p \in A$ implies $p \notin \overline{A\backslash p}$; so by Proposition 5.2.4, $p \in \overline{E\backslash A}^*$. This is true for all elements $p \in A$ if and only if the set $E\backslash A$ spans in the geometry G^*.

A subset $A \subseteq E$ is a basis for G if and only if A is both independent in G and spans G. By the preceding paragraph, this is true if and only if the complementary set $E\backslash A$ both spans and is independent in G^*, that is, if and only if $E\backslash A$ is a basis for G^*.

A subset $A \subseteq E$ is a circuit in G if and only if it is a minimal dependent set in G. By the first paragraph of this proof, that is true if and only if the complementary set $E\backslash A$ is a maximal nonspanning subset of G^*, that is, if and only if $E\backslash A$ is a copoint (a hyperplane) in the geometry G^*. This is true if and only if the original set A is a bond in the orthogonal geometry G^*. $\qquad\square$

Look again at Figure 5.1. The subset *bcde* in the fourth part is a bond in *G* because it is the complement of the copoint (i.e., line) *af*. Thus, *bcde* appears as a circuit in the orthogonal geometry. Every spanning subset in the second part must contain both *d* and *e* and at least two of the three points a, b, c. Thus, the only independent subsets of the orthogonal geometry are the empty set and the three singletons a, b, and c. The geometry G^* is thus one triple point $a = b = c$, together with two "loops" d, e in the closure of the empty set.

This concludes our introduction to the combinatorial notion of orthogonality. Before we can show how this abstracts the usual geometric or linear algebra notion, we shall need some theorems concerning vector geometries.

5.3. VECTOR GEOMETRIES AND FUNCTION-SPACE GEOMETRIES

A *vector geometry* is any subgeometry of a projective geometry *W* of finite dimension over a field *F*. It is the purpose of this section to explain the connection between vector geometries and function-space geometries, geometries defined by a vector space of functions from a set into a field. Function-space geometries figure prominently in the literature of combinatorial geometries: "those associated with matrices" in Whitney (1935), "chain groups" in Tutte (1956), and "function-space geometries" in Crapo and Rota (1970). We describe these geometries in terms of strong maps from free geometries into projective geometries.

A *strong map* ϕ from a geometric lattice *G* to a geometric lattice *H* is a function from *G* to *H* that preserves joins (including $0 = \vee \phi$) and preserves the relation $\downarrow$ ("covers or is equal to"). Strong maps will be studied in detail in Chapter 7. For our purposes in this section, it suffices to remark that every strong map gives rise to a *quotient geometry* whose flats are those flats of *G* that are maximal with any given image under σ. That is, $x \in G$ is a flat of the quotient geometry G/σ if and only if for every $y \in G$, $x < y$ implies $\sigma(x) < \sigma(y)$. This quotient can also be seen as the set of closed flats relative to a *Galois connection* between *G* and H^{opp}, because the map σ has an *adjoint* τ defined by

$$\tau(z) = V\{x \in G; \sigma(x) \leq z\}$$

for all flats $z \in H$, and satisfying

$$\sigma(x) \leq z \quad \text{if and only if } x \leq \tau(z)$$

for all flats $x \in G$, $z \in H$. For any such Galois connection between complete lattices, the composite [here $\tau(\sigma)$] of the two maps is a closure operator on the lattice *G*, and the set of closed elements of *G* itself forms a complete lattice.

A *function-space geometry* $G(E, V)$ is a geometry on a set E of points, determined by a finite-dimensional vector space of functions from the set E into a field F, as follows. For any subset $A \subseteq E$, the *hull* $\sigma(A)$ is a subspace of V,

$$\sigma(A) = \{\alpha \in V; \alpha(x) = 0 \text{ for every } x \in A\},$$

and for any subspace $S \subseteq V$, the *kernel* $k(S)$ is a subset of E:

$$k(S) = \{x \in E; \alpha(x) = 0 \text{ for every } \alpha \in S\}.$$

The mappings σ and k are *adjoint* to one another: For all subsets $A \subseteq E$ and all subspaces $S \subseteq V$,

$$S \subseteq \sigma(A) \quad \text{if and only if} \quad A \subseteq k(S),$$

these two statements meaning simply that $\alpha(x) = 0$ for all elements $x \in A$ and all functions $\alpha \in S$. The maps σ and k thus form a Galois connection between the Boolean algebra $\mathscr{B}(E)$ and the lattice $\mathscr{S}(V)$ of subspaces of the vector space V. It follows that both mappings carry joins to meets, and that the composites $k(\sigma)$, $\sigma(k)$ are closure operators on $\mathscr{B}(E)$ and $\mathscr{S}(V)$, respectively.

Especially significant for our purposes is the fact that σ also inverts, but otherwise *preserves*, the relation "covers or is equal to." Taking $\omega: \mathscr{S}(V) \to \mathscr{S}^{\text{opp}}(V)$ to be the map that inverts the partial order on the lattice $\mathscr{S}(V)$, we find that the composite $\sigma^{\text{opp}} = \omega(\sigma): \mathscr{B}(E) \to \mathscr{S}^{\text{opp}}(V)$ is a *strong map*. Consequently, the set of subsets of E closed with respect to the closure operator $k(\sigma)$ forms a geometric lattice, the *quotient* of the strong map σ^{opp}. These factors are summed up in the following propositions.

5.3.1. Proposition. *For any set E and any finite-dimensional subspace V of the vector space F^E of functions from the set E into a field F, the mapping $\sigma: \mathscr{B}(E) \to \mathscr{S}(V)$ from the Boolean algebra of subsets of E to the lattice of subspaces of V defined, for all subsets $A \subseteq E$, by*

$$\sigma(A) = \{\alpha \in V; \alpha(x) = 0 \text{ (for every } x \in A)\}$$

takes suprema in $\mathscr{B}(E)$ to infima in $\mathscr{S}(V)$, and inverts the relation $\downarrow$ ("covers or is equal to").

Proof. The mapping σ is well defined, because $\sigma(A)$ is a linear subspace of V. For any family $\{A_\mu\}$ of subsets of E, with $A = \bigcup_\mu A_\mu$,

$$\sigma(A) = \left\{\alpha \in V; \alpha(x) = 0 \text{ (for every } x \in \bigcup_\mu A_\mu)\right\}$$

$$= \bigcap_\mu \{\alpha \in V; \alpha(x) = 0 \text{ (for every } x \in A_\mu)\}$$

$$= \bigcap_\mu \sigma(A_\mu).$$

Assume that B covers A in $\mathcal{B}(X)$, so $B = A \cup p$ for some element $p \notin A$. Say $\sigma(B) \subset Q \subseteq \sigma(A)$ for some subspace Q of V. Then there is a function $\alpha \in Q$ with $\alpha(p) \neq 0$ and $\alpha(x) = 0$ for all $x \in A$. Let β be any function in $\sigma(A)$. Define a function γ by

$$\gamma(x) = \alpha(p)\beta(x) - \beta(p)\alpha(x).$$

Then $\gamma \in \sigma(B)$ because α and β are in $\sigma(A)$ and $\gamma(p) = 0$. Because $\alpha(p) \neq 0$, β is in the subspace spanned by α and γ, both of which are in Q. Thus, $\sigma(A) = Q$, and $\sigma(A)$ covers or is equal to $\sigma(B)$ in $\mathcal{S}(V)$. $\square$

We call the closed subsets of E *algebraic sets*.

5.3.2. Proposition. *For any set E and any n-dimensional subspace $V \subseteq F^E$, the lattice $G(E, V)$ of algebraic sets is geometric of rank n and is representable as a subgeometry of the projective geometry of rank n over F. The rank $r(A)$ of a subset $A \subseteq E$ is given by the formula*

$$r(A) = n - \lambda[\sigma(A)],$$

where $\lambda[\sigma(A)]$ is the rank of the subspace $\sigma(A) \subseteq V$ of functions equal to zero on A.

Proof. By Proposition 5.3.1, the map σ^{opp} is join-preserving and ($\downarrow$)-preserving. Further, it has the finite-basis property: For all subsets $A \subseteq E$, there is a finite subset $A_f \subseteq A$ such that $\sigma^{\text{opp}}(A_f) = \sigma^{\text{opp}}(A)$, because chains of algebraic sets in $\mathcal{B}(E)$ are bounded in length by n. For any maximal chain

$$\phi = A_0 \subset A_1 \subset \cdots \subset A_m = \bar{A}$$

of algebraic sets, if p_i is any point in the difference set $A_i \backslash A_{i-1}$ $(i = 1, \ldots, m)$, it follows that $A_i = \overline{A_{i-1} \cup p_i}$, so $\bar{A} = \overline{p_1 \cup \cdots \cup p_m}$, and σ^{opp} has the finite-basis property.

Thus, σ^{opp} is a strong map from $\mathcal{B}(E)$ to $\mathcal{S}^{\text{opp}}(V)$. The lattice of algebraic subsets of E, that is, the lattice of flats of the function-space geometry $G(E, V)$, is a geometric quotient of the free geometry $\mathcal{B}(E)$ and is isomorphic to the image of the strong map σ^{opp}, a subgeometry of the n-dimensional projective geometry $\mathcal{S}^{\text{opp}}(V)$. The restriction of σ^{opp} to the quotient lattice is *rank-preserving*, so for any subset $A \subseteq E$,

$$r(A) = \lambda[\sigma^{\text{opp}}(A)]$$
$$= n - \lambda[\sigma(A)].$$

The geometry $G(E, V)$ has rank n because $\sigma(E) = \{0\}$. $\square$

In place of the strong map σ^{opp} we might equally well have used the strong map $\sigma^*: \mathcal{B}(E) \to \mathcal{S}(V^*)$, which maps each subset $A \subseteq E$ to the subspace $\sigma^*(A)$ of linear functionals that are zero on all functions that are zero on A.

5.3.3. Proposition. *Under the strong map* σ^*: $\mathcal{B}(E) \to \mathcal{S}(V^*)$ *the image* $\sigma^*(A)$ *of any subset* $A \subseteq E$ *is the subspace spanned by the evaluations* π_p *for* $p \in A$, *where* $\pi_p(\alpha) = \alpha(p)$. *Thus, the function-space geometry* $G(E, V)$ *is isomorphic to the subgeometry of point evaluations in* V^*.

Proof. The lattices $\mathcal{S}^{\text{opp}}(V)$ and $\mathcal{S}(V^*)$ are isomorphic under the mapping τ, where, for any subspace $W \subseteq V$, $\tau(W)$ is the subspace consisting of all multiples $k\pi_p$ of the evaluation π_p.

Any multiple $k\pi_p$ of the functional π_p is in $\sigma^*(p)$. For the converse, observe that the kernel $\{\alpha \in V; g(\alpha) = 0\}$ of any linear functional $g \in V^*$ either is equal to V or is a hyperplane in V, and that any two linear functionals with the same kernel are multiples of one another. Furthermore, if f, g are linear functionals for which $\ker f \subseteq \ker g$, then g is a multiple of f (perhaps equal to 0). Consequently, if $g \in V^*$ is in $\sigma^*(p)$, then, because the kernel of g contains the hyperplane (or entire space) $S = \{\alpha \in V; \alpha(p) = 0\} = \ker \pi_p$, g is some multiple $k\pi_p$ of π_p. The exact value of k can be computed at any function $\alpha \in V$ for which $\alpha(p) \neq 0$, by the formula $k = g(\alpha)/\alpha(p)$. $\square$

As an important and very general example of function-space geometries, consider any basis E for a vector space T over a field F and any subspace $V \subseteq T$. Any vector $\alpha \in V$ can be expressed uniquely as a linear combination of elements of the basis E, with coefficients in F, and can thus be represented as a *function* $\alpha(x)$ from E into the field F.

Indeed, any function-space geometry on a finite set can be visualized in this way. Given a vector space V of functions from a finite set E into a field F, take E to be the basis of a vector space W, take each function $\alpha \in V$ to be a linear combination of elements of E, and identify V with a subspace of the space W. In this way, every function-space geometry on a finite set is represented by a subspace of a vector space in which a definite basis is prescribed.

As a further and equally general example along these lines, consider any $n \times m$ matrix M over a field F. The n rows of this matrix form a vector geometry that is invariant with respect to multiplication of M on the right by an $m \times m$ nonsingular matrix. That is, any column-equivalent matrix yields the same row geometry. All that counts is the *column space* V and its representation with respect to the standard basis E (the basis that relates the matrix M to its column space V). The *flats* of this geometry are the zerosets of vectors in the column space, together with all intersections of such zero-sets. For any sufficiently rich field, such as the real numbers, we can form a *general* linear combination of any given finite set of vectors and can thus produce a *single vector* whose zero-set is equal to the intersection of the zero-sets of the given vectors. In such a case, the flats of the row geometry are simply all zero-sets of vectors in the column space.

The rows of the matrix M give the values of the column "functions," and thus determine the evaluation functionals π_p, $p \in E$ of Proposition 5.3.3.

Consequently, the function-space geometry $G(E, V)$ is isomorphic to the *geometry of rows* of M, a straightforward vector geometry.

This matrix example is equally as general as the examples which precede it. Given any subspace V of a finite-dimensional vector space T with prescribed basis E, choose any matrix M whose columns are vectors in T that *span* the subspace V. The rows of this matrix M correspond to the elements of the set E and, as above, the row geometry of M is the function-space geometry $G(E, V)$.

5.4. ORTHOGONALITY OF VECTOR GEOMETRIES

Perhaps the most direct introduction to orthogonality of vector geometries is in terms of matrices. Given any $n \times k$ matrix M with row geometry G over a field F, we can construct (in many ways) a matrix N each of whose columns is orthogonal to all the columns of M, and such that the columns of M and N, taken together, span the entire n-dimensional vector space F^n. The row geometries of any such pair of matrices are orthogonal; that is, $H = G^*$, where H is the row geometry of the matrix N. These facts, as well as those sketched in the following introductory paragraphs, will be proved later.

For a function-space geometry, or for the geometry given by a subspace in a vector space with fixed basis, the construction of the orthogonal geometry is equally simple. Given a function-space geometry $G(E, V)$ on a finite set E, we can construct the space $V^\perp$ of functions g that are orthogonal to all functions $f \in V$ over E, in the sense that

$$\sum_{x \in E} f(x)g(x) = 0 \qquad \text{(for every } f \in V\text{)}.$$

Then the orthogonal geometry $G^*(E, V)$ is isomorphic to the function-space geometry $G(E, V^\perp)$.

If, instead, we are given a subspace V in a finite-dimensional vector space T with prescribed basis E, we construct the annihilator V^* of V in the dual space T^* of linear functionals on T and prescribe the dual basis E^* for T^*. Then the geometries $G(E, V)$ and $G(E^*, V^*)$ are orthogonal to one another.

5.4.1. Proposition. *If M and N are both matrices with n rows over a field F, such that each column of M is orthogonal to every column of N, and such that the columns of M and N, taken together, span F^n, then the row geometries of M and N are orthogonal.*

Proof. Let G and H be the row geometries of M and N, respectively. Let A, $\{p\}$, B be a partition of the n-element set E of rows of M (and of N) into three (disjoint) parts. Then the row p is in the G-closure of rows A if and only if row p is expressible as a linear combination of the rows A, that is, if there is an n-vector α such that $\alpha_p \neq 0$ while $\alpha_i = 0$ for all rows $i \in B$ and $\sum_{i \in E} \alpha_i M_{ij} = 0$ for all columns j. This n-vector is thus orthogonal to all the columns of M. Because the columns of M and N together span all of F^n, the column space of N not only is contained in but also is equal to the orthogonal complement of the column space of M. Thus, the vector α, being in this orthogonal complementary space, is in the column space of the matrix N. So far, we see that $p \in \bar{A}^G$ if and only if there is an n-vector α in the column space of N, with $\alpha_i = 0$ for all $i \in B$, and $\alpha_p \neq 0$.

Any such vector α is expressible as a linear combination of the columns of N, using coefficients that form an m-vector γ (m being the number of columns of N). This vector γ is orthogonal to the rows B of N and has a nonzero inner product with row p of N. The existence of such a vector γ is a necessary and sufficient condition that the row p of N be outside the space spanned by the rows B of N. Thus it is a necessary and sufficient condition that $p \notin \bar{B}^H$. This completes the proof that $p \in \bar{A}^G$ if and only if $p \notin \bar{B}^H$. So the row geometries G and H of these matrices are orthogonal by Proposition 5.2.5. $\square$

Let us look more closely at this example of orthogonal geometries. In particular, we look at vectors with minimal nonempty support in the column spaces of the matrices M and N. Such vectors are obtained by *elimination* between column vectors, carefully choosing linear combinations of columns so as to obtain zero entries in certain positions (rows). The minimal nonempty supports obtained by elimination among the columns of the matrix M are the *bonds* of the row geometry G.

On the other hand, every column of the matrix M gives a *dependence* between the rows of the matrix N. Indeed, the column space of M is exactly the space of dependences between the rows of N. The minimal nonempty supports of such vectors are precisely the *circuits* of the geometry H of rows of N. These observations merely underline a fact we proved in Theorem 5.2.6 for all pairs of orthogonal geometries: If $H = G^*$, then all the bonds of G are the circuits of H, and conversely.

As a special case of such pairs of matrices yielding orthogonal row geometries, consider an $n \times m$ matrix $M = I : A$ that consists of an $m \times m$ identity matrix I placed above an arbitrary $(n - m) \times m$ matrix A. A companion matrix having the orthogonal row geometry is easily constructed by placing the negative transpose $-A^{\text{tr}}$, an $m \times (n - m)$ matrix, above an $(n - m) \times (n - m)$ identity matrix J, to form an $n \times (n - m)$ matrix $N = -A^{\text{tr}} : J$.

5.4.2. Proposition. *The matrices $I : A$ and $-A^{\text{tr}} : J$ described in the preceding paragraph satisfy the hypotheses of Proposition 5.4.1 and thus have orthogonal row geometries.*

Proof. The column spaces of M and N are orthogonal because the inner product of the ith column of M with the jth column of N is simply

$$-A_{ij}^{\mathrm{tr}} + A_{ji} = -A_{ji} + A_{ji} = 0.$$

Because the column spaces of M and N are orthogonal and of ranks m and $n - m$, respectively, they are also complementary, and together they span all of F^n.   □

5.4.3. Proposition. *For any function-space geometry $G(E, V)$ on a finite set E, the orthogonal geometry $G^*(E, V)$ is isomorphic to the function-space geometry $G(E, V^\perp)$, $V^\perp$ being the space of functions orthogonal to all of V relative to the inner product*

$$(f, g) = \sum_{x \in E} f(x)g(x).$$

Proof. If we represent the function-space geometries $G(E, V)$ and $G(E, V^\perp)$ by matrices with a set E of rows, their columns spanning the vector spaces V and $V^\perp$, respectively, then the two matrices have orthogonal complementary column spaces, and thus orthogonal row geometries, by Proposition 5.4.1.   □

5.4.4. Proposition. *Given a prescribed basis E for a finite-dimensional vector space T, and a subspace $V \subseteq T$, the associated geometry $G(E, V)$ has orthogonal geometry $G^*(E, V)$ isomorphic to $G(E^*, V^*)$, where E^* is the dual basis for the dual space T^*, and V^* is the annihilator of V, a subspace of T^*.*

Proof. Any linear functional annihilating the vector space V can be expressed as a linear combination of the coordinate-evaluation functionals associated with the basis E, and thus as an n-tuple orthogonal to all the n-tuple representations of vectors in V relative to the basis E. This is the situation in Proposition 5.4.3, and the same proof applies.   □

5.5. ORTHOGONALITY OF SIMPLICIAL GEOMETRIES

A nontrivial example of pairs of orthogonal geometries arises in the study of simplicial geometries. As was mentioned in Section 1.3.C, the *simplicial geometry* S_k^n is a geometry whose points are the k-element subsets of an n-element set X, a geometry coordinatized by the rows of its *boundary matrix*. Simplicial geometries will be treated in detail in a later volume. We use notation from Section 1.3.C.

What we shall show presently is that the simplicial geometries S_k^n and S_{n-k}^n can be coordinatized by the rows of two matrices, the column spaces of which are orthogonal complements of one another in the vector space F^m, where $m = \binom{n}{k}$. It then follows from Proposition 5.4.1 that the row geometries are orthogonal, and

$$S^n_{n-k} \simeq (S^n_k)^*.$$

This theorem, to be proved later, is equivalent to the Alexander duality theorem applied to a simplex, namely, that for any subset $A \subseteq \binom{X}{k}$,

$$\beta_{k-1}\left[\binom{X}{k} \setminus A\right] = \beta_{n-k-2}(A'),$$

these being Betti numbers for the complexes $\binom{X}{k} \setminus A$ of k-element subsets not in the family A, and A' of all complements in X of subsets in the family A.

To simplify the proof of this orthogonality theorem, we introduce the following notation: Let a finite set X be listed in a fixed linear order. For any two *disjoint* subsets A, B of X, compare the linear order induced on the subset $A \cup B$ with the linear order (A, B) obtained by listing A first, then B, each in the induced order. The sign (1 or -1) of this permutation $A \cup B \to (A, B)$ we denote

$$\text{sign}(A, B).$$

5.5.1. Proposition. *The simplicial geometries S^n_k and S^n_{n-k} are orthogonal.*

Proof. The points of S^n_k are the k-element subsets of an n-element set X. For each such point $D \subseteq X$, we take the complementary set $X \setminus D$ as the corresponding point in the geometry S^n_{n-k}. We coordinatize S^n_k by the usual matrix M; so the entry in row C, column A, of M is equal to zero unless $A \subseteq C$, in which case the entry is equal to $\text{sign}(p, A)$, where p is the single element in the difference set $C \setminus A$.

If we were to coordinatize S^n_{n-k} in the usual fashion, the columns of the resulting matrix would not be orthogonal to the columns of M. Instead, we multiply each row C of the usual matrix for S^n_{n-k} by $\text{sign}(C, X \setminus C)$. Thus, the entry in row $X \setminus C$, column B, of the resulting matrix N is equal to zero unless $B \subseteq X \setminus C$, in which case the entry is equal to

$$\text{sign}(q, B)\, \text{sign}(X \setminus C, C),$$

where q is the unique element in the difference set $(X \setminus C) \setminus B$.

For any fixed $(k - 1)$-element subset $A \subseteq X$ and any fixed $(n - k - 1)$-element subset $B \subseteq X$, consider the associated columns in the matrices M and N, respectively. The entries in these columns are somewhere simultaneously nonzero only when the sets A and B are *disjoint*, in which case $A \cup B = X \setminus \{p, q\}$ for some pair p, q of elements of X. The only two rows in which these entries are simultaneously nonzero are $A \cup p$ and $A \cup q$ in

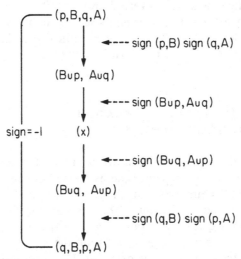

sign = -1

FIGURE 5.2. A factorization of a permutation.

TABLE 5.1. Orthogonal Simplicial Geometries

	The matrix M for the simplicial geometry S_2^5						The matrix N for the simplicial geometry S_3^5; each row C has been multiplied by $\text{sign}(C, X\setminus C)$									
	1	2	3	4	5			12	13	14	15	23	24	25	34	35	45
12	1	−1					345								1	−1	1
13	1		−1				245						−1	1			−1
14	1			−1			235					1		−1		1	
15	1				−1		234					−1	1		−1		
23		1	−1				145			1	−1						1
24		1		−1			135		−1		1					−1	
25		1			−1		134		1	−1					1		
34			1	−1			125	1					−1		1		
35			1		−1		124	−1		1						−1	
45				1	−1		123	1	−1			1					

M, corresponding to rows $B \cup q$ and $B \cup p$, respectively, in N. The inner product of these columns A in M and B in N is thus equal to

$$\text{sign}(p, A)\,\text{sign}(q, B)\,\text{sign}(B \cup q, A \cup p)$$
$$+ \text{sign}(q, A)\,\text{sign}(p, B)\,\text{sign}(B \cup p, A \cup q).$$

This *sum* of products is equal to zero if and only if the *product* of the six (± 1)-factors is equal to -1. This is easily seen to be the case by factoring the odd permutation $(p, A, q, B) \to (q, A, p, B)$ as in Figure 5.2. □

The example $(S_2^5)^* \simeq S_3^5$ is shown in Table 5.1.

5.6. ORTHOGONALITY OF PLANAR GRAPHIC GEOMETRIES

A fundamentally different example of orthogonality between simplicial geometries arises in the theory of planar graphs. Every 3-connected planar graph (and, thus, by the theorem of Steinitz, the edge-skeleton of any convex polyhedron) has a unique *dual graph*. This proposition is proved in the following chapter in greater generality, as the hypothesis of 3-connectedness is omitted. In this more general situation the dual graph is no longer unique (although the corresponding orthogonal matroid still is unique). The proof of this important result is given both in this chapter, where it is easier to follow because of the stronger hypothesis, and in the next chapter in its full generality.

5.6.1. Proposition. *The geometries of a 3-connected planar graph and of its dual graph are orthogonal.*

Proof. An edge e connecting vertices a and b in a 3-connected planar graph H is in the H-closure of a set $A \subseteq E$ of edges of H if and only if there is an A-path, that is, a sequence of edges in A, that links the vertices a and b. (*Note:* If e is a loop, the empty sequence suffices to prove that e is in the closure of the empty set.) Let $A, \{e\}, B$ be any three-part partition of the set E of edges of H. Let $G(H)$, $G^*(H)$, and $G(H^*)$ denote the geometry of the graph H, its orthogonal geometry, and the geometry of the dual graph H^*, respectively. Then the edge e is in the $G^*(H)$-closure of the set A if and only if e is not in the $G(H)$-closure of the set B, if and only if there is no B-path from a to b in H. Let s and t be the dual vertices connected by the dual edge e^*. H and H^* being dual graphs, there is no B-path from a to b in H if and only if there is an A-path from s to t in H^*, if and only if e is in the $G(H^*)$-closure of the set A. Thus, the $G^*(H)$- and $G(H^*)$-closures of any set A are equal, and the geometries $G^*(H)$ and $G(H^*)$ are isomorphic. □

An example of the orthogonality of geometries of dual graphs is shown in Figure 5.3 (although this example is not 3-connected). Note how this example differs from those discussed in the preceding section. In the latter, we find an orthogonality between the geometry S_2^n of the complete graph K_n on n vertices and a geometry S_{n-2}^n of $(n-2)$-element sets.

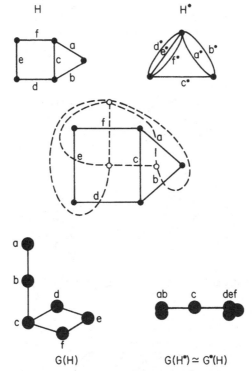

FIGURE 5.3.   Dual graphs and their orthogonal matroids.

5.7. RESEARCH PROBLEM: ORTHOGONALITY BETWEEN OTHER PAIRS OF SIMPLICIAL GEOMETRIES

There is certainly more to be known about orthogonality of simplicial geometries. For one thing, the orthogonality between geometries of dual graphs is not a special case of the orthogonality between the complete simplicial geometries S_k^n and S_{n-k}^n. One possibility is that there is an orthogonality theorem for more general sorts of relative duality, generalizing the planar-graph case of duality relative to the 2-sphere. As an indication that this may be true, we give the following example of orthogonality between simplicial geometries for complexes dual with respect to the 3-sphere.

In this example we deal with topological vertices, arcs, faces, and regions of space, rather than simply with sets of vertices. The geometries corresponding to skeletons in each dimension can then be defined in terms of Betti numbers. For instance, a circuit of two-dimensional faces is simply a minimal set of faces having $\beta_2 = 1$. In the examples we have studied, the

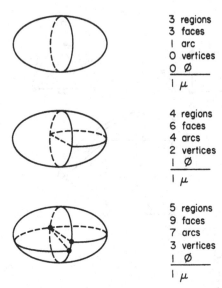

3 regions
3 faces
1 arc
0 vertices
0 ∅
———
1 μ

4 regions
6 faces
4 arcs
2 vertices
1 ∅
———
1 μ

5 regions
9 faces
7 arcs
3 vertices
1 ∅
———
1 μ

FIGURE 5.4. Simplicial decomposition of a 3-sphere.

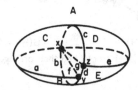

FIGURE 5.5. Parts of the decomposition T.

only difference between the geometry of a complex and that of a subdivision is that between two geometries with the same orthogonal geometry, but different orthogonal matroids.

Consider a real three-dimensional sphere (i.e., a real 3-space with a single point at infinity) cut into two regions by a two-dimensional sphere, then interior faces added one by one, each separating into two some single interior region. Figure 5.4 shows such a construction.

This simplicial decomposition of the 3-sphere we denote by T. We form an orthogonal decomposition T^* by placing a *-vertex within each region of T, drawing a *-arc across each face of T, thereby connecting each adjacent pair of *-vertices, and for each arc q of T, sketching a *-face between the *-arcs corresponding to those faces meeting the arc q. Each vertex of the decomposition T will then be surrounded by *-faces and will correspond to a region of the decomposition T^*.

We label the various parts of the decomposition T as in Figure 5.5. The construction of the *-arc from C^* to E^* and the construction of the

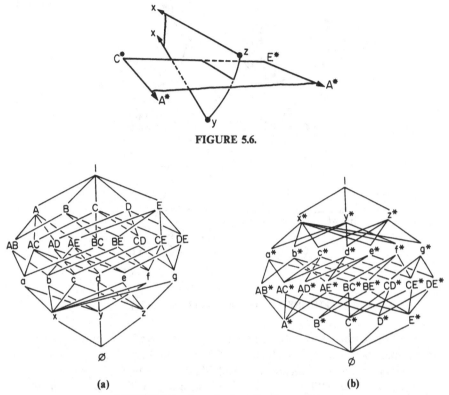

FIGURE 5.6.

FIGURE 5.7. An order-inverting isomorphism.

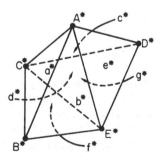

FIGURE 5.8. An orthogonal decomposition of the 3-sphere.

triangular *-face orthogonal to the arc *d* are shown in cutaway section in Figure 5.6. If we order the empty set, vertices, arcs, faces, and regions by inclusion, for both decompositions *T* and *T**, we see (Figure 5.7) that passage from *T* to *T** induces an order-inverting isomorphism between their incidence schemes. These schemes are not lattices.

The orthogonal decomposition *T** of the 3-sphere is shown in Figure 5.8. The faces are labeled with lowercase letters. The region *x** is the exterior

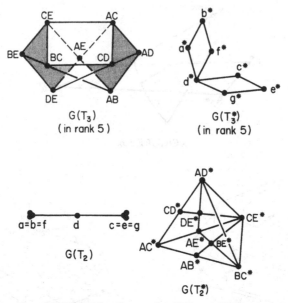

FIGURE 5.9. Geometries $G(T_k)$ and $G(T_k^*)$.

region. The region y^* is within the tetrahedron $A^*B^*C^*E^*$, and the region z^* occupies the other tetrahedron $A^*C^*D^*E^*$. The face d^* is $A^*C^*E^*$.

Finally, the geometries $G(T_k)$ and $G(T_k^*)$ are drawn in Figure 5.9. The relations $G(T_k^*) \simeq G^*(T_{n-k})$, for $k = 2, 3$ and $n = 5$ in this case, are a geometric consequence of the Poincaré duality theorem.

5.8. THE ORTHOGONAL OF A STRUCTURE GEOMETRY

Let a 3-connected planar graph H be realized by a plane drawing. This need not be a *planar* drawing; we admit the possibility that edges cross each other in the plane representation. Consider all possible spatial preimages H' of H under projection from a fixed point q (at infinity, if you like), such that the *faces* of the planar graph H are flat (coplanar sets of vertices) in the spatial preimage H'. These are the *polyhedral* preimages of H.

An edge e is *articulated* in a polyhedral preimage H' if and only if the two faces incident with the edge e are not coplanar in H'. As we shall prove in a chapter on structure geometries in a later volume, a set A of edges in H is *dependent* in the structure geometry $S(H)$ if and only if some subset $C \subseteq A$ is the set of articulated edges of some polyhedral preimage H' of H.

For any such plane representation of a graph H, we define the *folding geometry* $F(H)$ of H as follows: We say that an edge e is in the $F(H)$-closure

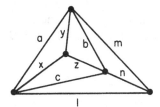

FIGURE 5.10.   Plane projection of a triangular bipyramid.

$\bar{A}$ of a set A of edges if and only if, for every polyhedral preimage H' of H, if no A-edges are articulated in H', then the edge e is not articulated either. That is, the flats of the folding geometry $F(H)$ are all (intersections of) sets of *nonarticulated edges* for the various polyhedral preimages H' of H. We place the words "intersections of" in parentheses because they can be omitted. We are working over the real field and can thus form general linear combinations of polyhedral preimages. In this way we can represent an arbitrary union of sets of articulated edges as the set of articulated edges of a *single* polyhedral preimage.

5.8.1. Proposition. *For any representation H of a 3-connected planar graph in the real plane, its structure geometry $S(H)$ and its folding geometry $F(H)$ are orthogonal.*

Proof. An edge e is in the orthogonal (S^*) closure of a set A of edges if and only if the edge e is not in the S-closure of the set $B = E \backslash (A \cup e)$, that is, if and only if there is no polyhedral preimage H' of H in which the edge e is articulated, and other than this, only edges in B are articulated. That is, the edge e is in the S^*-closure of the set A if and only if for every polyhedral preimage in which the edges in A are not articulated, e is not articulated either. Thus, the S^*- and F-closures agree for all subsets $A \subseteq E$, and $S^*(H) \simeq F(H)$. $\square$

For example, consider the plane projection H of a triangular bipyramid, as in Figure 5.10. If any one of the three edges at a trivalent vertex is not articulated, the vertex star must be coplanar, and the other two edges at that vertex are not articulated either. Thus, the sets lmn and xyz have nullity 2 in the folding geometry $F(H)$. If one of the edges a, b, or c is not articulated, the polyhedral preimage can be at best a rectangular pyramid (with base $lmxy$ if edge a is not articulated). If neither the edge a nor the edge b is articulated, then the preimage must be in a single plane. The set ab forms a *basis* for the geometry $F(H)$. This being the case, the complementary set $clmnxyz$ should form a basis for the structure geometry $S(H)$. To verify this, observe that xzc and cnl form two triangles sharing an edge, and the remaining vertex is attached to this minimally rigid structure by two bars y and m. This is minimally rigid (in general position) in the plane, and thus it forms a basis for the structure geometry $S(H)$ whenever H is realized in general position in the plane.

REFERENCES

Crapo, H., and Rota, G.-C. (1970). *On the Foundations of Combinatorial Theory: Combinatorial Geometries*, preliminary edition. M.I.T. Press, Cambridge, Mass.
Tutte, W. T. (1956). A class of abelian groups. *Canad. J. Math.* **8**: 13–28.
Whitney, H. (1935). On the abstract properties of linear dependence. *Amer. J. Math.* **57**: 509–33.

CHAPTER 6

Graphs and Series-Parallel Networks

James Oxley

The two fundamental examples in Whitney's seminal paper on independence theory (Whitney 1935) were vector matroids and graphic matroids. It is therefore not surprising that so many aspects of matroid theory are extensions and developments of concepts that were originally introduced in vector spaces or in graphs. In this chapter we shall study in detail the class of graphic matroids. The first section will consider the polygon and bond matroids of a graph, that is, those matroids on the edge-set of a graph Γ whose circuits are respectively the circuits and the bonds of Γ. The main result of Section 6.1 characterizes those graphs whose bond matroids are graphic. The polygon and bond matroids are not the only matroids that can be defined on the edge-set of a graph. Several other such matroids will be considered in detail in a chapter on matroidal families in a later volume.

In Section 6.2 we shall consider the concept of connectivity in matroid theory, indicating its relationship to various notions of connectivity for graphs. These ideas will be used in Section 6.3, where the main result of this chapter is proved. This result, due to Whitney (1933), characterizes precisely when two graphs have isomorphic polygon matroids. In the final section we shall investigate two graph-theoretic operations having their origins in electrical-network theory. One of the important aspects of these operations in the present context is the fact that they have been very successfully generalized to matroids. A detailed discussion of these matroid operations will be given in Chapter 7.

It will be assumed in this chapter that the reader has some familiarity with graphs and with the basic terminology of graph theory. Where a graph-theoretic concept is left undefined here, we refer the reader to Bondy and Murty (1976). All graphs considered in this chapter will be assumed to be finite. Consideration of independence structures on infinite graphs will be postponed until a chapter in a later volume, where matroids on infinite sets will be thoroughly investigated.

6.1. POLYGON MATROIDS, BOND MATROIDS, AND PLANAR GRAPHS

It was noted in Chapters 1 and 2 that if Γ is an arbitrary graph having edge-set E, then we can obtain a matroid $M_\Gamma(E)$ on E by taking the edge-sets of forests in Γ to be the independent sets of the matroid. We call $M_\Gamma(E)$ the *polygon matroid* of Γ. An arbitrary matroid is *graphic* if it is isomorphic to the polygon matroid of some graph. Evidently, the circuits of the matroid $M_\Gamma(E)$ are precisely the edge-sets of the circuits (polygons) in the graph Γ. Thus, M_Γ is a geometry if and only if Γ has no loops or parallel edges. If Δ is the graph in Figure 6.1, then the circuits of $M_\Delta(E)$ are $\{1\}, \{2,3,4,5\}$, $\{3,6,7\}$, and $\{2,4,5,6,7\}$.

No proof was given in the earlier chapters of the fact that $M_\Gamma(E)$ is indeed a matroid. We shall remedy this by showing directly that the edge-sets of circuits in Γ obey the matroid circuit axioms. Before doing this, however, we shall sketch an alternative proof that has as a consequence the fact that $M_\Gamma(E)$ is coordinatizable over all fields.

Take a graph Γ and arbitrarily assign a direction to each edge of Γ. Now form the matrix A_Γ as follows: Index the columns by the edges of Γ and the rows by the vertices of Γ; the entry in the (i,j) position of the matrix is zero unless the edge j meets the vertex i, and j is not a loop. In that case the (i,j) entry is $+1$ if j begins at i, and -1 if j ends at i. An example of the formation of such a matrix is given in Figure 6.2.

Now, for an arbitrary field F, let $A_\Gamma(F)$ denote the matrix obtained by reducing each entry of A_Γ modulo F. This leaves A_Γ unchanged unless F

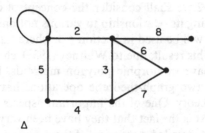

FIGURE 6.1. A graph.

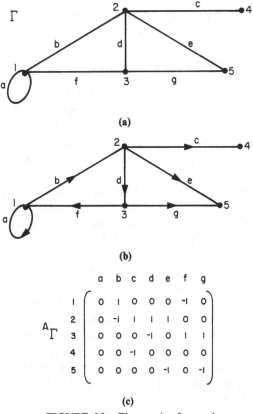

FIGURE 6.2.   The matrix of a graph.

has characteristic 2, in which case each negative entry changes sign. We leave the reader to verify that the edge-sets of circuits of Γ are precisely the circuits of the vector matroid of $A_\Gamma(F)$. From this it follows that $M_\Gamma(E)$ is a matroid, and furthermore that $M_\Gamma(E)$ is coordinatizable over all fields.

We now give a direct graph-theoretic proof of the fact that $M_\Gamma(E)$ is a matroid.

6.1.1. Proposition. *Let Γ be an arbitrary graph. Then the set $\mathscr{C}$ of edge-sets of circuits in Γ obeys the matroid circuit axioms.*

Proof. Clearly the empty set is not in $\mathscr{C}$. Moreover, Γ does not have two distinct circuits with the edge-set of one contained in the edge-set of the other.

To complete the proof we need to check that $\mathscr{C}$ satisfies the circuit-elimination axiom (c3) (see Chapter 2). Here and frequently throughout this chapter we shall adopt the standard practice of not distinguishing between various subgraphs of Γ and their edge-sets. Assume that C_1 and C_2 are distinct circuits of Γ, and suppose that x is an edge of both C_1 and C_2 having endpoints u and v. To construct a circuit in Γ that uses only edges of $(C_1 \cup C_2) - x$, we proceed as follows: Beginning at the vertex u, we follow the path $C_2 - x$ until the first time a vertex w is reached at which the next edge in $C_2 - x$ is not in C_1. Because $C_2 \neq C_1$, such a vertex does exist and may or may not equal u. Now we continue to follow $C_2 - x$ in the same direction until the next time we encounter a vertex z in C_1. Because, in particular, v is in C_1, there must be such a vertex z. Both $C_1 - x$ and $C_2 - x$ contain exactly one path that joins w and z, and putting these two paths together gives a circuit C_3 that has the required properties. $\square$

The circuits in a polygon matroid M_Γ are easily recognized from the graph Γ. What about the other fundamental subsets? Can you find the 11 bases of M_Δ, where Δ is the graph in Figure 6.1? This graph is *connected*; that is, if u and v are distinct vertices of Δ, they are joined by a path. It is straightforward to check that for an arbitrary connected graph Γ, the bases of M_Γ are the edge-sets of the spanning trees in Γ. More generally, suppose that Γ is the disjoint union of k connected graphs $\Gamma_1, \Gamma_2, \ldots, \Gamma_k$. (We then call $\Gamma_1, \Gamma_2, \ldots, \Gamma_k$ the *connected components* or just the *components* of Γ.) In such a case, the bases of M_Γ are the spanning forests of Γ, that is, all sets of the form $\bigcup_{i=1}^{k} T_i$, where T_i is the edge-set of a spanning tree of Γ_i. In Figure 6.1, the edge 8 is an isthmus of Δ and appears in every basis of M_Δ; the edge 1 is a loop and occurs in no basis. In an arbitrary matroid M, an *isthmus* is an element that appears in every basis, and a *loop* is an element that does not appear in any basis.

By an easy induction argument it can be shown that the number of vertices in a tree is 1 more than the number of edges in the tree. From this it follows that if Γ is a connected graph having vertex-set $V(\Gamma)$, then a spanning tree of Γ has $|V(\Gamma)| - 1$ edges; so M_Γ has rank $|V(\Gamma)| - 1$. In general:

6.1.2. Proposition. *If Γ has $\omega(\Gamma)$ connected components, then M_Γ has rank $|V(\Gamma)| - \omega(\Gamma)$.*

Note that if we choose a single vertex from each component of Γ and form a new graph Γ' by identifying all these vertices, then $M_\Gamma \cong M_{\Gamma'}$. Hence, every graphic matroid is isomorphic to the polygon matroid of a connected graph. A solution to the problem of determining the precise relationship between two graphs that have isomorphic polygon matroids will be presented in Section 6.3.

If the graph Γ has vertex-set $V(\Gamma)$ and edge-set $E(\Gamma)$ and $V' \subseteq V(\Gamma)$, we shall denote by $\Gamma[V']$ the subgraph of Γ *induced* by V'. This subgraph has vertex-set V' and edge-set consisting of all members of $E(\Gamma)$ having both endpoints in V'. For a subset E' of $E(\Gamma)$, the subgraph $\Gamma[E']$ of Γ *induced*

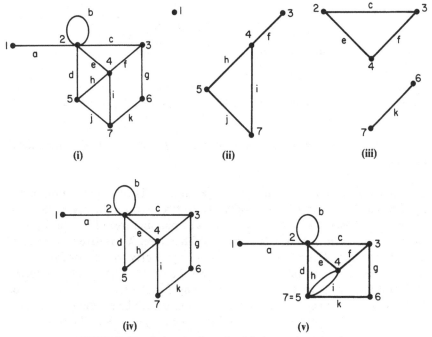

FIGURE 6.3. Induced subgraphs, deletion, and contraction.

by E' has edge-set E' and vertex-set consisting of all endpoints of edges in E'. Thus, if Γ is the graph in Figure 6.3(i), the subgraphs induced by $\{1,3,4,5,7\}$ and $\{c,e,f,k\}$ are shown in (ii) and (iii).

The graphs in (iv) and (v) are obtained by respectively deleting and contracting the edge j from Γ. In general, if Γ is a graph having edge-set S and x is an edge of Γ with endpoints u and v, then $\Gamma - x$ is the subgraph of Γ obtained by *deleting* x, that is, $\Gamma - x$ has vertex-set $V(\Gamma)$ and edge-set $S - x$; the graph Γ/x is obtained from Γ by *contracting* x, that is, by identifying u and v and then deleting x. These operations can be extended in the obvious way so that instead of deleting or contracting single edges from Γ, we can delete or contract an arbitrary set X of edges, denoting the graphs thus obtained by $\Gamma - X$ and Γ/X, respectively. The important matroid generalizations of these operations will be looked at in the next chapter.

In a graph Γ, a set of edges whose deletion from Γ increases the number of connected components is called an *edge cut* of Γ. A minimal edge cut is called a *bond*. We recall from Chapter 5 that the bonds of an arbitrary matroid M are the minimal sets meeting every basis of M. The next result shows that, just as with the term *circuit*, the matroid use of the term *bond* agrees with its use for graphs.

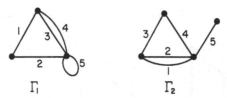

FIGURE 6.4. Dual graphs (the bond matroid of Γ_1 is the polygon matroid of Γ_2).

6.1.3. Proposition. X *is a bond in* M_Γ *if and only if* X *is a minimal edge cut in* Γ.

Proof. Suppose that X is a minimal edge cut in Γ. Then Γ has a component Γ_i such that the deletion of X breaks Γ_i into two components. Now let B be an arbitrary basis of M_Γ. Then B contains a spanning tree of Γ_i. But $\Gamma_i - X$ has two components; hence, $X \cap B \neq \varnothing$. Therefore, X meets every basis of M_Γ. The fact that X is a minimal set with this property will follow from what we show next, namely, that a set Z of edges that meets every spanning forest of Γ contains an edge cut. Now a spanning forest of $\Gamma - Z$ is a basis of $M_{\Gamma-Z}$ and an independent set of M_Γ. It can therefore be extended to a basis of M_Γ. But Z meets every such basis; hence, $r(M_{\Gamma-Z}) < r(M_\Gamma)$. Thus, by Proposition 6.1.2,

$$|V(\Gamma - Z)| - \omega(\Gamma - Z) < |V(\Gamma)| - \omega(\Gamma).$$

But $V(\Gamma) = V(\Gamma - Z)$. Therefore, $\omega(\Gamma - Z) > \omega(\Gamma)$, and so Z contains an edge cut of Γ and hence contains a minimal edge cut of Γ. It follows easily that the bonds in M_Γ coincide with the minimal edge cuts in Γ, and so the proposition is proved.   □

Because the circuits of M_Γ^* are precisely the bonds of M_Γ, we call M_Γ^* the *bond matroid* of Γ. An arbitrary matroid that is isomorphic to the bond matroid of some graph is called *cographic*. As an example, consider the graphs Γ_1 and Γ_2 in Figure 6.4. It is not difficult to check that if $E = \{1, 2, 3, 4, 5\}$, then the identity map on E induces an isomorphism between $M_{\Gamma_1}^*(E)$ and $M_{\Gamma_2}(E)$. Hence, in this example, the bond matroid of Γ_1 is graphic, or, equivalently, the polygon matroid of Γ_1 is cographic. The remainder of this section will be concerned with finding general conditions on a graph Γ under which its bond matroid is graphic. As we shall see, this occurs precisely when Γ is planar. We now recall some definitions that will be needed.

A *plane graph* is a graph drawn in the Euclidean plane so that the vertices are points of the plane, the edges are Jordan curves, and two distinct edges do not intersect except possibly at their endpoints. A graph Γ is *planar* if it is isomorphic to a plane graph Γ'. Under such circumstances, Γ' is called a *planar embedding* of Γ. The five-vertex graph Γ shown in Figure 6.5 is not

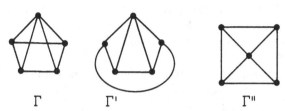

FIGURE 6.5. A graph and two planar embeddings.

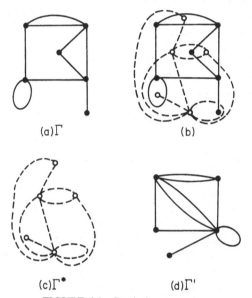

FIGURE 6.6. Dual planar graphs.

a plane graph. However, because it is isomorphic to the plane graph Γ', it is planar. The graph Γ'' is a different planar embedding of Γ.

A plane graph Γ partitions the Euclidean plane into regions. Such regions, called *faces* of Γ, can be formally defined as follows: Suppose that $\mathcal{P}$ is the set of points of the plane that are not vertices of Γ and that do not lie on edges of Γ. Two points x and y of $\mathcal{P}$ are in the same face of Γ if there is a Jordan curve joining x and y, all points of which lie in $\mathcal{P}$. A plane graph has exactly one unbounded face; it is called the *infinite face*.

To prove that M_Γ^* is graphic when Γ is planar, we construct a graph whose polygon matroid is isomorphic to M_Γ^*. We illustrate this construction by reference to the example in Figure 6.6. Unless one feels the need for a more formal description of the construction, which is provided later, one is advised to study the example and then move on to the next lemma.

In Figure 6.6, parts (a) and (c), the plane graph Γ and its geometric dual Γ^* are shown. In (b), Γ and Γ^* have been superimposed, showing that

Γ^* is formed by taking a vertex in each face of Γ and joining two such vertices when the corresponding faces share an edge. The edges of Γ^* are drawn so as to cross the corresponding edges of Γ. The graph Γ' in (d) is a different planar embedding of Γ^*.

   Given a plane graph Γ, the construction of the graph Γ^*, the *geometric dual* of Γ, is formally described as follows: Choose a single point v_F in each face F of Γ. These points are to be the vertices of Γ^*. If the set of edges bounding both of the faces F and F' is $\{e_1, e_2, \ldots, e_k\}$, then we join v_F and $v_{F'}$ by k edges $e'_1, e'_2, \ldots, e'_k$, where e'_i crosses e_i but no other edge of Γ. The only points common to two distinct edges of Γ^* can be their endpoints. If the edge e lies on the boundary of a single face F, we add a loop e' at v_F that crosses e but no other edge of Γ or Γ^*.

   Evidently the graph Γ^* just constructed is a plane graph having Γ as its geometric dual. For an arbitrary planar graph Γ, a geometric dual of Γ is the geometric dual of some planar embedding of Γ. We observe that because Γ may have several different planar embeddings, it may have several different geometric duals.

   6.1.4. Lemma. *If Γ^* is a geometric dual of the planar graph Γ, then* $M_\Gamma^*(S) \cong M_{\Gamma^*}(T)$.

Proof. Suppose that Γ^* is the geometric dual of the planar embedding Γ_0 of Γ. The construction of Γ^* from Γ_0 determines a bijection α from the edge-set S of Γ_0 to the edge-set T of Γ^*. We shall show that α induces an isomorphism between $M_{\Gamma_0}^*(S)$ and $M_{\Gamma^*}(T)$. As $\Gamma_0 \cong \Gamma$, the required result then follows.

   Let C be a circuit in $M_{\Gamma_0}(S)$. We want to show that $\alpha(C)$ is a bond in Γ^*. Now C forms a closed Jordan curve in the plane, and each edge in $\alpha(C)$ has one endpoint inside and the other endpoint outside this closed curve. Thus, $\alpha(C)$ contains an edge cut in Γ^*. The fact that $\alpha(C)$ is a minimal edge cut will follow from what we shall show next, namely, that if X is a minimal edge cut in Γ^*, then $\alpha^{-1}(X)$ contains a circuit of Γ_0. Evidently $\Gamma^* - X$ has two components Γ_1^* and Γ_2^*. Moreover, every edge in X has one endpoint in Γ_1^* and the other endpoint in Γ_2^*. If $|X| = 1$, then it is clear from the construction of Γ^* that the single edge in $\alpha^{-1}(X)$ is a loop in Γ_0. If $|X| > 1$ and F is a face of Γ^* having some edge x of X on its boundary, then, because x is not an isthmus of Γ^*, the boundary of F contains a circuit C_x containing the edge x. Let the endpoints of x be u and v. Then, because X is a minimal edge cut of Γ^*, the vertices u and v must be in the same component of $\Gamma^* - (X - x)$ but in different components of $\Gamma^* - X$. Therefore, $|X \cap C_x| \geq 2$. Hence, if a face of Γ^* meets an edge of X, it meets more than one such edge. But the faces of Γ^* correspond to vertices of Γ_0. Therefore, every vertex of Γ_0 meeting an edge of $\alpha^{-1}(X)$ meets at least two such edges. Because a graph in which every vertex has degree at least 2 contains a circuit, the induced graph $\Gamma_0[\alpha^{-1}(X)]$ contains a circuit in Γ_0. The lemma now follows without difficulty.   □

To complete the characterization of those graphs whose bond matroids are graphic, we now wish to establish that if M_Γ^* is a graphic matroid, then Γ is a planar graph. To prove this result we shall need Kuratowski's well-known characterization of planar graphs, a proof of which can be found in Bondy and Murty (1976, Section 9.5); see also Thomassen (1981). Recall that the graph Δ is *homeomorphic from* the graph Γ if Δ can be obtained from Γ by replacing edges of the latter by paths.

6.1.5. Kuratowski's Theorem. *A graph is nonplanar if and only if it contains a subgraph homeomorphic from K_5 or $K_{3,3}$ (Kuratowski 1930).*

A geometric dual Γ' of the graph Γ has the property that there is a bijection ψ from the edge-set of Γ to the edge-set of Γ' such that X is a circuit in Γ if and only if $\psi(X)$ is a bond in Γ'. Suppose now that Γ' is an arbitrary graph for which such a bijection exists. Then Γ' is called an *abstract dual* of Γ. Thus, every geometric dual is an abstract dual, although the converse of this statement is not true (see Exercises 6.10 and 6.24). The concept of an abstract dual will facilitate the proof of the main result of this section: Theorem 6.1.7. Before presenting this, we shall require one further preliminary that follows easily using matroid orthogonality.

6.1.6. Lemma. *If Γ' is an abstract dual of Γ, then Γ is an abstract dual of Γ'.*

Proof. This is left as Exercise 6.14.   □

6.1.7. Theorem. *The following statements are equivalent for a graph* Γ:

(i) Γ *is planar.*
(ii) M_Γ^* *is a graphic matroid.*
(iii) Γ *has an abstract dual.*

Proof. The fact that (i) implies (ii) follows immediately from Lemma 6.1.4. To see that (ii) implies (iii), note that if M_Γ^* is the polygon matroid of a graph Δ, then, by Lemma 6.1.6, Δ is an abstract dual of Γ. It remains to show that (iii) implies (i), and, following Parsons (1971), we break the proof of this up into several steps, the first two of which are easy preliminaries designed to allow us to use Kuratowski's Theorem.

6.1.8. Lemma. *If Γ_2 is a subgraph of the graph Γ_1 and Γ_1 has an abstract dual, then so does Γ_2.*

Proof. The general result will follow if we can show that it holds in the special case in which Γ_2 is obtained by deleting a single edge, say x, from Γ_1. Let Γ_1' be an abstract dual of Γ_1, and let x' be the edge of Γ_1' corresponding to x. Then Γ_1'/x' is an abstract dual of Γ_2, for the circuits of Γ_2 are just the circuits of Γ_1 not containing x, and the bonds of Γ_1'/x' are just the bonds of Γ_1' not containing x'.   □

6.1.9. Lemma. *If Γ_1 has an abstract dual Γ_1' and Γ_1 is homeomorphic from Γ_2, then Γ_2 has an abstract dual.*

Proof. We shall prove the lemma in the case in which Γ_1 is obtained from Γ_2 by replacing an edge x by a path of length 2 having edges y and z. The general result will then follow without difficulty. Let y' and z' be the edges of Γ_1' corresponding to y and z, respectively. Then y' and z' are parallel in Γ_1' unless x is an isthmus in Γ_2, in which case y' and z' are loops in Γ_1'. In either case, we easily check that Γ_1'/z' is an abstract dual of Γ_2.   □

We shall require one further preliminary to enable us to use Kuratowski's Theorem.

6.1.10. Lemma. *Neither K_5 nor $K_{3,3}$ has an abstract dual.*

Proof. We shall prove that K_5 does not have an abstract dual, leaving the argument for $K_{3,3}$ as an exercise. Suppose that Γ is an abstract dual of K_5. Then, by Lemma 6.1.6, K_5 is an abstract dual of Γ. The bonds of K_5 are of just two types separating either a single vertex or a single edge from the rest of the graph. Those of the first type have four edges, and those of the second type have six edges. Therefore, all circuits of Γ have either four or six edges. Hence, Γ is simple and bipartite, and $|V(\Gamma)| \geq 6$.

Because all circuits of K_5 contain at least three edges, every vertex of Γ has degree at least 3. Now the sum of the vertex degrees in Γ equals $2|E(\Gamma)|$. Hence, $2|E(\Gamma)| \geq 3|V(\Gamma)|$. Because $|E(\Gamma)| = |E(K_5)| = 10$, it follows that $|V(\Gamma)| \leq 6$.

We conclude that $|V(\Gamma)| = 6$, and so, because Γ is simple and bipartite, $|E(\Gamma)| \leq 9$, which is a contradiction.   □

To complete the proof of Theorem 6.1.7, suppose that Γ is a nonplanar graph having an abstract dual. By Kuratowski's Theorem, Γ has a subgraph Γ_1 homeomorphic from K_5 or $K_{3,3}$. But, by Lemma 6.1.8, Γ_1 has an abstract dual, and hence, by Lemma 6.1.9, either K_5 or $K_{3,3}$ has an abstract dual, contrary to Lemma 6.1.10.   □

6.2. CONNECTIVITY FOR GRAPHS AND MATROIDS

The main result of this chapter, which characterizes when two graphs have isomorphic polygon matroids, will be proved in the next section. In this section we shall develop the properties of graph connectivity that will be needed in the proof. In addition, we shall introduce the fundamental structural property of connectivity or nonseparability for matroids. This concept will be looked at further in the next chapter. In this section we shall concentrate on those properties of matroid connectivity that will be needed later in this chapter.

It was noted in the preceding section that if M is a graphic matroid, then $M \cong M_\Gamma$ for some *connected* graph Γ. It follows that there is no matroid concept that corresponds directly to the idea of connectedness for graphs. However, there are several extensions of the latter, and, as we shall see, matroid connectedness essentially corresponds to one of these.

In the same way that edges can be deleted from a graph, vertices can also be deleted. When this is done, not only the vertices themselves are removed but also all edges that meet these vertices. A subset X of the vertex-set of a graph Γ is called a *vertex cut* if the graph $\Gamma - X$ obtained by deleting the vertices of X from Γ has more components than Γ. If the vertex cut X contains a single vertex v, then v is called a *cut-vertex* of Γ.

Consider the graphs Δ_1 and Δ_2 in Figure 6.7. Neither has a cut-vertex. However, Δ_1 has a two-element vertex cut, whereas the smallest vertex cut in Δ_2 contains three elements. In general, if k is a positive integer, a graph Γ having at least $k + 1$ vertices is called *k-connected* if it is connected and has no vertex cut with fewer than k vertices. Hence, if Γ has at least two vertices, it is 1-connected if and only if it is connected. Thus, Δ_1 is 2-connnected but not 3-connected, whereas Δ_2 is 3-connected but not 4-connected. Notice that $M_{\Delta_1} \cong M_{\Delta_3}$, where Δ_3 is the graph shown in Figure 6.8. On the other hand, the reader is urged to check that if Δ_4 is a graph whose polygon matroid is isomorphic to M_{Δ_2}, then $\Delta_4 \cong \Delta_2$. These two examples give further hints as to the form of Whitney's theorem.

We have already noted that for $k = 1$, the notion of k-connectedness does not extend to matroids. Consider now the case in which $k = 2$. The

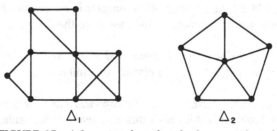

Δ_1 $\qquad\qquad\qquad$ Δ_2

FIGURE 6.7.   A 2-connected graph and a 3-connected graph.

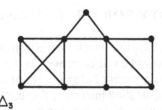

Δ_3

FIGURE 6.8. A graph with the same polygon matroid as Δ_1.

definition of 2-connectedness for graphs, relying as it does on deleting vertices, does not obviously extend to matroids. However, the following alternative characterization of 2-connectedness does.

6.2.1. Proposition. *Let* Γ *be a loopless graph without isolated vertices, and suppose that* $|V(\Gamma)| \geq 3$. *Then* Γ *is 2-connected if and only if for every pair of distinct edges* e *and* f *of* Γ, *there is a circuit containing both.*

This proposition can be proved by a direct graph-theoretic argument. We prefer to give a slightly different proof involving matroids, and we postpone this until later in the section.

We now indicate how the notion of 2-connectedness for graphs can be extended to matroids. For each element x of a matroid $M(E)$, let $\kappa_x = \{x\} \cup \{y: M$ has a circuit containing both x and $y\}$, and define the relation κ on E by $x \kappa y$ if and only if $x \in \kappa_y$.

6.2.2. Proposition. *For all matroids* $M(E)$, κ *is an equivalence relation on* E.

Proof. This is just Proposition 3.4.6, according to Proposition 3.4.5. □

6.2.3. Corollary. *If* $T \subseteq E$ *and* T *is a* κ-equivalence class, then for all elements t of T, $\kappa_t = T$.

The κ-equivalence classes are called the *(connected) components* of $M(E)$. Clearly every loop of $M(E)$ is a component; so is every isthmus. If E is a component of $M(E)$, then we call $M(E)$ *connected* or *nonseparable*; otherwise $M(E)$ is *disconnected* or *separable*. Thus, we have the following:

6.2.4. Proposition. $M(E)$ *is connected if and only if for every pair of distinct elements of* $M(E)$ *there is a circuit containing both.*

We note that the term *connected component* has now been defined for both graphs and matroids. Although this may seem undesirable, it does conform with standard usage. To highlight the difference here, consider the

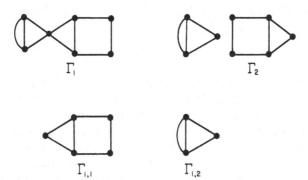

FIGURE 6.9.   Connectedness in matroids and in graphs.

graphs in Figure 6.9. The connected components of the matroids M_{Γ_1} and M_{Γ_2} are isomorphic to the matroids $M_{\Gamma_{1,1}}$ and $M_{\Gamma_{1,2}}$. On the other hand, the components of the graph Γ_2 are $\Gamma_{1,1}$ and $\Gamma_{1,2}$, whereas the graph Γ_1 has a single component.

The next result is no more than a restatement of Proposition 6.2.1. Its proof is therefore also a proof of Proposition 6.2.1.

6.2.5. Proposition. *Let Γ be a loopless graph without isolated vertices, and suppose that $|V(\Gamma)| \geq 3$. Then M_Γ is a connected matroid if and only if Γ is a 2-connected graph.*

Proof. It is straightforward to show, using Proposition 6.2.4, that Γ is 2-connected if M_Γ is connected. We leave this as an exercise.

Now suppose that Γ is 2-connected, and let Δ be the subgraph of Γ induced by some arbitrarily chosen component of M_Γ. We can assume that $\Delta \neq \Gamma$, for otherwise the required result holds. Therefore, Γ has an edge xy with exactly one endpoint, say x, in Δ. Now x is not a cut-vertex of Γ; so $\Gamma - x$ contains a path P_{yz} that joins y to some vertex z of $\Delta - x$ but is otherwise disjoint from $\Delta - x$. Because the graph Δ is certainly connected, it contains a path P_{zx} joining z and x. The union of P_{yz}, P_{zx}, and the edge xy is a circuit in Γ that contains xy together with at least one edge of Δ. But Δ is a component of M_Γ, and hence we obtain the contradiction that $xy \in E(\Delta)$. $\square$

There are many alternative characterizations of matroid connectivity. We shall consider just one. A number of others will be looked at in the exercises and in the next chapter. If $M(E)$ is a matroid and $T \subseteq E$, then T is a *separator* of M if T is a union of components of M. Clearly, if T is a separator of $M(E)$, so is $E - T$. Moreover, both E and $\varnothing$ are separators of $M(E)$. All other separators are called *nontrivial separators*. These are characterized in the following result.

6.2.6. Proposition. *Let S_1 be a proper nonempty subset of E. Then S_1 is a nontrivial separator of $M(E)$ if and only if*

$$r(S_1) + r(E - S_1) = r(E). \tag{6.1}$$

Proof. Suppose that S_1 satisfies equation (6.1), and let $S_2 = E - S_1$. Assume that S_1 is not a separator of $M(E)$. Then M has a circuit C intersecting both S_1 and S_2. Now, for $i = 1, 2, C \cap S_i$ is an independent subset of S_i and can therefore be extended to a basis B_i for S_i. Because B_i spans S_i, it follows that $B_1 \cup B_2$ spans $M(E)$. But $B_1 \cup B_2$ contains the circuit C; hence, $r(E) < |B_1 \cup B_2| = |B_1| + |B_2| = r(S_1) + r(S_2)$, a contradiction to equation (6.1). We conclude that if S_1 satisfies equation (6.1), it is a nontrivial separator of M. The proof of the converse is similar and is left to the reader.
□

The following appealing property of matroid connectivity is routine to verify using the preceding proposition and the formula for the rank function of M^* (see Chapter 5).

6.2.7. Corollary. *M is connected if and only if M^* is connected.*

Proof. The proof is left as Exercise 6.24.
□

6.3. WHITNEY'S 2-ISOMORPHISM THEOREM

In this section we shall prove Whitney's theorem characterizing when two graphs have isomorphic polygon matroids. Our proof, due to Truemper (1980), is considerably shorter than the original.

If Γ is a graph, then evidently adding vertices to Γ with no incident edges will not alter the polygon matroid of the graph. For this reason, throughout this section we shall consider only graphs having no isolated vertices. The following operations on a graph Γ have appeared, at least implicitly, in earlier sections. An example of each is given in Figures 6.10 and 6.11.

(a) *Vertex identification.* Let v and v' be vertices of distinct components of Γ. We modify Γ by identifying v and v' as a new vertex $\bar{v}$.

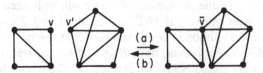

FIGURE 6.10. Vertex identification and vertex splitting.

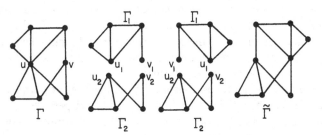

FIGURE 6.11. A twisting.

(b) *Vertex splitting.* This is the inverse operation of vertex identification, so that a graph can be split only at a cut-vertex.

(c) *Twisting.* Suppose that the graph Γ is obtained from the disjoint graphs Γ_1 and Γ_2 by identifying the vertices u_1 of Γ_1 and u_2 of Γ_2 as the vertex u of Γ, and identifying the vertices v_1 of Γ_1 and v_2 of Γ_2 as the vertex v of Γ. In a *twisting* $\tilde{\Gamma}$ of Γ *about* $\{u, v\}$, we identify, instead, u_1 with v_2 and v_1 with u_2. We call Γ_1 and Γ_2 the *pieces* of the twisting.

The graph Γ is *2-isomorphic* to the graph Δ if Δ can be transformed into a graph isomorphic to Γ by a sequence of operations of types (a), (b), and (c). Most of the rest of this section will be devoted to proving the next theorem. We shall require two more definitions. A connected graph is called a *block* provided that its polygon matroid is connected. Hence, a block consists of a single vertex, a single loop, or a connected loopless graph with no cut-vertices. A *block of a graph* Γ is a subgraph of Γ that is a block and that is maximal with this property. Evidently two distinct blocks of Γ have at most one common vertex. We note that, by Proposition 6.2.5, every block with at least three vertices is 2-connected.

6.3.1 Whitney's 2-isomorphism theorem. Let Γ and Δ be graphs having no isolated vertices. Then M_Γ and M_Δ are isomorphic if and only if Γ and Δ are 2-isomorphic (Whitney 1933).

Proof. It is clear that none of the operations (a), (b), or (c) alters the edge-sets of the circuits of a graph. Hence, if Γ and Δ are 2-isomorphic, their polygon matroids are isomorphic. For the converse, we argue by induction on $|V(\Gamma)|$. Let $\hat{\Gamma}$ and $\hat{\Delta}$ be the graphs consisting of the disjoint unions of the blocks of Γ and Δ, respectively. Because $M_{\hat{\Gamma}} \cong M_\Gamma \cong M_\Delta \cong M_{\hat{\Delta}}$, it follows that there is a bijection from the set of blocks of $\hat{\Gamma}$ to the set of blocks of $\hat{\Delta}$ so that corresponding blocks have isomorphic polygon matroids. Now Γ and $\hat{\Gamma}$ are 2-isomorphic, as are Δ and $\hat{\Delta}$. Therefore, as 2-isomorphism is clearly an equivalence relation, the required result will follow if we can show that $\hat{\Gamma}$ and $\hat{\Delta}$ are 2-isomorphic. This will be done by proving that if Γ is a block and Δ is a graph for which M_Γ and M_Δ are isomorphic, then Δ can be transformed into a graph isomorphic to Γ by a sequence of twistings. We shall therefore assume, for the remainder of the proof, that Γ is a block. Clearly, if Γ has fewer than four vertices, the required result holds. Hence, suppose Γ has at least four vertices. Then Γ is 2-connected.

The following lemma establishes the required result in the case that Γ is 3-connected. The proof, due to Edmonds (see Truemper 1980) and Greene (1971), uses the elementary observation that if Φ is a 2-connected graph, then the set of edges meeting a fixed vertex of Φ is a bond in M_Φ. Such a bond is called a *vertex bond* of M_Φ.   □

6.3.2. Lemma. *Let Γ and Δ be loopless graphs without isolated vertices. If M_Γ and M_Δ are isomorphic and Γ is 3-connected, then Γ and Δ are isomorphic graphs.*

Proof. If Ω is a loopless block, a hyperplane of M_Ω is connected precisely when the corresponding bond is a vertex bond. Moreover, if the number of connected hyperplanes of M_Ω is $|V(\Omega)|$, then, up to relabeling, M_Ω uniquely determines the vertex-edge incidence matrix of Ω, and hence uniquely determines Ω itself.

Because Γ is 3-connected, the complement in $E(\Gamma)$ of a vertex bond is a connected hyperplane. Thus, M_Γ has $|V(\Gamma)|$ connected hyperplanes. But $M_\Gamma \cong M_\Delta$, and M_Γ is connected and loopless, and so Δ is a loopless block. Because, by Proposition 6.1.2,

$$|V(\Gamma)| - 1 = r(M_\Gamma) = r(M_\Delta) = |V(\Delta)| - 1,$$

we have $|V(\Gamma)| = |V(\Delta)|$. Therefore, M_Δ has precisely $|V(\Delta)|$ connected hyperplanes, and so $\Gamma \cong \Delta$, and the lemma is proved.   □

The proof of Whitney's theorem will require two further lemmas, the first of which is Tutte's characterization (1966) of the structure of those 2-connected graphs that are not 3-connected.

Suppose $k \geq 2$. A connected graph Γ is called a *generalized circuit* with *parts* $\Gamma_1, \Gamma_2, \ldots, \Gamma_k$ if the following conditions hold:

(i)   Each Γ_i is a connected subgraph of Γ having a nonempty edge-set, and if $k = 2$, both Γ_1 and Γ_2 have at least three vertices.

(ii) The edge-sets of $\Gamma_1, \Gamma_2, \ldots, \Gamma_k$ partition the edge-set of Γ, and each Γ_i shares exactly two vertices, its *contact vertices*, with $\bigcup_{j \neq i} \Gamma_j$.

(iii) If each Γ_i is replaced by an edge joining its contact vertices, the resulting graph is a circuit.

6.3.3. Lemma. *Let Γ be a block having at least four vertices, and suppose that Γ is not 3-connected. Then Γ has a representation as a generalized circuit, each part of which is a block.*

Proof. If $\{u, v\}$ is a vertex cut of Γ, choose a component Δ_1 of $\Gamma - \{u, v\}$ and let $\Delta_2 = \Gamma - \{u, v\} - \Delta_1$. For $i = 1, 2$, let Γ_i be the subgraph of Γ induced by $V(\Delta_i) \cup \{u, v\}$. Then Γ is a generalized circuit with parts Γ_1 and Γ_2. If Γ_1 is not a block, it is the union of two connected subgraphs $\Gamma_{1,1}$ and $\Gamma_{1,2}$ having at least two vertices each but only one vertex x in common. Because Γ is a block, x is distinct from both u and v, and one of u and v is in $\Gamma_{1,1}$, with the other in $\Gamma_{1,2}$. Therefore, Γ is a generalized circuit with parts $\Gamma_{1,1}$, $\Gamma_{1,2}$, and Γ_2. If necessary, we repeat the foregoing process until a representation of Γ is obtained in which each part is a block. $\square$

The core of the proof of Whitney's theorem is contained in the next result.

6.3.4. Lemma. *Suppose that Γ is a block having a representation as a generalized circuit, the parts $\Gamma_1, \Gamma_2, \ldots, \Gamma_k$ of which are blocks. Let Δ be a graph for which there is an isomorphism θ from M_Γ to M_Δ. Then Δ is a generalized circuit with parts $\Delta_1, \Delta_2, \ldots, \Delta_k$, where, for all i, Δ_i is the subgraph of Δ induced by $\theta[E(\Gamma_i)]$.*

Proof. For an arbitrary element j of $\{1, 2, \ldots, k\}$, let $\bar{\Gamma}_j = \bigcup_{i \neq j} \Gamma_i$ and $\bar{\Delta}_j = \bigcup_{i \neq j} \Delta_i$. The main part of the proof of this lemma involves showing that

$$|V(\Delta_j) \cap V(\bar{\Delta}_j)| = 2. \tag{6.2}$$

Now $\Gamma, \Gamma_1, \Gamma_2, \ldots, \Gamma_k$ are blocks. Hence, all of $\Delta, \Delta_1, \Delta_2, \ldots, \Delta_k$ are blocks. Because Γ has a circuit meeting each of $E(\Gamma_1), E(\Gamma_2), \ldots, E(\Gamma_{k-1})$ and $E(\Gamma_k)$, $\bar{\Delta}_j$ has an edge e having an endpoint x in Δ_j. Choose an edge f in Δ_j. Then, because Δ is a block, it has a circuit C_1 containing both e and f. Starting at x, move along C_1 using the edge e first, and stop at the first vertex y of Δ_j that is encountered. Let P be the path from x to y thus traversed. Because C_1 contains the edge f of Δ_j, the vertices x and y are distinct. Hence, $|V(\Delta_j) \cap V(\bar{\Delta}_j)| \geq 2$. We shall complete the proof of equation (6.2) by showing that

$$V(\Delta_j) \cap V(\bar{\Delta}_j) = \{x, y\}. \tag{6.3}$$

Let u be one of the contact vertices of Γ_j. Because Γ_j is a block, the set E_u of edges of Γ_j meeting u is a bond in Γ_j. Hence, $\theta(E_u)$ is a bond in Δ_j. Moreover, every circuit of Γ that contains an edge of Γ_j and an edge of $\bar{\Gamma}_j$ also contains an edge of E_u. Hence, we have proved:

6.3.5. Lemma. *Every circuit of Δ that contains an edge of Δ_j and an edge of $\bar{\Delta}_j$ also contains an edge of $\theta(E_u)$.*

Now $\Delta_j - \theta(E_u)$ has two components, one, say Δ_j^x, containing x, and the other, Δ_j^y, containing y. Assume that equation (6.3) does not hold. Then there is a vertex z in $[V(\Delta_j) \cap V(\bar{\Delta}_j)] - \{x, y\}$. Let g and h be edges of $\bar{\Delta}_j$ and Δ_j, respectively, meeting z, and suppose, without loss of generality, that $z \in \Delta_j^x$. Because Δ is a block, it has a circuit C_2 that contains both g and h. Starting at z and using the edge g first, move along C_2 stopping at the first vertex w of $\Delta_j \cup P$ that is encountered. Evidently $w \neq z$. Let Q be the path from z to w thus traversed.

We now distinguish three cases, showing that each leads to the conclusion that Δ has a circuit C_3 meeting both Δ_j and $\bar{\Delta}_j$ but containing no edge of $\theta(E_u)$. Because this contradicts Lemma 6.3.5, it will follow that equation (6.3), and hence equation (6.2), holds. The cases are as follows:

(i) $w \in V(P)$.
(ii) $w \in V(\Delta_j^x) - V(P)$.
(iii) $w \in V(\Delta_j^y) - V(P)$.

In case (i), C_3 is obtained by combining Q, that part of P from w to x, and an arbitrary path from x to z in Δ_j^x. In case (ii), we obtain C_3 by combining Q with an arbitrary path from w to z in Δ_j^x. Finally, in case (iii), C_3 is formed by combining P and Q with a path from x to z in Δ_j^x and a path from y to w in Δ_j^y.

Because we have now proved equation (6.2), it follows from the fact that Δ is a block that Δ is a generalized circuit having parts $\Delta_1, \Delta_2, \ldots, \Delta_k$. $\square$

Completion of the proof of Whitney's theorem is now relatively straightforward. Suppose that $\theta: E(\Gamma) \to E(\Delta)$ induces an isomorphism of the polygon matroids M_Γ and M_Δ. It was noted earlier that the required result follows in general provided it can be established in the case in which Γ is a block having at least four vertices.

By Lemma 6.3.2, if Γ is 3-connected, then $\Delta \cong \Gamma$, and the required result certainly holds. It follows that we can assume that Γ is not 3-connected. Therefore, by Lemma 6.3.3, Γ has a representation as a generalized circuit in which each of the parts $\Gamma_1, \Gamma_2, \ldots, \Gamma_k$ is a block. Hence, by Lemma 6.3.4, Δ has a representation as a generalized circuit, the parts of which are $\Delta_1, \Delta_2, \ldots, \Delta_k$, where $\Delta_i = \Delta[\theta[E(\Gamma_i)]]$.

For each i in $\{1, 2, \ldots, k\}$, consider the graphs $\Gamma_i + e_i$ and $\Delta_i + f_i$ that are obtained from Γ_i and Δ_i by adding edges e_i and f_i so that in each case the new edge joins the contact vertices. Evidently the isomorphism between M_{Γ_i} and M_{Δ_i} can be extended to an isomorphism between $M_{\Gamma_i + e_i}$ and $M_{\Delta_i + f_i}$ by mapping e_i to f_i. Now $|V(\Gamma_i + e_i)| = |V(\Gamma_i)| < |V(\Gamma)|$; hence, by the induction assumption, $\Delta_i + f_i$ can be transformed by a sequence of twistings into a graph isomorphic to $\Gamma_i + e_i$. Because every two-element vertex cut in $\Delta_i + f_i$ is also a vertex cut in Δ, we can perform the same sequence of twistings in Δ. After this has been done for all i in $\{1, 2, \ldots, k\}$, we obtain a generalized circuit $\hat{\Delta}$ with parts $\hat{\Delta}_1, \hat{\Delta}_2, \ldots, \hat{\Delta}_k$ such that $M_\Gamma \cong M_{\hat{\Delta}}$, and for each i there is an isomorphism θ_i from Γ_i to $\hat{\Delta}_i$ under which the contact vertices of Γ_i are mapped to the contact vertices of $\hat{\Delta}_i$.

Suppose that the cyclic order of the parts of Γ is $\Gamma_1, \Gamma_2, \ldots, \Gamma_k$, and the cyclic order of the parts of $\hat{\Delta}$ is $\hat{\Delta}_1, \hat{\Delta}_{\sigma(2)}, \ldots, \hat{\Delta}_{\sigma(k)}$ for some permutation σ of $\{2, 3, \ldots, k\}$. Then it is clear that by a sequence of twistings, $\hat{\Delta}$ can be transformed into a generalized circuit in which the parts, in cyclic order, are $\hat{\Delta}_1, \hat{\Delta}_2, \ldots, \hat{\Delta}_k$. If the resulting graph $\hat{\hat{\Delta}}$ is still not isomorphic to Γ, this last remaining difficulty can be overcome by twisting some of the parts of $\hat{\hat{\Delta}}$ about their contact vertices.

This completes the proof of Whitney's 2-isomorphism theorem.   □

We note that in transforming Δ into a graph isomorphic to Γ, no attempt was made to minimize the number of twistings. We leave to the exercises consideration of Truemper's extension of Whitney's theorem that determines a best-possible bound on the minimum number of twistings necessary.

The last result of this section uses Whitney's theorem to show that a 3-connected loopless planar graph has a unique abstract dual. The proof of this will require some further properties of planar graphs. In particular, we note that a graph is planar if and only if it can be embedded on the surface of a sphere so that distinct edges do not meet, except possibly at their endpoints. To see this, suppose that we have such an embedding Γ' of a graph Γ. Choose a point P on the surface of the sphere so that P is not a vertex of Γ' and is not on any edge. Let Q be the point on the surface diametrically opposite P. Now project Γ' from P onto the tangent plane to the sphere at Q. This operation, which is known as *stereographic projection*, gives a plane graph Γ'' isomorphic to Γ. To establish that every planar graph can be embedded on the surface of a sphere so that distinct edges do not cross, we need only to reverse the foregoing construction.

Notice that under stereographic projection the points on the sphere that are in the region containing P are mapped into points in the infinite face of Γ''. Using this idea, it is not difficult to show that for any edge e of a planar graph Γ, there is a planar embedding of Γ in which e is on the boundary of the infinite face.

With these preliminaries we are now ready to prove the next result.

6.3.6. Proposition. (*Whitney 1932b.*) *Let Γ be a 3-connected loopless planar graph. If Γ_1 and Γ_2 are abstract duals of Γ, each having no isolated vertices, then $\Gamma_1 \cong \Gamma_2$.*

Proof. Because Γ is 3-connected, it has no bonds with fewer than three members. Moreover, it is not difficult to check that at least one bond of Γ has more than three members. Therefore, Γ_1 is a simple graph having at least four vertices. In addition, because M_Γ is connected, Corollary 6.2.7 implies that M_{Γ_1} is connected, and so Γ_1 is a block.

Because $M_{\Gamma_1} \cong M_{\Gamma_2}$, we can assume that Γ_1 is not 3-connected, for otherwise the required result follows by Lemma 6.3.2. But Γ_1 is a block, and so, by Whitney's 2-isomorphism theorem, it can be transformed into a graph isomorphic to Γ_2 by a sequence of twistings. Evidently a twisting of a graph produces an isomorphic graph unless each piece of the twisting contains a circuit. Thus, either we obtain the required result that $\Gamma_1 \cong \Gamma_2$, or Γ_1 has a representation as a generalized circuit with two parts, $\Gamma_{1,1}$ and $\Gamma_{1,2}$, each of which contains a circuit. In the latter case, because Γ_1 is planar, it is clear that for i in $\{1, 2\}$, the graph $\Gamma_{1,i} + e_i$ obtained from $\Gamma_{1,i}$ by adding an edge e_i joining its contact vertices must be planar. It follows from our earlier observations concerning planar graphs that there is a planar embedding of $\Gamma_{1,i} + e_i$ in which e_i is on the boundary of the infinite face. Hence, $\Gamma_{1,i}$ has a planar embedding $\Gamma'_{1,i}$ in which the contact vertices lie on the infinite face. Putting these planar embeddings for $\Gamma_{1,1}$ and $\Gamma_{1,2}$ together, we obtain a planar embedding Γ'_1 for Γ_1. The graph obtained from Γ'_1 by replacing each part by a single edge is a circuit and so has exactly two faces. It follows that Γ'_1 has exactly two faces (say F_1 and F_2) that are bounded by edges from both $\Gamma'_{1,1}$ and $\Gamma'_{1,2}$. Because both $\Gamma'_{1,1}$ and $\Gamma'_{1,2}$ contain circuits, each has a face different from its infinite face. Let $F_{1,1}$ and $F_{1,2}$, respectively, be such faces. Clearly, each is a face of Γ'_1 distinct from F_1 and F_2. Now, by Lemma 6.3.2, the geometric dual of Γ'_1 is isomorphic to Γ and is therefore 3-connected. However, if we delete from this geometric dual the vertices corresponding to F_1 and F_2, the resulting graph is disconnected, because it does not contain a path joining the vertices corresponding to $F_{1,1}$ and $F_{1,2}$. This contradiction completes the proof of the proposition.   □

6.4. SERIES-PARALLEL NETWORKS

The operations of joining electrical components in series or in parallel are fundamental in electrical-network theory. In this section we shall investigate these operations for graphs. We shall show how, by beginning with a loop or an isthmus and using these operations, we can build up a class of planar graphs with the property that an abstract dual of a member of the class is also in the class. The network operations considered here have been successfully extended to matroids, and Chapter 7 will present a detailed discussion of these matroid operations.

We begin with some examples. The graph Γ_1 in Figure 6.12 is obtained from Γ_0 by adding the edge f in *parallel* with the edge e, whereas in Γ_2, f has been added in *series* with e. We call Γ_1 a *parallel extension* of Γ_0, and Γ_2 a *series extension* of Γ_0. For an arbitrary graph Γ, these operations are defined as follows: Δ is a parallel extension of Γ if Δ has a two-element circuit $\{e, f\}$ such that $\Delta - f \cong \Gamma$. If, instead, $\{e, f\}$ is a two-element bond of Δ and $\Delta/f \cong \Gamma$, then Δ is a *series extension* of Γ. Thus, for example, replacing

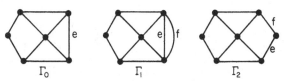

FIGURE 6.12.   Parallel and series extensions of a graph.

FIGURE 6.13.   A series extension.

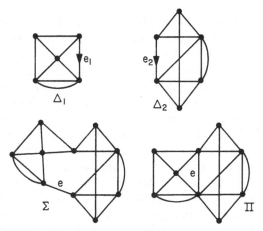

FIGURE 6.14.   Series and parallel connections of graphs.

an isthmus by a path of length 2 is not a series extension, although if the graph Ψ is homeomorphic from the graph Φ and Φ is 2-connected, then Ψ can be obtained from Φ by a sequence of series extensions. The example in Figure 6.13 shows that not every series extension consists of replacing an edge by a path of length 2. However, a consequence of Whitney's 2-isomorphism theorem is that if Δ_1 and Δ_2 are series extensions of the same graph, then their polygon matroids are isomorphic.

The operations of series extension and parallel extension for graphs are special cases of the operations of series connection and parallel connection of graphs. Examples of these operations are illustrated in Figure 6.14. The graphs Σ and Π are, respectively, the series and parallel connections of Δ_1 and Δ_2 relative to the directed edges e_1 and e_2. Evidently, if Δ_2 were the graph consisting of a pair of parallel edges, then the resulting series and parallel connections would be just a series extension and a parallel extension of Δ_1.

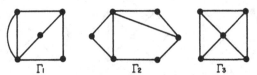

FIGURE 6.15. Γ_1 and Γ_2 are series-parallel networks; Γ_3 is not.

Notice that although both graphs Δ_1 and Δ_2 in Figure 6.14 are 3-connected, neither Σ nor Π is. Furthermore, $\Sigma - e$ and Π/e are not even 2-connected. These observations foreshadow several important properties of series and parallel connections of matroids. These matroid operations, which are of basic structural importance, will be looked at in detail in the next chapter.

Electrical networks that can be built up by adding resistors in series and in parallel arise frequently in electrical engineering. An attractive feature of such a network is that its joint resistance can be easily calculated (Duffin 1965). We now consider the graphs corresponding to such networks. A graph Γ without isolated vertices is called a *series-parallel network* if it can be obtained from a loop or an isthmus by a sequence of operations each of which is either a series or a parallel extension. Thus, as the reader can easily check, the graphs Γ_1 and Γ_2 in Figure 6.15 are series-parallel networks, but Γ_3 is not. Kuratowski's Theorem characterized planar graphs by a list of graphs from which no subgraph of a planar graph is homeomorphic. The main result of this section is a similar characterization of series-parallel networks. The reader may wish to try to derive this result. In that case, a detailed consideration of Γ_3 may prove helpful.

The definition of series-parallel networks strongly suggests that inductive arguments may be successfully applied in determining their properties. Indeed, each of the following properties of a series-parallel network Γ can be quickly obtained using such an argument.

6.4.1. Lemma. Γ *is a planar graph.*

6.4.2. Lemma. Γ *is a block.*

6.4.3. Lemma. If Γ *has at least three edges and* e *is an edge of* Γ*, then either* $\Gamma - e$ *or* Γ/e *is not a block.*

The next result is essentially an extension of Lemma 6.1.9.

6.4.4. Lemma. Let Γ *be a planar graph and* Δ *an abstract dual of* Γ*. Then, corresponding to every series extension* Γ' *of* Γ *there is a parallel extension* Δ' *of* Δ *that is an abstract dual of* Γ'*. Moreover, every parallel extension of* Δ *is an abstract dual of a series extension of* Γ*.*

Proof. Beacuse Δ is an abstract dual of Γ, there is a bijection ψ from $E(\Gamma)$ into $E(\Delta)$ with the property that X is a circuit in Γ if and only if $\psi(X)$ is a bond in Δ. Now suppose that the graph Γ' is obtained by adding a new edge f in series with the edge e of Γ. Then let Δ' be the graph obtained from Δ by adding a new edge g in parallel with $\psi(e)$. It is routine to verify that Δ' is an abstract dual of Γ', and the remainder of the proof is left as an exercise.

$\square$

From this lemma we immediately obtain the following:

6.4.5. Proposition. *An abstract dual of a series-parallel network is a series-parallel network.*

The next theorem is the main result of this section, and the rest of the section will be devoted to proving it.

6.4.6. Theorem. *A graph Γ with at least one edge is a series-parallel network if and only if it is a block having no subgraph homeomorphic from K_4 (Dirac 1952; Duffin 1965; Brylawski 1971).*

Proof. If Γ is a series-parallel network, a straightforward induction argument shows that it is a block with no subgraph homeomorphic from K_4. We leave the details to the reader.

The converse is also proved by induction on $|E(\Gamma)|$. The result clearly holds for $|E(\Gamma)| = 1$. Assume it true for $|E(\Gamma)| < n$, and let $|E(\Gamma)| = n$. If Γ has two elements in series or two elements in parallel, then the result follows easily by the induction assumption. We can therefore assume that Γ is simple and that every vertex of Γ has degree at least 3. The next lemma, due to Dirac (1952), completes the proof of the theorem.

A *chord* of a circuit in a graph is a path that is edge-disjoint from the circuit and that joins two nonneighboring vertices of the circuit. An *interior vertex* of the chord is a vertex of the chord other than one of its end vertices.

6.4.7. Lemma. *A simple 2-connected graph Γ in which the degree of every vertex is at least 3 has a subgraph homeomorphic from K_4.*

Proof. From among the circuits of Γ, choose one, say C, that has the greatest number of edges. Let $a_1, a_2, \ldots, a_m$ be the vertices of C in cyclic order. Then, as we shall now show:

6.4.8. Lemma. *Every vertex of C meets a chord of C.*

Proof. Because every vertex of Γ has degree at least 3, for each i in $\{1, 2, \ldots, m\}$, there is an edge e_i that meets a_i but is not in C. Because Γ is 2-connected, e_i is the first edge of a path P_i in Γ that joins a_i to some other vertex a_j of C

and that is otherwise disjoint from C. If this path is not a chord of C, then a_i and a_j are neighboring vertices of C. But then, either P_i has more than one edge, in which case the choice of C is contradicted, or P_i has exactly one edge, and the simplicity of Γ is contradicted. We conclude that Lemma 6.4.8 holds. $\qquad\qquad\qquad\qquad\qquad\qquad\qquad\qquad\qquad\qquad\qquad\qquad\qquad\qquad$ □

We now distinguish two cases:

(i) C has two chords joining different pairs of vertices and having a common interior vertex.
(ii) Whenever two chords of C have a common interior vertex, they have the same end vertices.

In the first case, let R_1 and R_2 be chords of C that have a common interior vertex but do not join the same pair of vertices. Suppose R_1 joins a_g and a_h, and R_2 joins a_s and a_t, where a_s, a_g, and a_h are distinct, although a_t may equal a_g or a_h. Beginning at a_s, follow R_2 toward a_t and stop at the first vertex b of R_1 that is encountered. Then the subgraph of Γ induced by C, R_1, and that part of R_2 joining a_s and b is homeomorphic from K_4.

In case (ii), consider the subgraph of Γ induced by C and its chords. In this subgraph, contract all but one edge of each chord. In the resulting graph Γ', each chord is a single edge that cuts C into two arcs. Take such an arc with the least number of edges and suppose that it joins a_1 and a_i. Now $i \neq 2$, and, by Lemma 6.4.8, C has a chord joining a_2 to a_j for some j. Because the edge a_2a_j cannot cut off a shorter arc than a_1a_i, we must have that $j > i$. It follows that the subgraph of Γ' induced by C and the edges a_1a_i and a_2a_j is homeomorphic from K_4.

This completes the proof of Lemma 6.4.7 and thereby the proof of Theorem 6.4.6. $\qquad\qquad\qquad\qquad\qquad\qquad\qquad\qquad\qquad\qquad\qquad\qquad\qquad\qquad$ □

EXERCISES

The more difficult exercises are marked with an asterisk.

Section 1

6.1. (Perfect 1981). Let $\mathcal{I}$ be the collection of edge-sets of forests in a graph Γ having edge-set E. Show directly that $\mathcal{I}$ is the set of independent sets of a matroid on E.

6.2. Let Γ be a graph.
  (i) Find the closure operator of M_Γ.
  (ii) Characterize those sets of edges that are flats in M_Γ.

6.3. (i) Give a direct argument to show that if $\mathcal{B}$ is the set of spanning trees in a connected graph Γ, then $\mathcal{B}$ is the set of bases of a matroid on $E(\Gamma)$.
  (ii) Repeat (i) with $\mathcal{B}^*$ in place of $\mathcal{B}$, where $\mathcal{B}^* = \{E(\Gamma) - B : B \in \mathcal{B}\}$.

FIGURE 6.16.

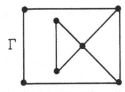

FIGURE 6.17.

6.4. If X is a set of edges in a graph Γ, what is the rank of X in M_Γ^*?

6.5. Find the largest integer m such that all matroids on fewer than m elements are graphic.

6.6. (i) Find all graphic matroids $M(E)$ such that $|E| \leq 6$ and $M(E) \cong M^*(E)$.

   (ii) Find an infinite family of graphic geometries M for which $M \cong M^*$.

   (iii) Find all graphic matroids $M(E)$ for which the identity map on E is an isomorphism between $M(E)$ and $M^*(E)$.

6.7. Show that each of the graphs in Figure 6.16 has its polygon matroid isomorphic to its bond matroid.

6.8. Find geometric duals of the graphs obtained from K_5 and $K_{3,3}$ by deleting a single edge of each.

6.9. (i) Find all simple graphs Δ with the property that whenever Δ is a subgraph of a simple graph Γ, the edge-set of Δ is a flat in M_Γ.

   (ii) Which of the graphs Δ in (i) have the stronger property that whenever Δ is a subgraph of an *arbitrary* graph Γ, the edge-set of Δ is a flat in M_Γ?

6.10. (D. R. Woodall, in Welsh 1976, Section 6.4) Construct a graph that is an abstract dual of the graph Γ in Figure 6.17 but is not a geometric dual of Γ.

6.11. (Wagner 1937a, 1937b; Harary and Tutte 1965) Prove that a graph is planar if and only if it contains no subgraph that is *contractible* to K_5 or $K_{3,3}$.

6.12. (Whitney 1932a; Ore 1967, Section 3.3) The graph Δ is a *Whitney dual* of the graph Γ if there is a bijection $\psi: E(\Gamma) \to E(\Delta)$ such that for all subsets Y of $E(\Gamma)$,

$$r(M_\Delta) - r(M_{\Delta - \psi(Y)}) = |Y| - r(M_{\Gamma[Y]})$$

where $r(N)$ denotes the rank of the matroid N.

(i)   Show that if Δ is a Whitney dual of Γ, then Γ is a Whitney dual of Δ.

(ii)   Determine the relationship between Whitney duals and geometric and abstract duals.

(iii) Prove that a graph is planar if and only if it has a Whitney dual.

6.13. For each positive integer n, find a planar graph Γ_n having at least n nonisomorphic geometric duals.

6.14. Prove that if Γ' is an abstract dual of Γ, then Γ is an abstract dual of Γ'.

Section 2

6.15. Find, up to isomorphism, all connected matroids on a set of four elements.

6.16. Give a purely graph-theoretic proof of Proposition 6.2.1.

*6.17. (Harary 1969, Theorem 3.3) Let Γ be a graph without isolated vertices, and suppose that $|V(\Gamma)| \geq 3$. Prove that the following statements are equivalent:

(i) Γ is 2-connected.

(ii)   If u and v are vertices of Γ, there is a circuit meeting both.

(iii) If v is a vertex and e is an edge of Γ, there is a circuit containing e and meeting v.

(iv)   If e and f are edges of Γ, there is a circuit containing both.

6.18. Prove that S_1 is a separator of a matroid $M(E)$ if and only if whenever I_1 is an independent subset of S_1 and I_2 is an independent subset of $E - S_1$, the set $I_1 \cup I_2$ is independent in M.

6.19. Show that no bond in an n-connected graph has fewer than n edges.

6.20. (i)   Prove that if x and y are distinct elements of a circuit C in a matroid M, then M has a bond C^* such that $C^* \cap C = \{x, y\}$.

(ii)   Use (i) to give a direct proof of the fact that S_1 is a component of $M(E)$ if and only if S_1 is a component of $M^*(E)$.

(iii) Prove that a matroid $M(E)$ is connected if and only if for every two distinct elements x and y of E, there is either a circuit or a bond containing both.

*6.21. (Dirac 1967; Plummer 1968; Oxley 1981a) Let Γ be a 2-connected graph such that for all edges e, the graph $\Gamma - e$ is not 2-connected.

(i)   Prove that Γ has a vertex of degree 2.

(ii)   Find a best-possible bound on the number of degree-2 vertices in Γ.

6.22. Let Γ be a 3-connected loopless graph and e, f, and g be distinct edges of Γ.

(i)   Show that Γ has either a circuit or a bond containing $\{e, f, g\}$.

(ii)   Will there always be a circuit containing $\{e, f, g\}$?

(iii) Can you find a weaker hypothesis on Γ under which (i) remains true?

*6.23. (Cunningham 1981; Inukai and Weinberg 1981; Oxley 1981b) Let n be an integer exceeding 1, Γ a connected graph, and r the rank function

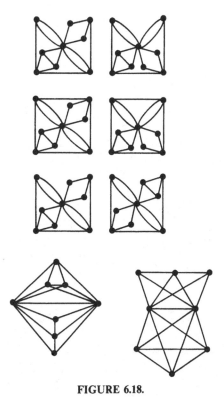

FIGURE 6.18.

of $M_\Gamma(E)$. Prove that Γ is n-connected if and only if there is no partition $\{X, Y\}$ of E such that for some positive integer k less than n,

$$r(X) + r(Y) - r(E) + 1 = k \quad \text{and} \quad \min\{r(X), r(Y)\} \geq k.$$

6.24. Prove that a matroid M is connected if and only if its dual M^* is connected.

Section 3

6.25. For each of the pairs of graphs in Figure 6.18, determine whether or not the polygon matroids are isomorphic.

*6.26. (Ore 1967, Theorem 3.4.1) Let Γ be a planar block and Γ^* a graph without isolated vertices. Prove that Γ^* is an abstract dual of Γ if and only if it is a geometric dual of Γ.

*6.27. (Truemper 1980) Suppose that Γ is a 2-connected graph having n vertices and Δ is a graph that is 2-isomorphic to Γ.

    (i) Prove that Γ can be transformed into a graph that is isomorphic to Δ in a sequence of at most $n - 2$ twistings.

    (ii) Show that for any integer N there are graphs Γ and Δ satisfying the foregoing conditions with $n \geq N$ such that at least $n - 2$ twistings are needed to transform Γ into a graph that is isomorphic to Δ.

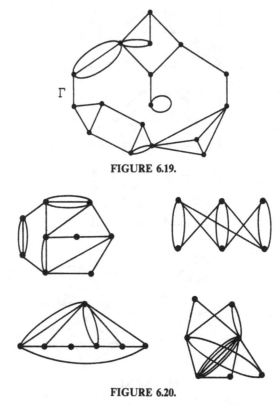

FIGURE 6.19.

FIGURE 6.20.

6.28. (Whitney 1932b) Let Γ be a 3-connected simple planar graph. Prove that every abstract dual of Γ is simple and 3-connected.

6.29. Find the number of nonisomorphic graphs Δ such that $M_\Delta \cong M_\Gamma$, where Γ is the graph in Figure 6.19.

Section 4

6.30. (i) Which of the graphs in Figure 6.20 is a series-parallel network?

(ii) For each of the graphs in Figure 6.20 that is not a series-parallel network, determine the least number of edges that must be deleted to give a series-parallel network.

6.31. Let Γ be a series-parallel network having no parallel edges. Determine the maximum number of chords possible in a circuit C of Γ.

6.32. The *chromatic number* $\chi(\Gamma)$ of a loopless graph Γ is the least number of colors needed to color the vertices of Γ so that if two vertices are adjacent they are colored differently.

(i) Show that if Γ is a series-parallel network without loops, then $\chi(\Gamma) \le 3$.

(ii) Suppose that the graph Γ_2 is obtained from a loopless graph Γ_1 by a sequence of series and parallel extensions. Show that if $\chi(\Gamma_1) \ge 3$, then $\chi(\Gamma_2) \le \chi(\Gamma_1)$.

(iii) Does (ii) remain true if $\chi(\Gamma_1) = 2$?

6.33. Let p be an edge of a block Γ for which $\Gamma - p$ has k blocks Γ_1, $\Gamma_2, \ldots, \Gamma_k$, where $k \geq 2$. If $\Gamma_0 = \Gamma[\{p\}]$, show that Γ is a generalized circuit with parts $\Gamma_0, \Gamma_1, \Gamma_2, \ldots, \Gamma_k$.

6.34. (Plummer 1968) Let Γ be a block without parallel edges and e an edge of Γ. Show that $\Gamma - e$ is a block if and only if e is a chord of some circuit in Γ.

6.35. (Dirac 1952) With chromatic number defined as in Exercise 6.32, prove that if Γ is a graph with chromatic number 4, then Γ has a subgraph homeomorphic from K_4. Is the converse of this true?

6.36. (Duffin 1965) Prove that if a graph Γ has a planar embedding in which every vertex lies on the boundary of the infinite face, then Γ is a series-parallel network. Is the converse of this true?

*6.37. (Duffin 1965) Let Γ be a block, and assign a direction to every edge of Γ. Prove that Γ is a series-parallel network if and only if whenever a circuit C of Γ meets a pair of edges, x and y, that are oppositely directed relative to C, then x and y are oppositely directed relative to every circuit containing them both.

REFERENCES

Bondy, J. A., and Murty, U.S.R. (1976). *Graph Theory with Applications*. Macmillan, London.

Brylawski, T. (1971). A combinatorial model for series-parallel networks. *Trans. Amer. Math. Soc.* **154**: 1–22.

Cunningham, W. H. (1981). On matroid connectivity. *J. Combin. Theory Ser. B* **30**: 94–9.

Dirac, G. A. (1952). A property of 4-chromatic graphs and some remarks on critical graphs. *J. London Math. Soc.* **27**: 85–92.

(1967). Minimally 2-connected graphs. *J. Reine Angew. Math.* **228**: 204–16.

Duffin, R. J. (1965). Topology of series-parallel networks. *J. Math. Anal. Appl.* **10**: 303–18.

Greene, C. (1971). *Lectures on combinatorial geometries* (with notes, footnotes, and other comments by Daniel Kennedy). N.S.F. advanced science seminar, Bowdoin College, Brunswick, Me.

Harary, F. (1969). *Graph Theory*. Addison-Wesley, Reading, Mass.

Harary, F., and Tutte, W. T. (1965). A dual form of Kuratowski's Theorem. *Canad. Math. Bull.* **8**: 17–20, 373.

Harary, F., and Welsh, D. (1969). Matroids versus graphs, in *The Many Facets of Graph Theory, Lecture Notes in Mathematics Vol. 110*, edited by G. Chartrand and S. F. Kapoor, pp. 155–69. Springer-Verlag, Berlin.

Inukai, T., and Weinberg, L. (1981). Whitney connectivity of matroids. *SIAM J. Alg. Disc. Methods* **2**: 108–20.

Kuratowski, C. (1930). Sur le problème des courbes gauches en Topologie. *Fund. Math.* **15**: 271–83.

Ore, O. (1967). *The Four-Color Problem*. Academic Press, New York.

Oxley, J. G. (1981a). On connectivity in matroids and graphs. *Trans. Amer. Math. Soc.* **265**: 47–58.

(1981b). On a matroid generalization of graph connectivity. *Math. Proc. Camb. Phil. Soc.* **90**: 207–14.

Parsons, T. D. (1971). On planar graphs. *Amer. Math. Monthly* **78**: 176–8.

Perfect, H. (1981). Independence theory and matroids. *Math. Gaz.* **65**: 103–11.

Plummer, M. D. (1968). On minimal blocks. *Trans. Amer. Math. Soc.* **134**: 85–94.

Thomassen, C. (1981). Kuratowski's Theorem. *J. Graph Theory* **5**: 225–41.

Truemper, K. (1980). On Whitney's 2-isomorphism theorem for graphs. *J. Graph Theory* **4**: 43–9.

Tutte, W. T. (1966). *Connectivity in Graphs.* University of Toronto Press.

Wagner, K. (1937a). Ueber eine Eigenschaft der ebenen Komplexe. *Math. Ann.* **114**: 570–90.

(1937b). Ueber eine Erweiterung eines Satzes von Kuratowski. *Deut. Math.* **2**: 280–5.

Welsh, D. J. A. (1976). *Matroid Theory.* Academic Press, London.

Whitney, H. (1932a). Non-separable and planar graphs. *Trans. Amer. Math. Soc.* **34**: 339–62.

(1932b). Congruent graphs and the connectivity of graphs. *Amer. J. Math.* **54**: 150–68.

(1933). 2-Isomorphic graphs. *Amer. J. Math.* **55**: 245–54.

(1935). On the abstract properties of linear dependence. *Amer. J. Math.* **57**: 509–33.

CHAPTER 7

Constructions

Thomas Brylawski

7.1. INTRODUCTION

In this chapter we shall give properties, examples, and applications of the more fundamental matroid operations and constructions. This chapter should go hand-in-hand with the chapter on orthogonal duality, because the latter operation is the most basic of all and generates many of the operations defined here. Most of the operations treated in the following were discovered first for other mathematical structures (duality for planar graphs, contraction in projective space, direct sum for vector subspaces, truncation in Euclidean space, etc.) and generalized to all matroids when it was recognized that many of their defining characteristics were essentially matroidal. Thus, in addition to defining each operation abstractly, we illustrate it for the four basic matroid classes of vector and affine matroids, transversal matroids, and graphical matroids.

Recall that a vector matroid (one coordinatizable over a field F) is one represented by the column vectors of a matrix where dependence is given by linear dependence over F. An affine matroid is one represented (analytically) by vectors over F where dependence is affine; that is, $\{\mathbf{v}_i: 1 \leq i \leq k\}$ is affinely dependent if $\sum_{i=1}^{k} a_i \mathbf{v}_i = \mathbf{0}$ for scalars $a_i \in F$, not all zero, with $\sum_{i=1}^{k} a_i = 0$. Synthetically, in an affine matroid, a subset of k points is dependent if it spans an affine flat (subspace not necessarily through the origin) of dimension less than $k - 1$ (three points on a line, four points on a plane, etc.). A matroid M is graphical when there is a graph whose edges correspond to the points of M and whose polygons give the circuits of M. Finally, a matroid $M(E)$ is transversal if it can be presented by a relation $R \subseteq E \times T$, where $I \subseteq E$ is independent if it can be matched into a subset of T (by a bijection contained in R). We illustrate transversal matroids by the bipartite graph defined by R, where independent sets are the E-vertices of some set of vertex-disjoint edges. We further assume (without loss of generality) that our representing matrices have full row rank (and occasionally are in row-echelon form in that a subset of columns corresponding to a basis of M forms an identity submatrix), that representing graphs are connected, and that, in transversal presentations, $|T| = r(M)$. Further description of the classes can be found in the Appendix of Matroid Cryptomorphisms.

The major operations treated vary from the conceptually simple one of deletion to the more difficult and less studied Dilworth truncation. They also include contraction, extension, principal extension, simplification, coextension, quotient, lift, truncated lift, matroid projectivity, truncation, (free, free principal) erection, bracing, direct sum, series and parallel connection, generalized parallel connection, matroid union, line completion, and others (see the index at the end of the chapter).

Most of these operations are characterized in terms of all commonly studied cryptomorphisms (although, for these descriptions, no proofs are given), and an algebraic or order structure is introduced, when possible, on families generated by some construction (such as rooted trees for series-parallel decompositions, the lattice of single-element extensions or of erections, and the order ideal of matroids in a hereditary class). Several operations on hereditary classes are defined, although, for the most part, there are many more questions raised than answered. A recurrent theme in the chapter is that when a specific operation is defined, we try to define its dual (by conjugating appropriately with the orthogonal duality operator) such as lift from quotient, matroid intersection from matroid union, Higgs lift from truncation, and so forth. In addition, whenever possible, we characterize "reverse" or inverse images of operations by describing the set of matroids that have a specific image under a specific operation such as lift from quotient, erection from truncation, extension from deletion, and so forth.

7.2. ISTHMUSES AND LOOPS

We begin with a discussion of two special types of points in a matroid. They not only give an example of the simplest kind of direct sum (Section 7.6) but also present slightly special cases when characterizing deletions (Section 7.3) and contractions (Section 7.4).

A *loop* (which corresponds to the zero vector in a linear matroid) is a point whose rank is zero. The name comes from graph theory, where it corresponds to an edge incident to only one vertex. In a transversal matroid, a point of E is a loop precisely when it is related to no point of T. Loops cannot exist in affine matroids (any single vector is affinely independent), and thus when they arise in "affine pictures," we must list them separately. More generally, loops are specifically excluded from combinatorial geometries (where, in addition, the rank of every doubleton is 2), so that special conventions must be made when treating them in essentially geometric examples (projective spaces, geometric lattices, simple graphs). However, even more awkward conventions would have to be made if we did not allow them in general, because they occur naturally in "matroid-type" examples such as vector spaces and their quotients and in duals of geometries.

The concept does not merit any deeper exploration, but, for completeness, we characterize when P is a loop for several cryptomorphic descriptions. The (one-line) proofs are left as exercises.

7.2.1. Proposition. A loop of a matroid $M(E)$ can be characterized in any of the following terms:

> *Closure: $p \in \bar{\varnothing}$.*
> *Rank: $r(p) = 0$.*
> *Closed sets: p is in every closed set.*
> *Hyperplanes: p is in every hyperplane.*
> *Bonds: p is in no bond.*
> *Spanning sets: If A spans, so does $A - p$.*
> *Bases: p is in no basis.*
> *Independent sets: p is not independent.*
> *Circuits: p is a circuit.*

One more way to think about loops (and several of the other elementary concepts in this chapter) is in terms of the basic-circuit incidence matrix of a matroid $M(E)$ with a given basis B. It is motivated by the submatrix A when M is a linear matroid represented by the $n \times K$ row-echelon matrix $[IA]$, where the first n columns form an identity matrix I_n corresponding to the basis $B = \{p_1, \ldots, p_n\}$. Then $a_{ij} \neq 0$ if and only if p_i is in the (unique) basic circuit contained in $B \cup \{p_j\}$. In general, the *basic-circuit incidence matrix* of M with respect to B, $A(M; B)$, is a $0 - 1$ matrix whose rows are indexed by B and whose columns are indexed by $E - B$, and where $a_{ij} := 1$ if p_i is in the circuit $C \subseteq B \cup \{p_j\}$. We introduce this concept because many properties of vector matroids have analogues in this more general context. Of course, a specific matroid is not characterized by one of its basic-circuit matrices unless something else is known about it (e.g., if M is binary, or if M is transversal and the degree of every vertex in B is 1). One can easily prove this fact and convince oneself, using Theorem 5.2.9, that for the orthogonal dual matroid M^*, $A(M^*; E - B) = A^t(M; B)$ (where t denotes matrix transpose). Now, $p_j \notin B$ is a loop if and only if c_j is the zero column in $A(M; B)$ (see Exercise 7.1).

We now introduce the only slightly more difficult concept of an isthmus. A point p is an *isthmus* of M if p is a loop of M^*. This gives an immediate characterization of an isthmus in terms of $A(M; B)$ (see Exercise 7.2). The term comes from graph theory, where isthmuses are one-edge cut-sets (and look like their geographic namesakes). In transversal matroids, they are not, in general, instantly recognizable, but $p \in E$ is an isthmus if, for example, it is related to a vertex $t \in T$ of degree 1 (see Exercise 7.3). A point p in affine space is an isthmus if M is a "cone" with "apex" p [i.e., $M - p$ is contained in an affine flat (translated hyperplane) that does not contain p]. In a linear matroid, p_j is an isthmus if it is not a loop, and if whenever the matrix A is row-reduced, turning the column c_j into a vector with a unique nonzero entry a_{ij}, then a_{ij} is the only nonzero entry in r_i, the ith row of A.

We leave the proof of the following as an exercise.

7.2.2. Proposition. *A point $p \in E$ is an isthmus of $M(E)$ if and only if it satisfies any of the following:*

> *Geometric lattice: $p \nleq \vee\{a: a$ is an atom of L, $a \neq p\}$. (Equivalently, there is a unique copoint $x \in L$ not above p, in which case L is the Cartesian product of $[0, x]$ and a two-element lattice.)*
> *Closure: If $p \in \bar{A}$, then $p \in A$.*
> *Rank: $r(E - p) = r(E) - 1$. (Equivalently, $r(A \cup p) = r(A) + 1$ for all $A \subseteq E - p$.)*
> *Closed sets: $E - p$ is closed.*
> *Hyperplanes: p is in all but one hyperplane (viz., $E - p$).*
> *Bonds: p is a bond.*
> *Bases: p is in every basis.*
> *Independent sets: $I \cup \{p\}$ is independent for every independent set I.*
> *Circuits: p is in no circuit.*

A combinatorial geometry each of whose n points is an isthmus is called a *Boolean algebra* or a *free matroid* and is denoted B_n.

7.3. DELETIONS, SUBMATROIDS, AND EXTENSIONS

The deletion operation for our four basic classes (and in most examples) is, when suitably interpreted, merely "erasure." It lies, however, at the heart of matroid theory. Indeed, matroids can be recursively characterized as those simplicial complexes (of independent sets) that are pure (i.e., all maximal simplices are equicardinal), and each of whose deletions is a matroid.

To make precise the notion of erasure, we can delete p from a linear matroid by deleting its column from the representing matrix, and we can delete it from a transversal matroid by deleting its vertex in E (along with its incident edges). We delete p from a graphical matroid by removing its representing edge (keeping the rest of the graph, including all the vertices, intact), and we delete p from an affine matroid by removing it but keeping the coordinates of all other points the same. This latter operation is viewed synthetically by removing p but keeping all the other dependences.

In general, we can delete more than once, obtaining the *submatroid* $M(E') := M(E) - \{p : p \in E - E'\}$. Here, the order of the deletions does not matter, because we can define $M(E')$ by restricting the rank function of M to subsets of E' so that $r_{M(E')}(A) = r_M(A)$ for all $A \subseteq E'$. (Submatroids are, in fact, often called *restrictions*.) When M is a combinatorial geometry, so is $M(E')$, and it can thus be called a subgeometry. In fact, the prototype for a combinatorial geometry is a subgeometry of a projective space.

Certain (minor) problems enter when isthmuses are deleted [and, in general, when, in $M(E')$, $r(E') < r(E)$], all arising from the fact that the rank of the matroid is lowered. In the applications given earlier, connected graphs do not remain connected; representing matrices with independent rows are no longer so; in transversal matroid presentations, maximal partial matchings do not cover T; and so forth. In light of this, and because the more general deletion can be handled in the following section, we make the convention that unless otherwise stated, $M - p := M(E - p)$ is defined only when p is not an isthmus, and the submatroid $M(E')$ is defined only for spanning subsets $E' \subseteq E$. The reader is invited to prove the cryptomorphic descriptions in the following proposition, where the characterizations with the symbol $\#$ are valid only when E' spans. [Note that here and in other cryptomorphic descriptions of operations $M \mapsto O(M)$, such as Propositions 7.3.3, 7.4.3, 7.4.9, 7.4.12, and so forth, each structure of the matroid $O(M)$ is described in the same terms as M.]

7.3.1. Proposition. *Let E' be a ($\#$ spanning) subset of the matroid $M(E)$. Then the submatroid $M(E')$ can be defined in any of the following cryptomorphic ways:*

> *Geometric lattice: The supremum subsemilattice of L generated by the atoms E'. ($\#$ This subsemilattice includes $1 \in L$, where 1 is the maximal element of the finite lattice L.)*
> *Closure: $\mathrm{cl}_{M(E')}(A) = \bar{A} \cap E'$.*
> *Rank: $r_{M(E')}(A) = r(A)$ for all $A \subseteq E'$.*
> *Closed sets: $\{K \cap E' : K \text{ is closed in } M\}$. ($\#$ E' is given uniquely in this description as $E \cap E'$.)*
> *Hyperplanes: Maximal sets among $\{H \cap E' : H \text{ is a hyperplane of } M \text{ that does not contain } E'\}$. ($\#$ We can eliminate the condition that $H \not\supseteq E'$.)*

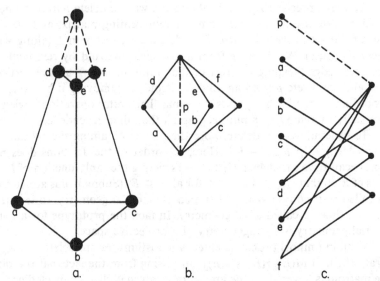

FIGURE 7.1. Affine (a), graphic (b), and transversal (c) representations of a matroid M and a deletion $M - p$.

> *Bonds: Minimal nonempty sets of the form $\{B \cap E': B \text{ is a bond of } M\}$.*
> *(# We can eliminate the word nonempty.)*
> *Spanning sets: # Spanning sets of M contained in E'.*
> *Bases: Maximal sets of the form $\{B \cap E': B \text{ is a basis of } M\}$. (# Bases*
>   *contained in E'.)*
> *Independent sets: Independent sets of M contained in (or, equivalently,*
>   *intersected with) E'.*
> *Circuits: Circuits of M contained in E'.*

To fix ideas, Figure 7.1 shows three representations of the same matroid deletion in rank 4 where $M - p$ is represented by erasing p and all dashed lines.

Even though the operation of deletion of a point (or subset) is quite elementary, when the reverse operation of adding points to a matroid is considered, the theory (known and conjectured) can be quite interesting. In particular, we say a matroid N is an *extension* of a matroid M by a subset E' if $N - E' = M$ and $r(M) = r(N)$. When $E' = \{p\}$, we call N a *single-element extension*, and in light of Proposition 7.3.1, we need only specify, for each flat K of M, whether or not K is a flat of N and whether or not $K \cup \{p\}$ is a flat of N. We can thus partition $\mathscr{K}$, the flats of M, into three types:

> $\mathscr{K}_1 = \{K: K \text{ and } K \cup \{p\} \text{ are both flats of } N\}$.
> $\mathscr{K}_2 = \{K: K \text{ is a flat of } N, \text{ but } K \cup \{p\} \text{ is not}\}$.
> $\mathscr{K}_3 = \{K: K \cup \{p\} \text{ is a flat of } N, \text{ but } K \text{ is not}\}$.

[Note that if it is known that p is a loop of N, then the structure of $M = N - p$ completely determines the structure of N, and we write this special case as $N = M \oplus p$. A similar result, with identical notation, occurs when p is an isthmus.]

Crapo (1965) characterized which partitions of $\mathcal{K}$ arise in single-element extensions, and later the theory was implemented into a computer program that found (and cataloged) all combinatorial geometries on eight or less points (Blackburn, Crapo, and Higgs 1973). (Clearly, any matroid is an extension of one of its bases.) It is an easy application of the closure-exchange axiom or the rank function that $\mathcal{K}_3$ is an order filter (i.e., if $A \in \mathcal{K}_3$ and $\bar{B} = B \supseteq A$, then $B \in \mathcal{K}_3$), $\mathcal{K}_1$ is an order ideal (closed subsets of sets in $\mathcal{K}_1$ remain in $\mathcal{K}_1$), and $A \in \mathcal{K}_2$ if and only if $A \notin \mathcal{K}_3$, but there is a $B \in \mathcal{K}_3$ that covers A. Thus, if N is a single-element extension of M, N is completely determined from M when it is known which closed sets of M are in $\mathcal{K}_3$. This is related to the theory of strong maps (see Chapter 8), but, for completeness, we list the characterization of which closed-set subfamilies $\mathcal{K}_3$, termed *modular cuts*, occur when constructing single-element extensions:

7.3.2. Proposition. *An order filter $\mathcal{M}$ of closed sets of $M(E)$ forms a modular cut if and only if $\mathcal{M}$ is nonempty ($E \in \mathcal{M}$) and if whenever A and $B \in \mathcal{M}$, with $r(A) + r(B) = r(A \cap B) + r(A \cup B)$, then $A \cap B \in \mathcal{M}$.*

We denote by $M +_{\mathcal{M}} p$ the extension of M via the modular cut $\mathcal{M}$. For the extension $(M - p) +_{\mathcal{M}} p$ in Figure 7.1, $\mathcal{M}$ consists of the flats $\{ad, be, cf, abde, acdf, bcef, abcdef\}$.

A geometrically appealing type of single-element extension is one with a modular cut consisting of a flat F and all closed sets that contain it. This construction, denoted $M +_F p$, is called the *principal extension* of M on F. Special cases occur when $F = E$ (the *free extension* $M + p$), when $F = \bar{\varnothing}$ (in which case p is a loop), and when F is an atom (in which case $F \cup p$ is a multiple point).

Note that we could consider modular cuts generated by more general subsets of M than flats, but the theory would reduce to the flat case by elementary reasoning (e.g., Exercise 7.8). In light of this, we could define principal extension on a subset A or on an independent subset I, but it is easily proved (and intuitively evident) that if $I \subseteq A \subseteq \bar{I}$, then $M +_I p = M +_A p = M +_{\bar{I}} p$.

For reference, we list the various cryptomorphic descriptions of principal extensions. Proofs can be obtained by suitably interpreting the foregoing $\mathcal{K}_1, \mathcal{K}_2$, and $\mathcal{K}_3$ for the case in which $\mathcal{K}_3$ is a principal order filter. Descriptions preceded by the symbol # apply to the free extension $M + p$.

7.3.3. Proposition. *The principal extension $M' = M +_F p$ (# and, in particular, the free extension $M'' = M + p$) can be described as follows (for* *all subsets $A \subseteq E$):*

Closure:

$$\mathrm{cl}_{M'}(A) = \begin{cases} \bar{A} \cup p & \text{if } \bar{A} \supseteq F, \\ \bar{A} & \text{otherwise} \end{cases}$$

(where $\bar{A}$ is the closure of A in M).

$$\mathrm{cl}_{M'}(A \cup p) = \begin{cases} \overline{A \cup F} \cup p & \text{if } r(A \cup F) - r(A) = 1, \\ \bar{A} \cup p & \text{otherwise} \end{cases}$$

[$\neq \mathrm{cl}_{M'}(A) = \bar{A}$ unless $\bar{A} = E$, in which case $\mathrm{cl}_{M''}(A) = E \cup p$; $\mathrm{cl}_{M''}(A \cup p) = \bar{A} \cup p$ unless $\bar{A}$ is a hyperplane, in which case $\mathrm{cl}_{M''}(A \cup p) = E \cup p$.]

Rank: $r_{M'}(A) = r(A)$, and

$$r_{M'}(A \cup p) = \begin{cases} r(A) & \text{if } r(A \cup F) = r(A), \\ r(A) + 1 & \text{otherwise.} \end{cases}$$

Closed sets: $\{K: K$ is closed in M, and $K \not\supseteq F\} \cup \{K \cup p: K$ is closed in M, and K is not covered by $\overline{K \cup F}\}$.

Hyperplanes: $\neq \{H: H$ is a hyperplane of $M\} \cup \{(H \cap H') \cup p: H \neq H'$ are hyperplanes of M, and $H \cap H'$ is maximal (i.e., a coline)$\}$.

Spanning sets: $\{S: S$ spans $M\} \cup \{S' \cup p: S' \cup q$ spans M for some $q \in F\}$.

Bases: $\{B: B$ is a basis of $M\} \cup \{(B - q) \cup p: B$ is a basis of M, and $q \in B \cap F\}$.

Independent sets: $\{I: I$ is independent in $M\} \cup \{I \cup p: \bar{I} \not\supseteq F\}$.

Circuits: $\{C: C$ is a circuit of $M\} \cup \{I \cup p: \bar{I} \supseteq F$, and I is minimal with respect to this property$\}$.

We now list some propositions showing the roles of extensions and principal extensions for various classes of matroids.

7.3.4. Proposition. *Every (finite) vector combinatorial geometry has a finite extension to a projective geometry.*

Proof. (After Rado 1957). By definition, a vector geometry G is coordinatized over a field F. We shall be done when we show that F can be chosen to be finite, because we can then take $\mathrm{PG}[r(G) - 1, F]$ for our extension. First, we show that F can be chosen as an algebraic extension over its prime subfield. This latter result can be considered as the matroid version of the Hilbert *nullstellensatz*. In particular, if G is coordinatized by the matrix A over F, let $\bar{F}$ be the algebraic closure of the prime field of F. Then the set of polynomials $\sum_{\sigma}(\pm \prod_i x_{i\sigma(i)}) = 0$ corresponding to all vanishing subdeterminants of A and the polynomials $x_\lambda \cdot \sum_{\sigma}(\pm \prod_i x_{i\sigma(i)}) - 1 = 0$ corresponding to all nonvanishing subdeterminants in A has a solution in F given by $x_{ij} = a_{ij} \in A$ and $x_\lambda = D_\lambda^{-1}$, where D_λ is the value of the appropriate nonvanishing determinant of A. Thus (a version of) the *nullstellensatz* (Atiyah and MacDonald 1969, Corollary 5.24) says that these polynomials can be simultaneously satisfied in $\bar{F}$, and hence in a field F' of finite degree over the prime field of F.

If this algebraic extension field F' has characteristic zero, we can multiply the columns of A so that each entry is an algebraic integer (e.g., if a_{ij} has minimal polynomial $b_n x^n + \cdots + b_0$, then, clearing denominators, we can assume that each b_i is an integer, and thus that $b_n a_{ij}$ is a root of $x^n + b_n b_{n-1} x^{n-1} + \cdots + b_n^{n-1} b_1 x + b_n^n b_0$, which is its monic minimal polynomial). Because the algebraic integers form a ring, repeated multiplication of columns by integers makes each a_{ij}, and hence each determinant, an algebraic integer. For each nonzero subdeterminant (and entry) D_λ, let $b_{\lambda,0}$ be the constant term in its minimal polynomial $x^n + b_{\lambda,n-1} x^{n-1} + \cdots + b_{\lambda,0}$. Choose a prime p that does not divide any of the $b_{\lambda,0}$'s. Now take the maximal ideal pR in the ring R of algebraic integers and consider the quotient ring R/pR that is a field $\hat{F}_p$ of characteristic p. The image of each a_{ij} under the ring homomorphism $R \to R/pR$ gives a matrix $\hat{A}$ over $\hat{F}_p$ in which determinants that are zero in A remain zero, and a determinant D_λ that is nonzero becomes a determinant $\hat{D}_\lambda$ that cannot be zero, because it solves the polynomial $x^n + \hat{b}_{\lambda,n-1} x^{n-1} + \cdots + \hat{b}_{\lambda,0}$, where $\hat{b}_{\lambda,i} = b_{\lambda,i} + Rp = b_{\lambda,i} \pmod{p}$, and $\hat{b}_{\lambda,0} \neq 0$. Thus, determinants vanish in $\hat{A}$ precisely when they do in A, and we can coordinatize G over the finite algebraic field $F_p(\ldots, \hat{a}_{ij}, \ldots)$.   □

In light of this latter theorem, we need never make, for example, a transcendental extension to represent a matroid, because an algebraic extension will always suffice. We shall, nevertheless, have occasion later to refer to the *transcendental extension* $F(\lambda_1, \ldots, \lambda_n)$, and we note that we do this merely for convenience, because the λ_i's play the role of coordinatizing elements that do not solve certain determinantal identities. Indeed, when F is sufficiently large, no extension need be made, because over fields of characteristic zero we can choose λ_i to be a sufficiently large rational integer N_i, and over sufficiently large fields of characteristic p we shall have an element x whose minimal polynomial over F_p has sufficiently high degree, and thus we can replace λ_i by an appropriate x^{N_i}. It is to impart a certain spirit, therefore, that we refer to transcendental extensions, and we leave it as a general research problem, for a given matroid or class of matroids, to determine how big an algebraic extension must be made or how big a "sufficiently large" field is needed to "finitize" a given transcendental construction.

As an example of supplying the technicalities for finitizing a transcendental construction, we prove the following proposition.

7.3.5. Proposition. *Let x be a flat of M of rank m, and let M be coordinatized over F. Then the principal extension $M +_x p$ can be coordinatized over an extension field of degree m, and, indeed, when F is finite and $M = \mathrm{PG}(n, F)$, no smaller field can coordinatize $M +_x p$.*

Proof. Extend a basis $p_1, \ldots, p_m$ for x to a basis for M, and represent M by A, where the first n columns form an identity matrix. Further, let α be a root of a minimal polynomial over F of degree m. Then the matrix $A' = [A\mathbf{v}]$ coordinatizes $M +_x p$, where $\mathbf{v}^t = (1, \alpha, \alpha^2, \ldots, \alpha^{m-1}, 0, \ldots, 0)$, because, by construction, the α^i's are linearly independent over F, and a subset of columns of A' is linearly independent if and only if it is indexed by an independent set specified in Proposition 7.3.3.

Conversely, if the matrix A coordinatizes $PG(n, F) +_x p$ over $\bar{F}$, and A is row-reduced as before, giving the coordinates $(1, \alpha_1, \alpha_2, \ldots, \alpha_{m-1}, 0, \ldots, 0)$ for $\mathbf{v}^t$, then any linear combination $\sum a_i \alpha_i$ ($a_i \in F$ not all zero) can be realized by a subdeterminant corresponding to a basis of x. [For example, when $a_0 = 1$, consider the basis consisting of $\mathbf{v}^t$ and the vectors $\mathbf{v}_j$ ($j = 2, \ldots, m$), where $v_{1,j} = -a_{j-1}, v_{j,j} = 1$, and $v_{i,j} = 0$ otherwise.] Thus, the elements $\{1, \alpha_1, \ldots, \alpha_{m-1}\}$ in $\bar{F}$ are linearly independent over F, and $\deg[\bar{F}: F] \geq m$.
□

The foregoing proposition suggests the following research problem: Find the *dual critical exponent* of a linear matroid M coordinatized over a field F, the minimal degree of $[\bar{F}: F]$, where $\bar{F}$ coordinatizes the free extension $M + p$. This is (projectively) dual to the critical problem in the following sense: An equivalent version of the critical problem of Crapo and Rota (1970) for a geometry $G \subseteq PG(n - 1, F)$ asks for the minimal degree $[\bar{F}: F]$ such that $G \subseteq AG(n - 1, \bar{F})$, that is, such that there exists a hyperplane H of $PG(n - 1, \bar{F})$ disjoint from G. When G is $PG(n - 1, F)$ and we projectively dualize, we get a point H^* of $PG(n - 1, \bar{F})$ that is not contained in any of the hyperplanes p^* spanned by points in $PG^*(n - 1, F) \simeq PG(n - 1, F)$. Thus, the free extension $PG(n - 1, F) + H^*$ is represented in $PG(n - 1, \bar{F})$. We call a matroid M *coaffine* over F if $M + p$ is linear over F (i.e., if its dual critical exponent is 1). Over sufficiently large fields, M is coaffine and $M + p$ is affine, so that in large fields free (or principal) extension is an affine operation. $M +_F p$ can be described synthetically as "putting a point freely on the flat F." Note that, unlike the critical exponent, the dual critical exponent is not a matroid invariant, and, in particular, the property of being coaffine depends on the coordinatization.

For example,

$$\begin{bmatrix} 1 & 0 & 0 & 1 & 1 & 1 \\ 0 & 1 & 0 & 1 & 2 & 3 \\ 0 & 0 & 1 & 1 & -3 & 2 \end{bmatrix}$$

represents a *complete arc* over F_7. That is, it has the uniform matroid structure $U_{3,6}$ (six points in the plane, no three on a line), and (as the reader can check) it is not coaffine over F_7. However, the same matroid can be given the coaffine coordinatization

$$\begin{bmatrix} 1 & 1 & 0 & 1 & 1 & 1 \\ 0 & -1 & 0 & 1 & 2 & 3 \\ 0 & 1 & 1 & 1 & -3 & 2 \end{bmatrix},$$

which admits the placing of

$$\begin{bmatrix} 1 \\ -2 \\ -3 \end{bmatrix} \quad \text{or} \quad \begin{bmatrix} 1 \\ -3 \\ 2 \end{bmatrix}$$

in free position.

Small fields (in particular, F_2) almost never coordinatize principal extensions by flats of rank greater than 1. The reader can generalize the following proposition to the binary case.

7.3.6. Proposition. *If M is represented by a 2-connected graph and $r(F) > 1$, then $M +_F p$ is not graphic.*

Proof. Assume that the graph G represents $M +_F p$, and consider the subgraph $G - p$, where p is an edge joining v_1 to v_2. This will be 2-connected (see Chapter 6 or Section 7.6 in this chapter), and thus there is a circuit C containing v_1 and v_2 as nonconsecutive vertices. Hence, there are two circuits $C_1 \cup p$ and $C_2 \cup p$, with $C_1 \cup C_2 = C$, so that C_1 and C_2 cannot both span F. Thus, a circuit is created in G that is not specified in Proposition 7.3.3. $\qquad \square$

7.3.7. Proposition. *Transversal matroids are not, in general, closed under principal extension by F, but $M +_F p$ is transversal if F is an intersection of flats of a transversal matroid M each of which, as a submatroid, has no isthmuses. On the other hand, every transversal matroid $M(e_1, \ldots, e_k)$ of rank n can be constructed by principal extension and deletion: $M = (B_n +_{F_1} p_1 + \cdots +_{F_k} p_k) - B_n$, where B_n is a Boolean algebra, and the principal extensions can be taken in any order.*

Proof. The matroid M consisting of three double points $\{p_i, p_i': i = 1, 2, 3\}$ on a line is not transversal. However, it is a principal extension $(M - p_i) +_{p_i} p_i$ (for any i) of the transversal matroid $M - p_i$.

If M is presented by the relation $R \subseteq E \times T$, and if $R(p_i) = \{t_{i_1}, \ldots, t_{i_m}\}$, then we can construct M by starting with a Boolean algebra $M^0 = B_n$ consisting of isthmuses $\{t_1', \ldots, t_n'\}$ and, at the ith stage, making the principal extension $M^i = M^{i-1} +_{F_i} p_i$, where $F_i = \overline{\{t_{i_1}', \ldots, t_{i_m}'\}}$, the closure being taken in M^{i-1}. Then $M = M^K - B_n$.

In this construction, any flat F_j of M (or M^K) that, as a submatroid, has no isthmuses must be of the form $\bar{B}^j$, where $B^j = \{t'_{j_1}, \ldots, t'_{j_m}\}$ for some subset B^j of B_n and must consist of the points $R^{-1}(\{t_{j_1}, \ldots, t_{j_m}\})$, along with $\{t'_{j_1}, \ldots, t'_{j_m}\}$. To freely extend M by the intersection $F_1 \cap \cdots \cap F_k$ (where k might be 0) of these flats, we add a vertex $p \in E$ and relate it to the set $T^1 \cap \cdots \cap T^k$ (or to T when $k = 0$), where $T^j = \{t_{j_1}, \ldots, t_{j_m}\}$.   □

7.4. CONTRACTIONS, MINORS, AND LIFTS

For any operation on matroids $O: \mathcal{M} \to \mathcal{M}$ that assigns to each matroid (or matroid isomorphism class) M a unique matroid M' we can formally define the *dual operation* O^* by conjugating with the orthogonal dual (perhaps the most characteristic matroid operation of all):

$$O^*(M) = [O(M^*)]^*.$$

We can extend this idea of (operational) duality to operations from, say, n-tuples of matroids to k-sets of matroids, from pointed matroids (i.e., "matroids with the point p") to matroids, and so forth. Note that when O is the operation "restrict to the subgeometry p," we have the observation that "isthmus is the dual of loop."

For future reference, we remind the reader of some of the cryptomorphic descriptions of the orthogonal dual (see Chapter 5 for proofs and motivations).

7.4.1. Proposition. *The orthogonal dual $M^*(E)$ can be described in terms of $M(E)$ by any of the following cryptomorphisms:*

Closure: $p \in \mathrm{cl}_{M^*}(A)$ *if either* $p \in A$ *or* $p \notin \overline{E - A - p}$.
Rank function: $r_{M^*}(A) = r(E - A) + |A| - r(E) = null(E) - null(E - A)$[*where null(A), the nullity of A, is equal to* $|A| - r(A)$]. *[Equivalently,* $corank_{M^*}(A) = null(E - A)$.*]*
Closed sets: Closed sets of M^* are complements of circuit unions of M.
Hyperplanes: Complements of circuits of M.
Bonds: Circuits of M.
Spanning sets: Complements of independent sets of M.
Bases: Complements of bases of M.
Independent sets: Complements of spanning sets of M.
Circuits: Bonds of M.

For our prototypal families we have the following:

7.4.2. Proposition. 1. *If M is a matroid of rank n and cardinality K that is a vector matroid over F, then M^* is a vector matroid with the matrix representation $[-A^t I_{K-n}]$ or $[A^t I_{K-n}]$, where $[I_n A]$ represents M. [Equiva-*

lently, if a matrix N represents M, then M* is coordinatized by any matrix whose row vectors span a subspace orthogonal (in F^K) to the row space of N.]

2.   *Duality is not an affine operation (a one-point geometry dualizes into a loop that is not affine).*

3.   (a)   *The dual of a transversal geometry $M \in \mathcal{T}$ is not, in general, transversal* [*the smallest example being the dual of the transversal matroid $M - p$ of Figure 7.1*]. $\mathcal{T}^* = \{M^*: M \in \mathcal{T}\}$ *is characterized as the class of strict gammoids: the class of matroids such that there exists a directed graph G with vertex set E and vertex subset B such that $I \subseteq E$ is an independent set of vertices if and only if there are vertex-disjoint directed paths in G joining each vertex of I to a subset of B (Ingleton and Piff 1973). Equivalently, $\mathcal{T}^*$ is the class of\ principal matroids (see remarks following Proposition 7.4.9).* [*Gammoids (see Exercise 7.9) are defined as subgeometries of strict gammoids, i.e., they do not have, as groundset, the full vertex set of G.*]

(b)   *If M is principal transversal, that is, if it has a transversal presentation of the form $R \subseteq (\bar{T} \cup S) \times T$, where $R_{\bar{T}}$ (R restricted to $\bar{T}$) is a bijection (matching) onto T, then M* is transversal and is presented by $R^* \subseteq (T \cup \bar{S}) \times S$, where $R_{\bar{T}}^* = R_S^{-1}$, and $R_{\bar{S}}^*$ is a bijection.*

4.   *The dual of a graphical matroid M (represented by the connected graph G) is graphical if and only if G is planar, in which case it is represented by G^*, the planar dual graph of G. Equivalently, M* is represented by the planar map of G, where countries play the role of vertices.*

We note the historical importance of Proposition 7.4.3, part 4: It was the attempt by Whitney to give a dual structure to a nonplanar graph that led to the invention of matroids.

To illustrate these ideas we shall dualize the representations of M in Figure 7.1, obtaining the rank-3 matroid M* in its various dual representations in Figure 7.2. In Figure 7.2b, the graph G(M) from Figure 7.1b is superimposed with dashed edges and hollow vertices.

Orthogonal matrix representations for the foregoing matroids M and M*, respectively, are given (over any field) by $N(M) \overset{*}{\leftrightarrow} N^{\perp}(M^*)$:

$$
N = \begin{matrix} & p & a & b & c & d & e & f \\ & \begin{bmatrix} 1 & 0 & 0 & 0 & 1 & 1 & 1 \\ 0 & 1 & 0 & 0 & -1 & 0 & 0 \\ 0 & 0 & 1 & 0 & 0 & -1 & 0 \\ 0 & 0 & 0 & 1 & 0 & 0 & -1 \end{bmatrix} \\ & \underbrace{}_{I_4} \ \underbrace{}_{A} \end{matrix} \overset{*}{\leftrightarrow} N^{\perp} = \begin{matrix} p & a & b & c & d & e & f \\ \begin{bmatrix} -1 & 1 & 0 & 0 & 1 & 0 & 0 \\ -1 & 0 & 1 & 0 & 0 & 1 & 0 \\ -1 & 0 & 0 & 1 & 0 & 0 & 1 \end{bmatrix} \\ \underbrace{}_{-A^t} \ \underbrace{}_{I_3} \end{matrix}
$$

In the context of dual operations, the dual of deletion (by a nonisthmus) is *contraction* (by a nonloop):

$$M/p := (M^* - p)^*.$$

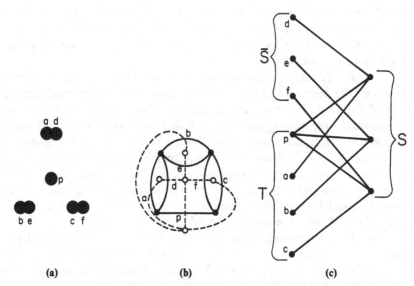

FIGURE 7.2. Affine, graphic, and transversal representations of M^*.

The dual notion of (spanning) submatroid is *contraction by a subset* $I \subseteq E$ (where I is independent):

$$M/I := \{[M^*(E)] - I\}^*.$$

Combining Propositions 7.3.1 and 7.4.1, we obtain the following cryptomorphic descriptions of contraction. As in Proposition 7.3.1, the (simpler) characterizations with the symbol $\#$ hold only when I is independent. Note that the case in which I is independent is not so special, because, in general, if I is a maximal independent subset of A, then points in $A - I$ are all loops of M/I, and $M/A = (M/I) - (A - I)$.

7.4.3. Proposition. *Let I be an ($\#$ independent) subset in the matroid $M(E)$. Then the contraction M/I is a matroid on the set $E - I$ and can be defined in any of the following cryptomorphic ways (where A is an arbitrary subset of $E - I$):*

> *Geometric lattice: The lattice of closed sets of M/I is isomorphic to the interval $[\bar{I}, 1]$ in the lattice L of M. However, as will be discussed later, M/I is not, in general, a geometry, even if M is.*
> *Closure:* $\mathrm{cl}_{M/I}(A) = \overline{A \cup I} - I.$
> *Rank:* $r_{M/I}(A) = r(A \cup I) - r(I) \, [\# = r(A \cup I) - |I|].$
> *Closed sets:* $\{K - I : K \supseteq I, \text{ and } K \text{ is closed in } M\}.$
> *Hyperplanes:* $\{H - I : H \supseteq I, \text{ and } H \text{ is a hyperplane of } M\}.$

Bonds: $\{B: B$ *is a bond of* M, *and* $B \cap I = \varnothing\}$.
Spanning sets: $\{S - I: (S \supseteq I$, *and)* S *spans* $M\}$.
Bases: Minimal sets of the form $\{B - I: B$ *is a basis of* $M\}$. (# $\{B - I$:
$B \supseteq I$, *and* B *is a basis of* $M\}$.)
Independent sets: $\{J - I: J$ *is independent in* M, *and* $J \cap I$ *is maximal*
[*i.e.,* $|J \cap I| = r(I)$]}. (# $\{J - I: J \supseteq I$, *and* J *is independent in* $M\}$.)
Circuits: Minimal nonempty sets of the form $\{C - I: C$ *is a circuit of* $M\}$
(# *when* I *is independent, the condition "nonempty" is redundant).*
Dependent sets: # $\{D - I: D$ *is a dependent subset of* $M\}$.

The contraction M/I is often not a combinatorial geometry even when M is. In fact, if M is a geometry, then M/A will have a loop unless A is closed and will be a geometry if and only if, further, the subsets $\{A \cup p: p \in E - A\}$ are all closed. For those who prefer to stay within the category of combinatorial geometries, we define the *simplification* $\bar{M}$ of a matroid M as the geometry formed by removing all loops and identifying any multiple point into a single point (atom) (in direct analogy with creating a simple graph from a multigraph, creating a projective space from a vector space, or identifying equal vectors in an affine space). For most reasonable classes $\mathscr{C}$ of matroids (certainly, for example, for those closed under deletion), if M is in $\mathscr{C}$, so is $\bar{M}$, but not necessarily conversely, as the counterexample of Proposition 7.3.7 shows for transversal matroids or a loop shows for affine matroids. We call the geometry $\overline{M/A}$ the *geometric contraction* of M by A, and we can identify it with the geometric-lattice interval $[x, 1]$, where x is the element of the geometric lattice of M corresponding to $\bar{A}$.

7.4.4. Proposition. *Let M be a matroid linear over F, with representing matrix N. Then M/A is linear, and its matrix is obtained as follows: First, choose a basis $I = \{p_1, \dots, p_k\}$ for the set A, row-reducing N so that I is represented by I', the first k columns of the identity matrix. Then $N(M/A)$ is the submatrix of N formed by deleting the first k rows of N and then deleting any column indexed by A (all of which necessarily have become zero-vectors).*

Proof. One method of proof is to realize that the foregoing construction is a step-by-step application of the dualize-submatroid-dualize definition of contraction using Proposition 7.4.2, part 1. Alternatively, we can show that $N(M/A)$ represents M/A directly by using the dependent-set characterization of contraction in Proposition 7.4.3. Indeed, a subset of columns of $E - A$ is dependent in $N(M/A)$ if and only if the corresponding columns of N along with I' are dependent. $\qquad \square$

One of the foregoing proofs used the duality between deletion and contraction to prove the given representation for M/A. Certainly, any class of matroids (such as those represented by planar graphs) that is closed under deletion and duality is closed under contraction, but the class of graphic matroids, although not closed under duality, is closed under contraction:

7.4.5. Proposition. 1. *Let G be a planar map with matroid $M(G)$ [bonds of $M(G)$ correspond to subsets of boundaries that enclose a simply connected region]. Then M/A has the planar-map representation $G - A$ (where the edges A are deleted from G).*

2. *If G is a connected graph with edge-set E, and $M(G)$ is its matroid, then M/A is represented by the connected graph G/A obtained as follows: The edges in A partition the vertices of G [the blocks are the connected components of the subgraph $G(A)$]. The graph G/A has vertex-set $\{v_1, \ldots, v_k\}$ corresponding to the blocks $\{B_1, \ldots, B_k\}$ of this partition, and an edge joins vertices v_i and v_j for every edge of $G - A$ joining a vertex in B_i to a vertex in B_j (where, perhaps, $i = j$). [In graph theory, this operation is variously described by "contraction" (of edges) or "coalescence" (of vertices).]*

Proof. The first part is a direct application of the dualize-delete-dualize definition. The second part follows easily from the dependent-set characterization of contraction, because we can assume that A is a forest of G. Then $A \cup E'$ contains a circuit in G if and only if there is a subset E'' of E' of the form $\{e_{i_1}, e_{i_2}, \ldots, e_{i_m}\}$, with e_{i_j} joining a vertex in B_{i_j} to a vertex in $B_{i_{j+1}}$ (where $B_{i_{m+1}} := B_{i_1}$).   □

Transversal matroids are not closed under contraction: The matroid M^* in Figure 7.2 is transversal, whereas M^*/p is not (see Proposition 7.3.7). The class of transversal matroids and their contractions forms the interesting class of gammoids [see Exercise 7.9 or Proposition 7.4.2, part 3(a), where a gammoid is a strict gammoid restricted to a subset of the vertices of G].

Contraction is not, in general, affine. This observation, in fact, probably formed the historical basis for projective geometry (!), because the geometric contraction of $AG(n, F)$ by a point is $PG(n - 1, F)$. The classic view of real projective geometry ("lines through the origin") motivates the following proposition, where part 1 is the synthetic version of the analytic part 2:

7.4.6. Proposition. 1. *If $M(E)$ is affine over F, then M/p is affine if and only if there exists a hyperplane in $AG[r(M) - 1, F]$ that contains p but no other point of M. Equivalently, M/p is affine if and only if $\bar{p} = p$ (i.e., p is not a multiple point) and there exists a hyperplane H in $AG[r(M) - 1, F]$ disjoint from p that is not parallel to any line $\{\overline{pq}: q \in M - p\}$.*

In this latter case, M/p has an affine representation in $H \simeq AG[r(M) - 2, F]$ consisting of $\{q' = \overline{pq} \cap H: q \in E - p\}$.

2. *If $M(E)$ has an affine representation over F, where $\mathbf{q}$ is the vector of affine coordinates for the point $q \in E$, then M/p has the matrix representation $N = \{\mathbf{q} - \mathbf{p}: q \in E - p\}$. Thus, for a vector matroid M represented by N, M and M/p are both affine if and only if some linear combination of the rows of N is nowhere zero, and some linear combination of the rows is zero exactly in the pth coordinate.*

Proof. The two conditions of part 1 are equivalent by taking parallel hyperplanes. We shall prove part 1 by synthetic means and leave the analytic proof of part 2 to the reader. It is an important consequence of "matroid-invariant" theory that if a matroid M is representable over F, then whether or not it is affine depends only on its matroid isomorphism type (not on its particular representation). [If $\chi(M; \lambda)$ is the characteristic polynomial of M and $|F| = q$, then M is affine if and only if $\chi(M; q) > 0$. See the chapter in a later volume on Möbius functions, or see Crapo and Rota (1970).] Thus, translating p to the origin and considering the projective space $P = PG[r(M) - 1, F]$, M/p is affine (in any representation) if and only if the projection of $E - p$ from p onto the "hyperplane at infinity" H_∞ misses a relative hyperplane $\bar{H}$ of H_∞. In this case, the hyperplane of P spanned by $\bar{H}$ and p, when restricted to $AG[r(M) - 1, F] = P - H_\infty$, contains no points of $E - p$.

Now consider a hyperplane H in $A = AG[r(M) - 1, F]$ that is not parallel to any line through p and with $p \notin H$. Then the following are equivalent:

The points $\{p, q_1, \ldots, q_m\} \subseteq E$ have affine dimension d.
The lines $\{\overline{pq_i}: 1 \le i \le m\}$ span a d-dimensional flat x in A.
The dimension of $x \cap H$ is $d - 1$.
The dimension of $\{q_i' := \overline{pq_i} \cap H: 1 \le i \le m\}$ is $d - 1$.

Thus, the rank description of M/p in Proposition 7.4.3 is satisfied.

$\square$

Combining single-element extension with contraction, we get an important construction: the *elementary quotient matroid* (with respect to a modular cut $\mathcal{M}$)

$$M \mapsto (M +_{\mathcal{M}} p)/p := T_{\mathcal{M}}(M).$$

The resulting matroid M' on the set E is characterized by $r(M') = r(M) - 1$ and the fact that if K is closed in M', then K is closed in M. Cryptomorphic versions of quotients (i.e., when M' is a quotient of M) come from letting $N = M +_{\mathcal{M}} p$ and comparing $N - p$ (Proposition 7.3.1) with N/p (Proposition 7.4.3). If we iterate this construction (k times), we obtain a *(nullity-k)* quotient. A nullity-k quotient is of the form $N - I \mapsto N/I$, with $|I| = r(I) = r(N) - r(N/I) = k$. [In Chapter 8 *quotient* is used to describe the (identity) *mapping* $N - I \mapsto N/I$, as it is for later c-quotient in this chapter.]

We have the following cryptomorphic characterizations of a quotient. Other properties and descriptions are to be found in Chapter 8 on strong maps, because M' is a quotient of M if and only if the identity map on E is a matroid strong map. For comparison we also list cryptomorphisms for weak-map images. Note that, in general, Proposition 7.4.7 describes when one matroid (M') is a quotient of another matroid (M). Some descriptions are stronger (such as that for the geometric lattice of a quotient) in that they further guarantee that M' is a matroid (lattice) whenever M is, or vice versa. The reader is invited to identify these stronger characterizations and to invent others.

7.4.7. Proposition. 1. *$M' (E)$ is a quotient of $M(E)$ if and only if any of the following cryptomorphic relationships hold:*

> *Geometric lattice: There is a supremum-preserving surjection f from $L(M)$ to $L(M')$ that takes an atom of $L(M)$ to an atom or the zero element of $L(M')$. (Equivalently, f preserves the zero element, suprema, and the property of "covers or equals.")*

> *Closure: If $p \in \mathrm{cl}_M(A)$, then $p \in \mathrm{cl}_{M'}(A)$ [i.e., $\mathrm{cl}_{M'}(A) \supseteq \mathrm{cl}_M(A)$]. (This is the characterization that most closely mimics the definition of a linear transformation.)*

> *Rank function: $r_M(A) - r_M(A') \geq r_{M'}(A) - r_{M'}(A')$ for all $A' \subseteq A \subseteq E$. [Equivalently, we have the recursive definition: $r_M(E) \geq r_{M'}(E)$, and any minor $(M'/I) - A$ of M' is a quotient of the corresponding minor $(M/I) - A$ of M.]*

> *Closed sets: Each closed set of M' is closed in M. [For an elementary quotient, the closed sets of M that do not remain closed in some M' form the set $\mathcal{K}_2$ of those closed sets covered by a (unique) flat in a modular cut (see Proposition 7.3.2).]*

> *Hyperplanes: Each hyperplane of M' is an intersection of (a nonempty subset of) hyperplanes of M. [For an elementary quotient, they are the hyperplanes of M contained in some linear subclass along with any coline covered by no member of the subclass (see Exercise 7.8).]*

> *Bonds: Each bond of M' is a union of bonds of M.*

> *Bases: For all bases B of M, $p \notin B$, there is a basis B' of M' with $B' \subseteq B$ and such that $\{q: B \cup p - q$ is a basis of $M\} \supseteq \{q': B' \cup p - q'$ is a basis of $M'\}$.*

> *Independent sets: $I \cup A \cup q$ is independent in M under the following three conditions: (1) $I \cup q$ is independent in M'; (2) $I \cup p$ is dependent in M' for all $p \in A$; (3) $I \cup A$ is independent in M.*

> *Circuits: Each circuit of M is a union of circuits of M'.*

Dual: M^* is a quotient of M'^*.

Weak map: M'/A is a weak-map image of M/A for all $A \subseteq E$ (cf. the foregoing recursive definition of rank function).

2. $M'(E)$ is a weak-map image of M [# where $r(M') = r(M)$] under any of the following conditions:

Closure: $\mathrm{cl}_{M'}(I) \supseteq \mathrm{cl}_M(I)$ for all independent subsets I in M'.

Rank function: $r_{M'}(A) \le r_M(A)$ for all $A \subseteq E$.

Closed sets: Each closed set of M is contained in some closed set of M' of lesser or equal rank.

Hyperplanes: # Each hyperplane of M is contained in some hyperplane of M'.

Bonds: # Each bond of M contains a bond of M'.

Spanning sets: # Each spanning set of M' is a spanning set of M.

Bases: # Each basis of M' is a basis of M.

Independent sets: Each independent set of M' is an independent set of M. [A subfamily $\mathscr{C} \subseteq \mathscr{I}$ of independent sets of M that become dependent in some matroid M' forms an order filter of $\mathscr{I}$ that contains $(I \cup \{p_1, p_2\}) - \{q\}$ whenever $I \in \mathscr{I} - \mathscr{C}$, $I \cup \{p_i\} \in \mathscr{C}$, and $q \in I$.]

Circuits: Each circuit of M contains a circuit of M'.

Dual: # M'^* is a weak-map image of M^*.

Truncation: M' is a (#) rank-preserving weak-map image of $T^{r(M)-r(M')}(M)$. (See Proposition 7.4.9.)

Because any matroid is a rank-preserving weak-map image of a uniform matroid, we see that this cannot be a linear, affine, or transversal operation. That it is a binary and a graphical operation is a major result due to Lucas (1975).

Strong maps are much too general to have their images preserve any interesting class of matroids, and so we must consider special classes. One such special case is to take a class $\mathscr{C}$ closed under deletion and contraction and consider only quotients of the form $N - I \mapsto N/I$ for $N \in \mathscr{C}$. We call these quotients $\mathscr{C}$-*quotients*. A quotient is not, in general, a $\mathscr{C}$-quotient, even if it is between two matroids in $\mathscr{C}$, as can be seen by letting I be a point and letting N be any excluded minor (see Table 7.1) for the class $\mathscr{C}$. Thus, for example, a triple point is a quotient of a three-point line, and both are binary (even graphical) matroids, but in this case, N is a four-point line, and so we do not have a binary (or graphical) quotient.

Even though affine matroids are not closed under contraction, we can define *affine quotients* as those of the form $N - I \mapsto N/I$, where both N and N/I are affine. Geometrically, nullity-k affine quotients $M \mapsto M'$ can be thought of as choosing an independent k-subset that spans an affine flat I disjoint from M in the ambient affine space A and sectioning the k-dimensional flat F_p spanned by I and the point $p \in M$ with a generally placed flat F in

TABLE 7.1. Examples of Hereditary Classes and Their Excluded Minors

$\mathscr{C}$ (hereditary class)	$\mathscr{E}$ (excluded minors)
1. $\mathscr{M}$: All matroids	None
2. $\mathscr{B}$: Boolean algebras	C_1: a loop
3. $\mathscr{B}'$: Boolean algebras with loops ("totally separable matroids")	C_2: the two-point circuit
4. $\bar{\mathscr{B}}$: Matroids whose simplifications are Boolean algebras	C_3: the three-point circuit
5. $\mathscr{M}_{\leq k}$: Matroids on at most k points	$\mathscr{M}_{k+1}$: all (isomorphism classes of) matroids on $k+1$ points
6. $\mathscr{U}$: Uniform matroids	$C_1 \oplus C_1^*$: the matroid consisting of a loop and an isthmus
7. $\bar{\mathscr{U}}$: Matroids that have uniform simplifications	$C_3 \oplus C_1^*$: the matroid consisting of a three-point circuit and an isthmus
8. $\mathscr{MG}$: Matchstick geometries (geometries cosisting of k lines and m isthmuses freely placed in rank $2k + m$)	C_4: the four-point circuit
9. $\overline{\mathscr{MG}}_k$: Matchstick geometries each of whose lines has k or fewer points	$\{C_4, L_{k+1}\}$, where L_{k+1} is the $(k+1)$-point line
10. $\mathscr{L}$: Matroids consisting of a line with isthmuses	$\{C_4, C_3 \oplus C_3\}$, where $C_3 \oplus C_3$ consists of two three-point lines freely placed in rank 4
11. $\mathscr{N}$: Nested matroids (those any two of whose isthmus-free flats are comparable) (Oxley et al. 1982)	$\{C_k^2: k \geq 2\}$, where C_k^2 consists of two disjoint, closed k-point circuits placed in rank k
12. $\bar{\mathscr{F}}_2$: Binary matroids (vector matroids over F_2)	L_4: the four-point line
13. $\bar{\mathscr{F}}_3$: Ternary matroids (vector matroids over F_3)	$\{F_7, F_7^*, U_{2,5}, U_{3,5}\}$, where F_7, the Fano plane, is PG(2, 2) and $U_{n,k}$ is the uniform matroid of rank n and size k
14. $\overline{\mathscr{SP}}$: Series-parallel networks (see Chapter 6 or Section 7.6)	$\{L_4, M(K_4)\}$, where $M(K_4)$ is the matroid of the complete graph K_4
15. $\bar{\mathscr{R}}$: Regular (or unimodular) matroids	$\{F_7, F_7^*, L_4\}$
16. $\bar{\mathscr{G}}$: Graphical matroids	$\{F_7, F_7^*, L_4, M^*(K_5), M^*(K_{3,3})\}$, where $K_{3,3}$ is the complete three-by-three bipartite graph
17. $\bar{\mathscr{G}}^*$: Cographical matroids	$\{F_7, F_7^*, L_4, M(K_5), M(K_{3,3})\}$
18. $\overline{\mathscr{PG}}$: Planar graphical matroids	$\{F_7, F_7^*, L_4, M(K_5), M^*(K_5), M(K_{3,3}), M^*(K_{3,3})\}$
19. $\mathscr{M}^{<r}$: Matroids of rank less than r	B_r: the Boolean algebra on r points
20. $\mathscr{M}^{<r} \cup \mathscr{M}^{<s^*}$: Matroids of rank less than r or nullity less than s	$\mathscr{M}_{r+s}^r$: matroids of rank r on $r + s$ points
21. $\bar{\mathscr{U}}'$: Uniform geometries with isthmuses	$\{C_3 \oplus C_3, P(C_3, C_3), T(C_3 \oplus B_2)\}$, where P is the parallel connection of Section 7.6 and $T(C_3 \oplus B_2)$ is a three-point line with two other points in rank 3

TABLE 7.1. (*Continued*)

$\mathscr{C}$ (hereditary class)	$\mathscr{E}$ (excluded minors)
22. $\overline{\mathscr{P}\mathscr{G}}$: Direct sums and 2-sums (see Section 7.6) of planar graphical matroids and copies of $M(K_5)$ (Hall 1943)	$\{F_7, F_7^*, L_4, M^*(K_5), M^*(K_{3,3}), M(K_{3,3})\}$
23. $\overline{\mathscr{R}}$: Direct sums and 2-sums of regular matroids and copies of F_7^* (Seymour 1980)	$\{L_4, F_7\}$
24. $\overline{\mathscr{F}}'_3$: Ternary, base-orderable geometries	$\{U_{2,5}, U_{3,5}, M(K_4)\}$

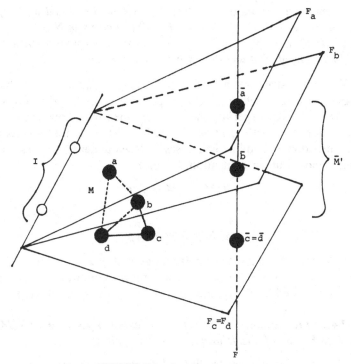

FIGURE 7.3. An affine quotient.

A of codimension k. The intersection $F \cap F_p$ is then a (zero-dimensional) point $\bar{p}$ of $\bar{M}'$ represented on F (Figure 7.3).

It is not hard to prove the following (where, for example, we use Proposition 7.4.6, part 1, and Proposition 7.4.8, part 1, to prove Proposition 7.4.8, part 4):

7.4.8. Proposition. 1. *M' is a graphical quotient of M if and only if some graph that represents M gives a representation for M' by identifying vertices. (See, e.g., Exercise 7.17).*

2. *Vector matroid quotients correspond to taking a quotient space: M is represented by a subset V' of vectors in a vector space V, and M' is represented by vectors $\{v + W: v \in V'\}$ in the quotient space V/W for some subspace W. In particular, every vectorial nullity-k quotient $M \mapsto M'$ can be realized by finding a matrix representation for M that becomes a matrix representation for M' on deletion of its first k rows.*

3. *If M' is a vector quotient of M, and M' is affine, then M is affine also. (However, a two-point line and a double point over F_2 show that such vector quotients need not be affine quotients. In fact, affine quotients do not exist over F_2!)*

4. *M' is a nullity-k affine quotient of M if and only if there exists an $n \times K$ matrix N that represents M whose last $n - k$ rows represent M', and, further, there exist vectors u, u', with $u_i = 0$ for all $i \leq k$ and $u'_i \neq 0$ for all $i \leq k$, such that $u \cdot v \neq 0$ and $u' \cdot v \neq 0$ for all columns v in N.*

5. *If M' is loopless and the field F is sufficiently large, then M' is a vectorial quotient of M if and only if it is an affine quotient.*

We now restrict our attention to elementary quotients in which the extension is principal, defining the *principal truncation* $T_F(M) := (M +_F p)/p$ [where $r(F) > 0$]. As a special case ($F = E$), when we follow free extension by contraction, we get the *truncation* of M, $T(M)$. Note that a class closed under contraction and adding isthmuses is closed under free extension if and only if it is closed under truncation, because

$$T(M) = (M + p)/p,$$

and

$$M + p = T(M \oplus p),$$

where $M \oplus p$ has M as a subgeometry along with p as an isthmus. We now characterize the principal truncation in several cryptomorphic ways. Characterizations preceded by the symbol # refer to ordinary truncation.

7.4.9. Proposition. *The principal truncation $T_F(M)$ [# or $T(M)$] can be described by any of the following cryptomorphisms:*

Geometric lattice: Delete all the elements y in $L(M)$ that are not above $x = F$, but are covered by some element $y' \geq x$.
Closure:

$$\text{cl}_{T_F(M)}(A) = \begin{cases} \overline{A \cup F} & \text{if } r(A \cup F) = r(A) + 1 \text{ (i.e., if } A \cup F \text{ covers } \overline{A}), \\ \overline{A} & \text{otherwise.} \end{cases}$$

Rank:

$$r_{T_F(M)}(A) = \begin{cases} r(A) - 1 & \text{if } r(A) = r(A \cup F), \\ r(A) & \text{otherwise.} \end{cases}$$

Closed sets: See the geometric-lattice description.

$\#\{K: K$ *is a closed set of* M *that is not a hyperplane*$\}$.

Hyperplanes: $\#$ *Colines of* M *(closed sets of corank 2).*

Spanning sets: $\{S: S \cup q$ *spans* M *for some* $q \in F\}$.

Bases: $\{B - p: B$ *is a basis of* M, *and* $p \in B \cap F\}$.

Independent sets: $\{I: I$ *is independent in* M, *and* $\bar{I} \not\supseteq F$ *(i.e.,* I *does not contain a basis for* F)$\}$.

Circuits: $\#$ $\{C: C$ *is a nonspanning circuit of* $M\} \cup \{B: B$ *is a basis of* $M\}$.

We can iterate principal truncation up to $r(F)$ times, where

$$T^i(M) := T[T^{i-1}(M)] \text{ and } T_F^i(M) := T_{T^{i-1}(F)}[T_F^{i-1}(M)].$$

If $i = r(F) - 1$, we get the *complete principal truncation*, $\bar{T}_F(M)$, wherein F becomes a (multiple) point. The iterated truncation $T^{K-n}(B_K)$ of a Boolean algebra is called the *uniform matroid* $U_{n,k}$ of rank n and size K.

The kth principal truncation $T_F^k(M)$ is the freest quotient M' of M such that $r_{M'}(F) = r_M(F) - k$. It was shown by Brown (1974) and by Dowling and Kelly (1974) that matroids obtained from Boolean algebras by a sequence of principal truncations form the class $\mathscr{T}^*$ of dual transversal matroids. Principal truncations have the intuitive affine description of the view of the matroid M when looking out from a generic point on the flat F. (Any picture, such as Figure 7.1a, of a rank-4 matroid is really a picture of its truncation!) One of their more technical properties (Brylawski 1975b) holds when F is a modular flat (see Exercise 7.11, part b).

In this case, $T(F)$ is a modular flat of $T_F(M)$, and we have the following identity for the characteristic polynomial: $\chi(M) \cdot \chi[T(F)] = \chi[T_F(M)] \cdot \chi(F)$. Iterating, we get $\chi(M) \cdot (\lambda - 1) = \chi[\bar{T}_F(M)] \cdot \chi(F)$. This latter identity shows an application to graph theory of the complete principal truncation of a graphical matroid even when $\bar{T}_F(M)$ is not graphic. In particular, if G is a graph with a k-clique F, then $\lambda(\lambda - 1) \ldots (\lambda - k + 1)$ divides the chromatic polynomial $c(G)$ of G, and the quotient $c(G)/\lambda(\lambda - 2)(\lambda - 3) \ldots (\lambda - k + 1)$ is the characteristic polynomial of the matroid $\bar{T}_F[M(G)]$.

Using the vector representations given earlier for principal extension (Proposition 7.3.5) and contraction (Proposition 7.4.4), we obtain the following vector transcendental representation for $T_F(M)$:

7.4.10. Proposition. *Let* $N = [I_n A]$ *represent the matroid* M *over the field* F, *and assume that the last* $k + 1$ *columns of* I_n, $(c_{n-k}, c_{n-k+1}, \ldots, c_n)$, *form a basis for the flat* x. *Then* $T_x(M)$ *is represented in the transcendental extension field* $F(\lambda_{n-k}, \ldots, \lambda_{n-1})$ *by the matrix* $N' = [I_{n-1} A']$, *where the first column of* A', c_n', *is the vector* $\lambda = (0, \ldots, 0, \lambda_{n-k}, \ldots, \lambda_{n-1})^t$, *and each other column of* A' *is given by* $c_j' = \bar{c}_j - a_{nj}\lambda$, *where* $\bar{c}_j$ *is the* jth *column of* A *with its last entry,* a_{nj}, *removed.*

As an immediate consequence of Exercise 7.12 we get that the set $\mathcal{Q}$ of all (labeled) elementary quotients of a geometry G forms a lattice, minus its zero element, under the weak order. When a zero element Z is adjoined, we call this lattice $L_{\mathcal{Q}}(G)$. Further, elementary quotients of the form $(G/q) \oplus q'$ (where q is a point of G and q' is a loop) are atoms in this lattice (because atoms are modular, and so their principal filter is a maximal modular cut). Further, the suprema of these atoms generate a subsemilattice isomorphic to $L(G)$, the geometric lattice of G. In particular, an element $x \in L(G)$ corresponds to the element $T_x(G)$ of $L_{\mathcal{Q}}(G)$. Several of the elements $T_{\mathcal{M}}(M)$ of $L_{\mathcal{Q}}(M)$ for the geometry M of Figure 7.1 are drawn in Figure 7.4 with their induced weak-map order shown and minimal sets of their defining modular cut $\mathcal{M}$ listed. Matroids that are principal truncations are identified as those having only one generator for $\mathcal{M}$. Note that the weak-map (lattice) ordering in $L_{\mathcal{Q}}(M)$ is given by $T_{\mathcal{M}}(M) \le T_{\mathcal{M}'}(M)$ if and only if every generator for $\mathcal{M}'$ contains some generator for $\mathcal{M}$. Looking at the diagrams for the quotients in $L_{\mathcal{Q}}$ we observe another fact: Elementary quotients are various perspective views of a matroid. The reader is encouraged to label the points in Figure 7.4 and to construct the appropriate single-element extensions of M (which give the perspective).

We note some more elementary properties of $L_{\mathcal{Q}}$. The unit element $1_{L_{\mathcal{Q}}(G)}$ is $T(G)$, and the coatoms are all of the form $T_H(G)$ for a hyperplane H of G. Further, $T_x(G) \wedge T_y(G) \ge T_{x \wedge y}(G)$ [where $T_{\varnothing}(G) = Z$], with equality if x and y form a modular pair [i.e., if $r(x) + r(y) = r(x \wedge y) + r(x \vee y)$], and, in any case, equals $T_{\mathcal{M}}(G)$, where $\mathcal{M}$ is the smallest modular cut containing x and y. (But note that even when x and y do not form a modular pair, $T_x \wedge T_y$ may equal $T_{x \wedge y}$, as in Exercise 7.6, part b.)

Using Exercise 7.12, part f, we see that over sufficiently large fields, matroids with compatible vectorial or affine quotients have a representable supremum in $L_{\mathcal{Q}}$. This fact has the following interpretation in the field of binocular perception: $M_l \vee M_r$ is the (two-dimensional) image perceived by the brain when M_l and M_r are the perspective views given by the two eyes.

Sequences of deletions and contractions are termed *minors* and can be realized in a number of ways stemming from the facts that $(M/p) - q$ equals $(M - q)/p$ [both have rank function $r'(A) = r_M(A \cup p) - r_M(p)$] for all $A \subseteq E - \{p, q\}$ and that, if p is an isthmus or a loop, then $M - p = M/p$. Thus, we can rearrange terms so that we do all contractions first, and so that we never delete an isthmus or contract a loop. This is the essence of Proposition 7.4.11, part 1, whose geometric version, part 2, is known as the scum theorem because it is used when checking whether or not a geometry G is not in a hereditary class $\mathcal{H}$ to guarantee that any "bad" geometric minor of $L(G)$ (i.e., excluded minor for $\mathcal{H}$) can always be found at the "top of the lattice."

7.4.11. Proposition. 1. *Let $M(E)$ be a matroid, and assume that M' is a minor of M on the set E' (i.e., M' results from M by a sequence of deletions*

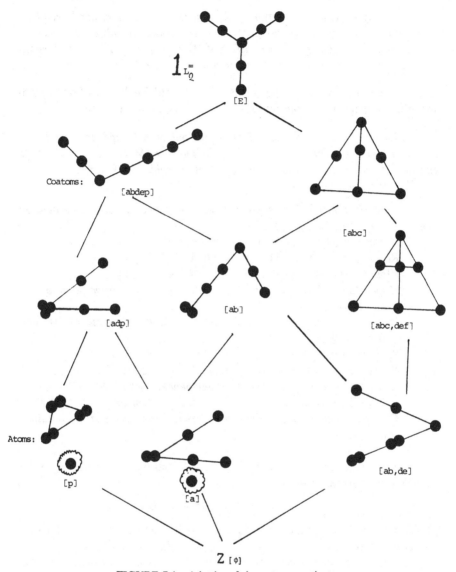

FIGURE 7.4. A lattice of elementary quotients.

and contractions of the points $E - E'$). Then there exists a subset $I \subseteq E - E'$ of cardinality $r(M) - r(M')$ that is independent in M such that

$$M' = (M/I)(E') = M/I - (E - E' - I) = M(E' \cup I)/I,$$

where E' spans M/I (or, equivalently, $E' \cup I$ spans M).

2. *Any sequence of deletions and geometric contractions can be realized by a minor of the form $(\overline{G/F})(A')$, where A' is a spanning subset of atoms of the geometry $\overline{G/F}$. The geometric lattice of $(\overline{G/F})(A')$ is isomorphic to a spanning supremum subsemilattice of the interval $[\overline{F}, 1]$ in $L(G)$.*

We can combine Propositions 7.3.1(#) and 7.4.3(#) to find cryptomorphic descriptions of the minor $(M/I)(E')$. We mention a few.

7.4.12. Proposition. *Let M be a matroid with independent set I and spanning set $I \cup E'$. Then the minor $M' = (M/I)(E') = (M/I) - \overline{E} = (M - \overline{E})/I$ can be defined in any of the following cryptomorphic ways, where $\overline{E} = E - I - E'$:*

> *Geometric lattice: The supremum subsemilattice of $[\overline{I}, 1]$ generated by the atoms $\{\overline{I \cup p}: p \in E' - \overline{I}\}$.*
> *Closure: $\mathrm{cl}_{M'}(A) = (\overline{A \cup I}) \cap E'$ for all $A \subseteq E'$.*
> *Rank: $r_{M'}(A) = r(A \cup I) - |I|$ for all $A \subseteq E'$.*
> *Closed sets: $\{K \cap E': K$ is closed in M, and $K \supseteq I\}$.*
> *Hyperplanes: Maximal sets among $\{H \cap E': H$ is a hyperplane of M, and $H \supseteq I\}$.*
> *Bonds: Minimal sets of the form $\{B - \overline{E}: B$ is a bond of M, and $B \cap I = \varnothing\}$.*
> *Spanning sets: $\{S - I: S$ spans M, and $(I \subseteq) S \subseteq I \cup E'\}$.*
> *Bases: $\{B - I: B$ is a basis of M, and $I \subseteq B \subseteq I \cup E'\}$.*
> *Independent sets: $\{J - I: J$ is an independent set of M, and $I \subseteq J \subseteq I \cup E'\}$. (Equivalently, $\{J \cap E': J$ is independent in M, and $I \subseteq J\}$.)*
> *Circuits: Minimal sets of the form $\{C - I: C$ is a circuit of M contained in $I \cup E'\}$.*

The operation of taking a minor is self-dual (i.e., commutes with the duality operator) because

$$[(M/A) - B]^* = (M^* - A)/B.$$

[This latter is an immediate consequence of the commutativity of deletion and contraction and the fact that $(M^* - p)^* = M/p$.] In fact, it is self-dual even in the independent-spanning sense described earlier. If I is an independent subset of M, and $I \cup E'$ spans M, then $E - I$ spans M^*, and $E - (I \cup E') = \overline{E}$ is independent in M^*, so that

$$[(M/I)(E')]^* = (M^*/\overline{E})(E'),$$

where I is independent in M, $\overline{E}$ is independent in M^*, and E' spans both M/I and $M^*/\overline{E}$.

We call a matroid class $\mathscr{C}$ closed under minors a *hereditary class*. Among the classes mentioned earlier, the hereditary classes include graphical matroids, planar graphical matroids, vector matroids (over a fixed field F), gammoids, oriented matroids, Boolean algebras, matroids with at most k points, and *uniform matroids* (those that are truncations of Boolean algebras). In addition, unions and intersections (finite or infinite) of hereditary families are themselves hereditary. Thus, for example, matroids representable over fields of fixed characteristic, being the union of families of matroids representable over fields with a common prime field, form a hereditary class.

Those not closed under minors include transversal matroids (and its dual class of strict gammoids) and affine matroids (except over infinite fields or over sets of fields $\mathscr{F}$ that include, for every finite $F \in \mathscr{F}$, a proper extension field of F, and when we formally allow loops to be adjoined).

Vector matroids give interesting hereditary classes. For example, the class of matroids coordinatizable over every field forms the class $\mathscr{R}$ of *regular* or *unimodular matroids* and is important in circuit theory and linear programming. As an easy consequence of Proposition 7.4.4 and the column-deletion linear representation of submatroids, we have that if M is vectorial with a representing matrix N in row-echelon form, then any submatrix of N represents a minor of M (with the possible addition of some loops coming from zero columns of the identity matrix). Such minors are called *visible minors*, and, in general, any minor M' is visible in some representing matrix N, where we need only be careful that if $M' = (M/I)(E')$, then the columns of N corresponding to I form a submatrix of the identity matrix.

Imitating the ideas of obstruction theory in topology and universal algebra, we have the notion of an *excluded minor* for a hereditary class $\mathscr{C}$. This is a matroid M that is not in $\mathscr{C}$, but each of whose proper minors (equivalently, each of whose single-element contractions and deletions) is in $\mathscr{C}$. A *geometric hereditary class* $\mathscr{C}$ is one such that a matroid M is in $\mathscr{C}$ if and only if its simplification is in $\mathscr{C}$ (a class closed under adjoining loops and parallel extension, in the language of Section 7.6). Most of the interesting hereditary classes are geometric (the only hereditary classes mentioned earlier that are not geometric are Boolean algebras, matroids with $\leq k$ points, and uniform matroids). As a theory cryptomorphic to that of geometric hereditary classes, we can intersect the class $\overline{\mathscr{C}}$ with the class $\mathscr{G}$ of all combinatorial geometries and consider only *geometric minors* $\overline{(G/A) - B}$. Here, the cryptomorphism associates any geometry G with the class of all matroids with the geometric lattice $L(G)$. The set $\mathscr{E}$ of excluded minors remains the same for $\overline{\mathscr{C}}$ in either theory.

Some of the (fairly easy to prove) properties of hereditary classes and their excluded minors appear next. Other properties will appear in the exercises and in Section 7.6.

7.4.13. Proposition. *Let $\mathscr{C}$ be a hereditary class, with $\mathscr{E}$ its set of excluded minors.*

1. *The class $\mathscr{C}^* = \{M: M^* \in \mathscr{C}\}$ is a hereditary class, with excluded minors $\mathscr{E}^* = \{M: M^* \in \mathscr{E}\}$.*

2. *If a geometry $G' = \overline{(G/A)} - B$ is a geometric minor, it can be realized by a sequence of deletions followed by (matroid) contractions such that G' is the resulting minor. If G is a geometry, all intermediate minors are geometries.*

3. *$\mathscr{C}$ is geometric if and only if all its excluded minors are geometries.*

4. *Any matroid not in $\mathscr{C}$ has, as a minor, a matroid isomorphic to an excluded minor in $\mathscr{E}$. If $\mathscr{C}$ is geometric, any geometry not in $\mathscr{C}$ has, in its geometric lattic, an upper interval that contains a spanning supremum subsemilattice isomorphic to the lattice of an excluded minor.*

5. *A class of matroids $\mathscr{M}'$ is the set of excluded minors for some hereditary subclass if and only if no member of $\mathscr{M}'$ is (isomorphic to) a minor of another member of $\mathscr{M}'$. In fact, such noncomparable families are cryptomorphic to hereditary families by the obvious relationship.*

6. *The class $\mathscr{C}$ is closed under the adjoining of loops if and only if all its excluded minors are loopless.*

Proof. 1. This is the hereditary-class restatement of the fact that M' is a minor of M if and only if M'^* is a minor of M^*.

2. Assume we have a minor of the form $G' = \overline{(G/I)}(E')$, where I is independent. Consider the atoms of the matroid $(G/I)(E')$. Each atom of G/I corresponds to a closed set of M of the form $\overline{p \cup I}$ for $p \notin \bar{I}$. Atoms of the matroid $(G/I)(E')$ are then those of the form $\overline{p \cup I}$ for $p \in E' - \bar{I}\}$. Further, the sets $\{\overline{p \cup I}: p \in E' - \bar{I}\}$ partition $E' - \bar{I}$, and so we can choose a representative from each block. Call this set of representatives A. Then $G' = G(A \cup I)/I$, and the property follows.

3. If $M \in \mathscr{E}$ is not a geometry, by definition M is not in $\mathscr{C}$, but its simplification $\bar{M}$ is a proper minor and so is in $\mathscr{C}$. The converse is also immediate. (See the proof of part 6.)

4. and 5. Obvious.

6. If $M \in \mathscr{E}$ has a loop p, then $M - p \in \mathscr{C}$, but M is not, and so $\mathscr{C}$ is not closed under adding loops. Conversely, if $\mathscr{C}$ is not closed under adding loops, let M be a smallest matroid in $\mathscr{C}$ such that $M \oplus p$ (p a loop) is not in $\mathscr{C}$. Then all proper minors M' of M are in $\mathscr{C}$, and, by assumption, so are all minors of the form $M' \oplus p$ (M' with a loop adjoined). Thus, $M \oplus p \in \mathscr{E}$. $\square$

Several interesting, well-known, or easy excluded-minor theorems are listed in Table 7.1. Geometric classes have bars.

One way to think about hereditary classes is in terms of the *minor ordering* on matroid isomorphism classes defined by $[M'] \leq [M]$ if some member of $[M']$ is a minor of some member of $[M]$. This gives a ranked poset (the rank of $[M]$ is the cardinality of any of its members), and a hereditary class is simply an order ideal in the poset. In analogy with matroid terminology, a hereditary class resembles the independent sets of a matroid, and excluded minors are analogous to circuits. Maximal members of $\mathscr{C} \cap \mathscr{M}^r$ termed *generators of rank r* correspond to bases, and an important study by Kahn and Kung (1982) identified those geometric classes with a single generator for each rank which are closed under direct sums (see Section 7.6).

We can also perform operations on hereditary classes. Examples include duality (Proposition 7.4.13, part 1), union ($\mathscr{C}_{20}$ in Table 7.1), and intersection. It is easy to see that the set of excluded minors of the intersection $\cap \mathscr{C}_\lambda$ of the hereditary class $\{\mathscr{C}_\lambda : \lambda \in \Lambda\}$ has, as excluded minors, minimal (in the minor ordering) members of the set $\bigcup_{\lambda \in \Lambda} \mathscr{E}(\mathscr{C}_\lambda)$. In our foregoing examples, $\overline{\mathscr{F}}_2 \cap \overline{\mathscr{F}}_3 = \overline{\mathscr{R}}$ and $\overline{\mathscr{G}} \cap \overline{\mathscr{G}^*} = \overline{\mathscr{P}\mathscr{G}}$. It is more difficult to describe excluded minors for hereditary class union. $\mathscr{E}(\mathscr{C} \cup \mathscr{C}')$ consists of minor-minimal matroids that contain both M and M' as minors, where $M \in \mathscr{E}(\mathscr{C})$ and $M' \in \mathscr{E}(\mathscr{C}')$. We leave it as a research problem for the reader to explore $\mathscr{E}(\mathscr{C} \cup \mathscr{C}')$ in general. For example, if $\mathscr{E}(\mathscr{C}) = \{M\}$ and $\mathscr{E}(\mathscr{C}') = \{M'\}$, is $\mathscr{E}(\mathscr{C} \cup \mathscr{C}')$ finite? (It is not bounded, as shown by the hereditary class $\mathscr{C}_{20} = \mathscr{M}^{<r} \cup \mathscr{M}^{<s*}$ of Table 7.1.) This problem is a generalization of one for graphs called the *intertwining problem*, and it was posed independently by Lovász and Ungar (Ungar 1978). Following their example, we call a minor-minimal matroid $I(M, M')$ that contains M and M' as minors an *intertwining* of M and M'.

A related research area is that of *well-quasi-ordering*. An as yet unsolved problem in graph theory is equivalent to stating that when restricted to graphic matroids, the minor ordering has no infinite antichain (equivalently, that $\mathscr{E}(\mathscr{C})$ can contain only a finite number of graphical matroids). This, of course, would imply the intertwining problem for graphs. A slightly weaker result has been proved in Robertson and Seymour (1985), namely, that if $\mathscr{E}(\mathscr{C})$ consists of graphical matroids, any one of which is planar, then $\mathscr{E}(\mathscr{C})$ consists of a finite number of matroids. There are many infinite antichains of matroids. A particularly easy one is the set $\mathscr{A}^* = \{A_3^*, A_4^*, \ldots, A_i^*, \ldots\}$, where A_i is a collection of $2i$ points in the plane $\{p_1, q_1, p_2, q_2, \ldots, p_i, q_i\}$ with the i three-point lines $p_j q_j p_{j+1}$ ($i + 1 := 1$). Each member of the antichain $\mathscr{A}^*$ is transversal and hence is representable over any infinite field. Further, by orthogonal duality, no contraction of any A_i has nullity greater than 3, and so there are no six-point lines as minors. As a research area, we can ask if there exist infinite antichains $\mathscr{A}'$ of matroids such that for some k and all $M \in \mathscr{A}'$, neither M nor M^* has a k-point line (this would imply that matroids representable over a fixed finite field are well-quasi-ordered).

In order to study intertwinings and well-quasi-ordering, we need first to explore the properties of a matroid M that has a fixed matroid M' as a

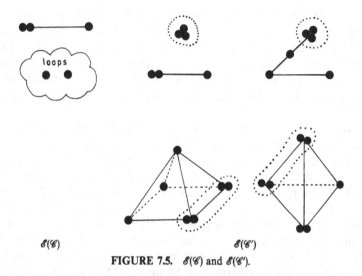

FIGURE 7.5. $\mathscr{E}(\mathscr{C})$ and $\mathscr{E}(\mathscr{C}')$.

minor. M will be an extension (which we have already discussed) of a *coextension* of M', where N is a coextension of M' if $N/I = M'$ for an independent subset I of N. To give another application of coextensions, we show how they are used to give the excluded minors for another operation on hereditary classes: adjoining loops (cf. the foregoing $\mathscr{C}_2 \cdot \mathscr{B}$ and $\mathscr{C}_3 \cdot \mathscr{B}'$, as well as Proposition 7.4.13, part 6). The straightforward proof is left to the reader.

7.4.14. Proposition. *Let $\mathscr{C}$ be a hereditary class, and form the hereditary class $\mathscr{C}' = \{M \oplus L : M \in \mathscr{C}, L \text{ a set of loops}\}$. The excluded minors for $\mathscr{C}'$ are (minimal members) of the set $\{N : N'/F = M\}$, where the loopless matroid M with k loops adjoined is in $\mathscr{E}(\mathscr{C})$, each point of the flat F is multiple, and the nullity of F equals k.*

For example, see Figure 7.5, where the dotted line encloses the flat F. Note that, although in this example each F, when simplified, is independent, F (and N) simplify to a three-point line in a member of $\mathscr{E}(\mathscr{M}^{<4^*})$.

The operation of single-element coextension is dual to that of extension:

$$M' = M \times_{\hat{\mathscr{M}}} p = (M^* +_{\mathscr{M}} p)^*,$$

where, by $\hat{\mathscr{M}}$, we desire a description and characterization, in terms of M, of the modular cut $\mathscr{M}$ of M^* that gives the extension. As we shall see, $\hat{\mathscr{M}}$ specifies which subsets A of E have the same rank in M' and M. Before going through arguments involving the orthogonal dual M^* (which tend to be more formal than insightful), we make some a priori observations about $\hat{\mathscr{M}}$:

0.   Knowing which subsets $A \subseteq E$ have their ranks unchanged in M

and M' characterizes a coextension, because, by definition, $M'/p = M$, so that every subset of the form $A \cup p$ has M'-rank equal to $r_M(A) + 1$ (because we never coextend by a loop). Further, for every subset $A \subseteq E$,

$$r_{M'}(A) \le r_{M'}(A \cup p) \, [= r_M(A) + 1] \le r_{M'}(A) + 1.$$

Hence, to fully describe the coextension, we need only distinguish those subsets (of E) that have their ranks increased by 1 and those that have their ranks unchanged.

    1.   Because $M = [(M' - p) +_{\mathscr{M}'} p]/p$ for some modular cut $\mathscr{M}'$ of $M' - p$, those subsets A not in $\hat{\mathscr{M}}$ are those that (when viewed in $M' - p$) are in $\mathscr{M}'$. In particular, $\hat{\mathscr{M}}$ is an order ideal of subsets, and we call it the *modular ideal* of the coextension.

    2.   If A and B are a modular pair of subsets of M [i.e., $r_M(A) + r_M(B) = r_M(A \cup B) + r_M(A \cap B)$] and both are in $\mathscr{M}$, then $r_{M'}(A) = r_M(A)$, $r_{M'}(B) = r_M(B)$, and $r_{M'}(A \cap B) = r_M(A \cap B)$. Hence, $r_M(A \cup B) = r_{M'}(A) + r_{M'}(B) - r_{M'}(A \cap B) \ge r_{M'}(A \cup B) \ge r_M(A \cup B)$. Thus, $A \cup B \in \hat{\mathscr{M}}$.

    3.   Independent subsets of M can never have their ranks increased in M'. More generally, if $A \in \hat{\mathscr{M}}$ and q is an isthmus of $M(A \cup q)$, then $A \cup q \in \hat{\mathscr{M}}$.

7.4.15. Proposition. *The subsets $\hat{\mathscr{M}}$, which have their ranks unchanged in M and M' in the coextension $M' = M \times_{\hat{\mathscr{M}}} p = (M^* +_{\mathscr{M}} p)^*$, are the complements of subsets whose closures are in the modular cut $\mathscr{M}$ of M^*. Such families $\hat{\mathscr{M}}$ are characterized in M by the following:*

    (1) *$\hat{\mathscr{M}}$ is a nonempty order ideal: If $B \in \hat{\mathscr{M}}$ and $B \supseteq A$, then $A \in \hat{\mathscr{M}}$.*

    (2) *If $A \in \hat{\mathscr{M}}$ and $B \in \hat{\mathscr{M}}$ are a modular pair of subsets of M, then $A \cup B \in \hat{\mathscr{M}}$.*

    (3) *If I is a subset that is independent of A in M [i.e., if $r(A \cup I) = r(A) + |I|$], and $A \in \hat{\mathscr{M}}$, then $A \cup I \in \hat{\mathscr{M}}$.*

    (3') *If p is not a loop of M, then $p \in \hat{\mathscr{M}}$.*

Proof. Conditions 3 and 3' are equivalent by condition 2. A subset A whose closure is in $\mathscr{M}$ has the property that in M^*, $r(A \cup p) = r(A)$, [i.e., $p \in \mathrm{cl}_{M^* +_{\mathscr{M}} p}(A)$]. By Proposition 7.4.1, this is equivalent to the property that in $(M^* +_{\mathscr{M}} p)^*$, $p \notin \mathrm{cl}(E - A)$; that is, $r_{M'}[(E - A) \cup p] = r_{M'}(E - A) + 1$. Equivalently, $r_M(E - A) = r_{M'}(E - A)$.

    Conditions 1, 2, and 3 for the modular ideal $\hat{\mathscr{M}}$ of M are cryptomorphic to the conditions for $\mathscr{M}$ being a modular cut of M^*, where the cryptomorphism takes a subset A of E in M^* to $E - A$ in M:

    1. $\{E - A: A$ is in a nonempty order filter$\}$ is a nonempty order ideal.

    2.   If A and B are a modular pair of subsets in M^*, then so are their closures, and hence $\mathrm{cl}_{M^*}(A \cap B) \in \mathscr{M}$. Thus, $E - (A \cap B) \in \hat{\mathscr{M}}$ if $E - A$ and $E - B$ both are. On the other hand, our cryptomorphism

preserves the property of being a modular pair of subsets by the following chain of equivalences:

$$r_{M^*}(A) + r_{M^*}(B) = r_{M^*}(A \cup B) + r_{M^*}(A \cap B),$$
$$\text{cor}_{M^*}(A) + \text{cor}_{M^*}(B) = \text{cor}_{M^*}(A \cup B) + \text{cor}_{M^*}(A \cap B),$$
$$\text{null}_M(E - A) + \text{null}_M(E - B) = \text{null}_M[E - (A \cup B)]$$
$$+ \text{null}_M(E - (A \cap B)),$$
$$r_M(E - A) + r_M(E - B) = r_M[(E - A) \cap (E - B)]$$
$$+ r_M[(E - A) \cup (E - B)].$$

3. $\mathcal{M}$ (as a modular cut of subsets in M^*) is closed under M^*-closure. This property is equivalent to the following for a subset A and a point $q \notin A$:

If $A \cup q \in \mathcal{M}$ and $q \in \text{cl}_{M^*}(A)$, then $A \in \mathcal{M}$.
If $(E - A) - q \in \hat{\mathcal{M}}$ and $q \notin \text{cl}_M[(E - A) - q]$, then $E - A \in \hat{\mathcal{M}}$.
$\hat{\mathcal{M}}$ is closed under adding points that increase M-rank. $\qquad\square$

Not unnaturally, we call the dual of a principal extension a *principal coextension*. Similarly, we have the *free coextension $M \times p$*. An application of the free coextension occurs in the theory of the characteristic polynomial (see the chapter on Möbius functions in the next volume), where it can be shown that the number of independent subsets of M of cardinality i is the (absolute) coefficient of λ^i in $\chi(M \times p, \lambda)/(\lambda - 1)$ (Brylawski 1977).

For a specific principal coextension we must decide what should play the role of the generator F in $M +_F p$. In general, for a subset $A \subseteq E$, $M +_A p = M +_{\bar{A}} p$ is the freest extension of M by p such that $r(A) = r(A \cup p)$. To designate a particular principal extension, we can make a maximal choice for A (necessarily a closed set and unique) or a minimal choice for A (necessarily an independent set and not unique in that any other independent set with the same closure gives the same principal extension).

Dually, we define $M \times_A p$, the *principal coextension of $M(E)$ by A* to be the freest coextension M' of M such that $E - A$ has the same rank in M as in M'. By elementary arguments in the theory of weak maps and orthogonal duality [primarily that if $r(M_1) = r(M_2)$ and if $M_1 \geq M_2$ in the weak order, then $M_1^* \geq M_2^*$ (see Proposition 7.4.8, part 2)], we have that $(M^* +_A p)^* = M \times_A p$, i.e., the freest way to preserve the rank of $E - A$ in a coextension of M dualizes to the freest way to have $r(A) = r(A \cup p)$ in an extension of M^*. The dual notion of a flat is the complement of a union of circuits, and the dual notion of an independent set is the complement of a spanning set. These arguments give us the following proposition:

7.4.16. Proposition. *For any principal coextension $M \times_A p$, there are appropriate choices A' and A'' with $A'' \subseteq A \subseteq A'$ such that $M \times_A p = M \times_{A'} p = M \times_{A''} p$, where $E - A'$ is a union of circuits and $E - A''$ is a spanning set.*

Further, if $M \times_A p = M \times_{A'} p$, where $E - A'$ is a circuit union, then $E - A'$ is uniquely specified as the cyclic part of $E - A$, the set $E - A$ with all of its (relative) isthmuses deleted.

If $M \times_A p = M \times_{A''} p$, with $A'' \subseteq A$ and $E - A''$ a spanning set, then $E - A''$ is of the form $I \cup (E - A)$, where I extends a basis of $E - A$ to a basis for M. In particular, the free coextension of M by p is given by $M \times_A p$ if and only if A contains a basis complement of M (i.e., if $E - A$ is independent so that its cyclic part is empty). The modular ideal $\mathcal{M}_A$ of $M \times_A p$ equals $\{A': A' \subseteq I \cup (E - A)$, where I extends some basis of $E - A$ to some basis for $M\}$.

Axiomatizing principal (or free) coextensions in terms of various cryptomorphisms is an elementary exercise in which we can orthogonally dualize each characterization of principal extension in Proposition 7.3.3. For example, to illustrate the proposition that follows, the reader will note that the matrix M of Figure 7.1 is the free coextension of M/p (a triangle of double points) by p. The reader is invited to prove the following proposition, which explores principal coextension for our basic classes.

7.4.17. Proposition. 1. *If the matrix N represents the matroid M over F, then $M \times_A p$ is representable in the extension field $F(\{\lambda_i: p_i \in A\})$ by the matrix*

$$N' = \begin{bmatrix} 1 & \mathbf{v} \\ 0 & N \end{bmatrix},$$

whose first column is indexed by p, where $\mathbf{0} = (0, 0, \ldots, 0)^t$, and where, for the row vector $\mathbf{v}$,

$$\mathbf{v}_i = \begin{cases} 0 & \text{if } p_i \notin A, \\ \lambda_i & \text{if } p_i \in A. \end{cases}$$

Without loss of generality we can assume that A is disjoint from a basis B of M and that N is of the form $[I_n N'']$, where the columns of I_n are indexed by B. Then N', as constructed in the previous paragraph, gives a representation of $M \times_A p$ whose first $n + 1$ columns form an identity submatrix.

2. *If M is a graphical matroid represented by the graph G, then $M \times_e p$, with $e \in M$, is a graphical matroid whose representing graph $G \times_e p$ is the graph G with the edge e series extended by an edge p (see Chapter 6 or Section 7.6). A graphical matroid M has a graphical free coextension if and only if every (maximal two-connected) block is a circuit C_i or an edge e_i, in which case $G \times p$ is a graph consisting of paths $\bar{C}_i$ all connected in parallel to the edge p along with isthmuses $\bar{e}_i$.*

3. *If M is affine and F (equivalently, the ambient space) is sufficiently large, we can represent $M \times_A p$ affinely by first constructing the cone over M with apex p (where M is affinely represented in a hyperplane H not containing*

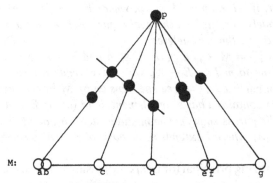

FIGURE 7.6. A coextension.

p). We then leave each point of E − A in the hyperplane H, but for each point $p_i ∈ A$, we replace it by a new point $\bar{p}_i$ freely placed on the line joining p_i with p.

In general, over large F, we can affinely represent the linearly representable matroid $M ×_{\mathscr{M}} p$ by sectioning lines from p to points in a subset $A ∈ \hat{\mathscr{M}}$ with affine flats whose affine dimension is $r_M(A) − 1$. Thus, when $\hat{\mathscr{M}}$ is the modular ideal generated by bcd and ef of the matroid M consisting of a line with doubletons ab and ef, we have (Figure 7.6) the coextension $M ×_{\hat{\mathscr{M}}} p$ (with solid points).

4. *If M is transversal, with presentation $R ⊆ E × T$, then $M ×_A p$ is transversal, with presentation $R' ⊆ (E ∪ p) × (T ∪ t)$, where $R' ∩ (E × T) = R$, $R'(p) = t$, and $R^{-1}(t) = A ∪ p$.*

Conversely, M can be constructed by a sequence of principal coextensions starting with a set of loops if and only if M has a principal transversal presentation [see Proposition 7.4.2, part 3(b)].

To complete our constructions of the previous two sections full circle, we note that the dual of an (elementary) quotient is also the inverse image of an (elementary) quotient: $M(E)$ is defined to be a *lift* of the matroid $M'(E)$ when M' is a quotient of M. Equivalently, M^* is a quotient of M'^*. Lifts are discussed in detail in Chapter 8, and the reader may, by now, have discovered their properties by dualizing properties for quotients or in the case of *elementary lifts* (where $M = \{M' ×_{\mathscr{M}} p\} − p = [(M'^* +_{\mathscr{M}} p)/p]^*\}$ by deleting the point p from a coextension. We need only mention that in our terminology, a *principal lift* $L_A(M)$ is one that comes from a principal coextension and is orthogonally dual to the principal truncation $T_A(M)$:

$$L_A(M) = T_A^*(M^*).$$

The dual of truncation is the *Higgs lift*: $L(M) = L_E[M(E)]$. Also, the concepts of $\mathscr{C}$-lift and $\mathscr{C}$-quotient coincide, because both demand that the matroids $N, N − I$, and N/I all be in the class $\mathscr{C}$. By a dualize-truncate-

TABLE 7.2. Relationships Among Constructions

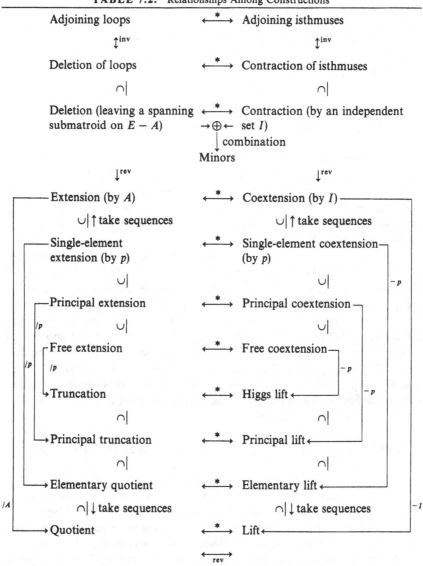

dualize argument, it is easy to see that Higgs lift is seldom graphical, but it does have a transversal construction (see Exercise 7.26, part a) as well as affine and linear constructions over sufficiently large fields, by Proposition 7.4.17.

Our constructions of the previous two sections are related in Table 7.2, where $\overset{*}{\leftrightarrow}$ denotes dual operations, and $O \xrightarrow{\text{rev}} O'$ denotes that the operations O and $\mathcal{O}'$ are reverse, in that if O is an operation, then $\mathcal{O}'$ is (in general) a family of operations such that if $O' \in \mathcal{O}'$, then $O \circ O'$ is the identity. Further, $O'' \circ O$ is the identity for some $O'' \in \mathcal{O}'$. The double arrow $O \xleftrightarrow{\text{inv}} O'$ indicates that the operations O and O' are inverses. An operation missing from the diagram is the reverse of truncation: erection. This will be treated in the following section.

7.5. TRUNCATIONS, LIFTS, AND MATROID BRACING

Understanding the truncation operator is essential for the "geometric" insight needed in many areas of matroid theory. We need only mention that "pictures" of matroids are almost invariably two-dimensional and thus are diagrams of the structure of $T^{n-3}(M)$, the matroid M truncated $r(M) - 3$ times to the plane. We begin with a construction combining truncation and its dual, the Higgs lift.

First, note that (restricted to matroids that have positive rank and nullity) truncation and Higgs lift commute: $TL(M) := T[L(M)] = L[T(M)]$. Both are matroids of rank $r(M)$ whose independent sets are given by the independent sets of M along with its nonspanning circuits. [Truncation increases nullity by 1 precisely for spanning sets, whereas the Higgs lift decreases (positive) nullity by 1.] Thus, $TL(M)$ is a lift of $T(M)$ and can be viewed affinely by placing a point p and a hyperplane H in free position with respect to M, projecting M from p onto H to obtain $T(M)$, and then freely lifting each point of $T(M)$ back on its line from p to H (so that H plays no role in the final construction). We call $TL(M)$ the *truncated lift* of M. Note that TL is a self-dual operator:

$$[TL(M^*)]^* = \{T[L(M^*)]\}^* = \{T[T(M)^*]\}^* = L[T(M)] = TL(M).$$

In imitation of planar projective geometry, we call two matroids *perspective* if they have a common quotient and *projective* if they can be linked by a sequence of such perspectivities. Unlike the vector prototype, however, matroid projectivity is practically unstructured. Whereas cross-ratios (Hirschfeld 1979) are preserved by linear projectivity in the plane, the only projective invariants in the matroid case are rank and size! An example of truncated lift as a perspectivity is shown in Figure 7.7.

7.5.1. Proposition. *Two matroids $M_1(E)$ and $M_2(E)$ are projective if and only if $r(M_1) = r(M_2)$. In fact, any matroid M of rank n on K points is projective with the uniform matroid $U_{n,K}$ by a sequence of $c(M)$ perspectivities where $c(M) := max_{A \subseteq E}\{min[cor(A), null(A)]\}$.*

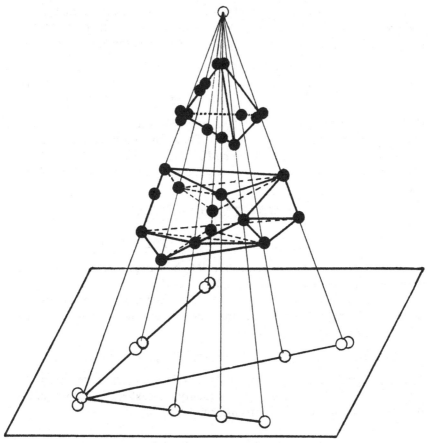

FIGURE 7.7. Truncated lift.

If we define the perspective difference $d(M_1, M_2)$ of M_1 and M_2 to be the minimum number of perspectivities needed to transform M_1 into M_2, then

$$c(M_1) + c(M_2) \geq d(M_1, M_2) \geq \bar{r}(M_1, M_2)$$

$$:= \tfrac{1}{2} max \left(\sum_{i=1}^{K} \left| r_{M_1}(p_1, \ldots, p_i) - r_{M_1}(p_1, \ldots, p_{i-1}) \right. \right.$$

$$\left. \left. - r_{M_2}(p_1, \ldots, p_i) + r_{M_2}(p_1, \ldots, p_{i-1}) \right| \right),$$

the maximum taken over all orderings on E, with equality in the first inequality if $c(M_i) = 0$ for $i = 1$ or 2.

Proof. $c(M) = 0$ if and only if M is a truncated Boolean algebra. For any subset $A \subseteq E$, if $null_M(A) > 0$, then $null_{L(M)}(A) = null_M(A) - 1$, so that if $cor(A)$

is also positive, then $\mathrm{null}_{TL(M)}(A) = \mathrm{null}_M(A) - 1$. Dually, if $\mathrm{null}_M(A) > 0$ and $\mathrm{cor}_M(A) > 0$, then $\mathrm{cor}_{TL(M)}(A) = \mathrm{cor}_M(A) - 1$, so that $c[TL(M)] = \max[0, c(M) - 1]$. Thus, $c(M)$ perspectivities give us the following chain:

$$
\begin{array}{ccccccc}
M & & TL(M) & & (TL)^2(M)\dots & & (TL)^c(M) = T^{K-n}(B_K) \\
& \searrow \ \nearrow & & \searrow \ \nearrow & & \searrow & \nearrow \\
& T(M) & & T^2L(M) & & T^cL^{c-1}(M) & & T^{K-n+1}(B_K)
\end{array} \tag{7.1}
$$

Conversely, for any quotient M' of M, $\mathrm{null}_M(A) \le \mathrm{null}_{M'}(A) \le \mathrm{null}_M(A) + 1$, and $\mathrm{cor}_M(A) - 1 \le \mathrm{cor}_{M'}(A) \le \mathrm{cor}_M(A)$. Thus, for two matroids M and M' with the same quotient, $|c(M) - c(M')| \le 1$, so that if $c(M_1) = 0$, then $d(M_1, M_2) = c(M_2)$ by the chain in equation (7.1). Combining two of these chains,

$$
\begin{array}{ccccc}
\dots \searrow & (TL)^{c_1}(M_1) = T^{K-n}(B_K) = (TL)^{c_2}(M_2) & \dots \\
\searrow & & \nearrow \\
T^{c_1}L^{c_1-1}(M) & & T^{c_2}L^{c_2-1}(M)
\end{array}
$$

we see that M_1 can be transformed into M_2 by a sequence of $c(M_1) + c(M_2)$ perspectivities. Thus, $d(M_1, M_2) \le c(M_1) + c(M_2)$.

The lower bound for $d(M_1, M_2)$ comes from the fact that for any quotient M' of M, $r_M(A) \ge r_{M'}(A)$, and that, for any sequence of points of E, there is exactly one choice of i such that

$$
r_M(p_1, \dots, p_i) - r_{M'}(p_1, \dots, p_i) - r_M(p_1, \dots, p_{i-1}) + r_{M'}(p_1, \dots, p_{i-1})
$$

is nonzero (and equals 1). Hence, if M and $\bar{M}$ are perspective, $\bar{r}(M_1, M) \ge \bar{r}(M_1, \bar{M}) - 1$, and, by the triangle inequality, at least $\bar{r}(M_1, M_2)$ perspectivities are needed to projectively transform M_1 to M_2. $\qquad \square$

One of the motivations for the study of extension was that because deletion "simplifies" a matroid until, eventually, a Boolean algebra is obtained, any matroid can be constructed via a sequence of single-element extensions. Truncation is another operation that will eventually yield a trivial, predictable matroid: B_K^*. We call a matroid M' an *erection* of M if $T(M') = M$. Then any matroid of rank n and cardinality K is constructible or *erectible* from K loops by a sequence of n erections. Whereas Blackburn, Crapo, and Higgs (1973) used extension to generate all eight-point matroids, perhaps erection would be a more efficient process [an isomorphism check among erections of rank n only involves checking which erections of the same rank-$(n-1)$ matroid (isomorphism class) are isomorphic, whereas extensions of two nonisomorphic matroids could well be isomorphic]. Knuth (1975) was probably the first to exploit the fact that erection was a good way to construct matroids, and he gave an algorithm using multiple erection to construct "random" matroids.

Essential to the study of erections is a description of the hyperplane family of a matroid given the closed-set structure of its other flats. The first description in Proposition 7.5.2, part (a), was given by Crapo (1970), and part (b), implicit in Knuth's study, was proved by Cravetz (1978).

7.5.2. Proposition. (a) *A family $\mathscr{H}$ is the set of hyperplanes for a matroid M such that $T(M) = M'$ if and only if the following hold:*

1. $\mathscr{H} \neq \{E\}$.

2. *$\mathscr{H}$ partitions the bases of M' in that if B is a basis of M', then there is a unique $H \in \mathscr{H}$, with $B \subseteq H$, and each member of $\mathscr{H}$ contains some basis.*

3. *Each member of $\mathscr{H}$ is $[r(M') - 1]$-closed in that if $|A| \leq r(M') - 1$ and $A \subseteq H \in \mathscr{H}$, then $\mathrm{cl}_{M'}(A) \subseteq H$.*

(b) *Equivalently, $\mathscr{H}$ is the family of hyperplanes of an erection M of M' if and only if conditions 1 and 2 hold, as well as the following:*

3'. *For all hyperplanes H' of M' and $p \in E - H'$, there is a set $H \in \mathscr{H}$ with $H' \cup p \subseteq H$.*

Further, the set of all erections of M' is lattice-ordered (except for the zero element, which we take to be the trivial erection, M' itself) by the weak order in which $E(M_1) \leq E(M_2)$ if and only if every hyperplane of $E(M_2)$ is contained in some hyperplane of $E(M_1)$. (**)

Proof. Both will follow from our own characterization in Proposition 7.5.3, part 3. For now, we shall show that conditions 3 and 3' are equivalent (in light of condition 2). Let the family $\mathscr{H}$ satisfy conditions 2 and 3, and let $H' \cup p$ be a one-point cover of a hyperplane H' of M'. Let B' be a basis for H'. Then $B' \cup p$ is a basis for M', and by condition 2 it is contained in some $H \in \mathscr{H}$. By condition 3, H also contains the closure of B' and hence contains $H' \cup p$.

Conversely, assume that $\mathscr{H}$ satisfies conditions 2 and 3', that $|A| < r(M')$, and that $A \subseteq H \in \mathscr{H}$. Let I be a basis for A. By condition 2, H contains a basis of M'. Therefore, by basis exchange, it contains a basis B' of M' that contains I. Let $q \in B' - I$. Then some member $\hat{H}$ of $\mathscr{H}$ contains the one-point hyperplane cover $\overline{B' - q} \cup q$. But this $\hat{H}$ also contains B', so that $\hat{H} = H$. Thus, $\bar{A} = \bar{I} \subseteq \overline{B' - q} \subseteq H$. $\square$

We note that the ordering (**) of Proposition 7.5.2 is, indeed, the weak order, because $r_{M_1}(A) = r_{M_2}(A)$ for all A with $r_{M'}(A) < r(M')$, and, for spanning subsets S of M', $r_{M_1}(S) = r(M')$ or $r(M') + 1$, depending on whether or not S' is in a hyperplane of M (i.e., a member of $\mathscr{H}$). This lattice order gives us $E(M') = \vee \{M: T(M) = M'\}$, the *free erection* of M', where, as before, $E(M') = M'$ if M' has no nontrivial erections. Las Vergnas (1975) gave an algorithm that computes the hyperplanes of the free erection using the Crapo characterization [Proposition 7.5.2, part (a)], and Knuth (1975) gave an algorithm using Proposition 7.5.2, part (b). Knuth's algorithm, more generally, computes a freest matroid $E_{\mathscr{C}}(M')$ that erects M' and, in addition, retains, as circuits, a given set $\mathscr{C}$ of spanning circuits of M'. In fact, the set $\mathscr{E}_{\mathscr{C}}(M') :=$ $\{M'\} \cup \{M: T(M) = M'$, and M contains C as a circuit for all $C \in \mathscr{C}\}$ is lattice-ordered by the weak-map order. Crapo probably knew this. It is implicit in Knuth's algorithm and was proved by Nguyen (1979) for the case in which there is a matroid M'' that is a weak preimage of M' and whose dependent sets intersect the spanning circuits of M' in $\mathscr{C}$. All of the foregoing facts follow rather easily from the observation that M is a lift of $T(M)$: $M = [T(M) \times_{\hat{M}} p] - p$, where the point p is in free position with respect to M (i.e., lies on no proper flat spanned by points of M). In this case, $\mathscr{M}$, the subsets of M that have equal rank in $T(M)$ and M (see Proposition 7.4.15), must include all nonspanning circuits. Conversely, any modular ideal $\hat{\mathscr{M}}$ of M' that contains all of M's nonspanning circuits $\mathscr{N}$, along with a fixed set $\mathscr{C}$ of spanning circuits, gives an erection M (perhaps the trivial one) of M', where M includes the members of $\mathscr{C}$ among its circuits. The hyperplanes of M, being maximal subsets of E that have equal rank in M and $T(M)$, are the maximal members of $\hat{\mathscr{M}}$.

All modular ideals of M' form a lattice L (in the weak order) isomorphic to the lattice $L_{\mathscr{Q}}(M^*)$ of elementary quotients of M^* (see the remarks accompanying Figure 7.5), because the intersection of two modular ideals is a modular ideal. Further, the set $\{\hat{\mathscr{M}}_{\mathscr{N} \cup \mathscr{C}}\}$ of all modular ideals that contain $\mathscr{N} \cup \mathscr{C}$ is also closed under intersection and gives a supremum subsemilattice of L when the zero element Z (corresponding to the trivial erection) is adjoined. Thus, all erections of M in which the members of $\mathscr{C}$ are dependent are lattice-ordered under the weak order. The 1 in this lattice (which may degenerate so that $1 = Z$), $E_{\mathscr{C}}(M)$, can be computed in any of the equivalent ways outlined below, the first being the Knuth construction, and the second adapted from the work of Crapo, Las Vergnas, and Nguyen cited earlier.

7.5.3. Proposition. *Let M' be a matroid and let $\mathscr{C}$ be a subset of its spanning circuits. Then there is a freest matroid $M = E_{\mathscr{C}}(M')$ ($\#$ or no matroid) such that $T(M) = M'$, and M contains, as circuits, the members of $\mathscr{C}$. The hyperplanes of $E_{\mathscr{C}}(M')$ can be computed in any of the following ways (as the family $\mathscr{F}_N$):*

1. $\mathscr{F}_0 := \{H \cup p: H \text{ is a hyperplane of } M', p \in E - H\} \cup \mathscr{C}$

    $\vdots$

    $\mathscr{F}_{i+1} := \text{maximal members of}$
    $\{F_1 \cup F_2: F_1, F_2 \in \mathscr{F}_i, r_{M'}(F_1 \cap F_2) = r(M')\}$

    $\vdots$

    terminate when $\mathscr{F}_N = \mathscr{F}_{N+1}$ ($\#$ or when $\mathscr{F}_N = \{E\}$).

2. $\mathscr{F}_0 := \{B\colon B \text{ is a basis of } M'\} \cup \mathscr{C}$

$\vdots$

$\mathscr{F}'_i := \{c(F)\colon F \in \mathscr{F}_i, \text{ where } c(F), \text{ the } [r(M') - 1]\text{-closure of } F, \text{ is } $ the smallest set containing F with the property that if $A \subseteq c(F)$ and $r(A) < r(M')$, then $\bar{A} \subseteq c(F)\}$

$\mathscr{F}_{i+1} := \text{maximal members of}$
$$\{F_1 \cup F_2\colon F_1, F_2 \in \mathscr{F}'_i, r_{M'}(F_1 \cap F_2) = r(M')\}$$

$\vdots$

terminate when $\mathscr{F}_N = \mathscr{F}'_N = \mathscr{F}_{N+1}$ (# or when $\mathscr{F}_N = \{E\}$).

3. *Take the maximal members of the smallest modular ideal $\hat{\mathscr{M}}$ that contains the members of $\mathscr{I} \cup \mathscr{C} \cup \mathscr{N}$, where $\mathscr{N}$ is the set of nonspanning circuits of M' and $\mathscr{I}$ is the set of independent sets of $\mathscr{M}'$. Following Proposition 7.4.15 $\hat{\mathscr{M}}$ can be constructed as follows:*

$\hat{\mathscr{M}}_0 := \mathscr{C} \cup \mathscr{N} \cup \mathscr{I}$

$\vdots$

$\hat{\mathscr{M}}'_i := \{A\colon A \subseteq F \in \hat{\mathscr{M}}_i\}$
$\hat{\mathscr{M}}_{i+1} := \{A \cup B\colon A, B \in \hat{\mathscr{M}}'_i, r(A) + r(B) = r(A \cup B) + r(A \cap B)\}$

$\vdots$

terminate when $\hat{\mathscr{M}}_N = \hat{\mathscr{M}}'_N = \hat{\mathscr{M}}_{N+1}$ (# or when $\hat{\mathscr{M}}_N = 2^E$).

Taking Proposition 7.5.3, part 3, as our fundamental algorithm (constructing freest modular ideals for lifts), we can modify it to solve other erection-type problems. For example, we can define *principal erection* to be the reverse of principal truncation: M' is a principal erection of M, with respect to a flat F of M, if $T_F(M') = M$. The following proposition gives necessary and sufficient conditions for the existence of a principal erection, as well as an algorithm that produces the freest one, $E_F(M)$, if any principal erection exists.

7.5.4. Proposition. 1. *Let $M(E)$ be a matroid with independent-set family $\mathscr{I}$, and let F be a flat of M. Then there exists some matroid M'' whose principal truncation $T_F(M'')$ equals M if and only if there is a modular ideal $\hat{\mathscr{M}}'$ of M that does not contain F, but contains $\mathscr{C}(F)$, those circuits whose closure is properly contained in F.*
2. *If some matroid has M as its principal truncation, then there is a freest one, $E_F(M)$, termed the free principal erection of M. It is determined by the modular ideal $\hat{\mathscr{M}}_{\mathscr{C}(F)}(M)$ generated in M by $\mathscr{C}(F)$. $\hat{\mathscr{M}}_{\mathscr{C}(F)}$ (# unless it does*

not exist) is given by $\hat{\mathcal{M}}_N$ *in the following algorithm:*

$$\hat{\mathcal{M}}_0 = \mathcal{C}(F) \cup \mathcal{I}$$

$$\vdots$$

$$\hat{\mathcal{M}}'_i = \{A': A' \subseteq A \in \hat{\mathcal{M}}_i\}$$

$$\hat{\mathcal{M}}_{i+1} = \{A \cup B: A, B \in \hat{\mathcal{M}}'_i, r(A) + r(B) = r(A \cup B) + r(A \cap B)\}$$

$$\vdots$$

terminate when $\hat{\mathcal{M}}_N = \hat{\mathcal{M}}'_N = \hat{\mathcal{M}}_{N+1}$ (# *or when* $F \in \hat{\mathcal{M}}_N$).

Then $E_F(M)$ *exists if and only if* $F \notin \hat{\mathcal{M}}_N$, *in which case* $E_F(M)$ *is defined by its rank function:*

$$r_{E_F(M)}(A) = \begin{cases} r_M(A) & \text{if } A \in \hat{\mathcal{M}}_N, \\ r_M(A) + 1 & \text{otherwise.} \end{cases}$$

3. *A sufficient condition for the existence of* $E_F(M)$ *is that* M *is a principal extension of* $M - p$ *by* $F - p$: $M = (M - p +_{F-p} p)$ *for some* p.

4. *A sufficient condition for the nonexistence of* $E_F(M)$ *is that* F *is the union of a modular pair of subsets* A *and* B, *each of lower rank.*

5. *A matroid* M *is characterized by the family of pairs* $\{(E_i, r_i): E_i$ *is an essential flat of rank* $r_i\}$. F *is essential if* $M(F)$ *is erectible.*

Proof. 1. Let $M = T_F(M'')$. Then M'' is defined by a modular ideal $\hat{\mathcal{M}}'$ of M, because M'' is a lift: $(M \times_{\hat{\mathcal{M}}'} p) - p$. The modular ideal $\hat{\mathcal{M}}'$ that consists of all $A \subseteq E$ such that $r_{M''}(A) = r_M(A)$ includes, by Proposition 7.4.9, all subsets A except those such that $r_{M''}(A) = r_{M''}(A \cup F)$ [i.e., those such that $\text{cl}_{M''}(A) = \text{cl}_{M''}(A \cup F)$]. Because there is a strong map from M'' to M, if $\text{cl}_{M''}(A) = \text{cl}_{M''}(A \cup F)$, then $\text{cl}_M(A) = \text{cl}_M(A \cup F)$. Thus, if C is a circuit of M such that $\bar{C} \subsetneq F$, then $\text{cl}_{M''}(C) \neq \text{cl}_{M''}(C \cup F)$, and C is in $\hat{\mathcal{M}}'$. Further, $r_M(F) = r_{M''}(F) - 1$, so that the conditions of Proposition 7.5.4, part 1, are necessary.

Conversely, assume that $\hat{\mathcal{M}}'$ is given as in part 1. By Proposition 7.4.15, $\hat{\mathcal{M}}'$ gives a coextension $M' = M \times_{\hat{\mathcal{M}}'} p$ of M. We shall be done when we show that $M' = (M' - p) +_{\mathcal{M}} p$ is the principal extension of $M' - p$ by $F - p$. Because $F \notin \hat{\mathcal{M}}'$, all closed sets that contain F are in $\mathcal{M}$. If there is some closed set G' in $\mathcal{M}$ that does not contain F, let G be a minimal such set, and let I be a basis of G. Then $(I \cup p)/p$ is dependent in M, and by the minimality of G, it must be a circuit. But then

$$r_M(C) = r_{M'}(C) - 1 < r_{M'}(C \cup F) - 1 = r_M(C \cup F),$$

so that $\text{cl}_M(C) \subsetneq \text{cl}_M(C \cup F)$, and, by construction, $C \in \hat{\mathcal{M}}'$. This contradicts the fact that $C \in \mathcal{M}$. Thus, $\mathcal{M}$ is the principal modular cut generated by F, and $M = T_F(M' - p)$.

2.   By construction, $\hat{\mathscr{M}}_{\mathscr{C}(F)}$ is the smallest modular ideal that contains $\mathscr{C}(F)$. If it does not contain F, the associated lift is clearly $E_F(M)$. If there is any modular ideal $\hat{\mathscr{M}}'$ that contains $\mathscr{C}(F)$ but does not contain F (giving, by Proposition 7.5.4, part 1, a principal erection of M), then $\hat{\mathscr{M}}' \cap \hat{\mathscr{M}}_{\mathscr{C}(F)}$ is a modular ideal that contains $\mathscr{C}(F)$ but not F. Thus, $\hat{\mathscr{M}}' \cap \hat{\mathscr{M}}_{\mathscr{C}(F)} = \hat{\mathscr{M}}_{\mathscr{C}(F)}$, and $F \notin \hat{\mathscr{M}}_{\mathscr{C}(F)}$.

3.   By part 2, we need only show that there is some principal erection of M. But we easily check that if $M = (M - p) +_F p$, then $M' = (M - p) \oplus p$ is such a principal erection, where $(M - p) \oplus p$ means we adjoin p as an isthmus to $M - p$ (see the remarks after Proposition 7.3.1).

4.   Let A and B be a modular pair whose union is F and each of whose ranks is less than F. Then A and B are both in $\hat{\mathscr{M}}_{\mathscr{C}(F)}$, and thus $A \cup B = F$ is in $\hat{\mathscr{M}}_{\mathscr{C}(F)}$. Hence, no principal erection is possible.

5.   We iterate the erection algorithm Proposition 7.5.3, where, at each level, each spanning circuit of E_i is put in $\mathscr{C}$. $\square$

We illustrate some of the arguments of this and the previous section by dualizing the matroids in Figure 7.4, thereby constructing some of the elements in the lattice of elementary lifts of M^* (see Table 7.2). The maximal cyclic elements of the associated modular ideal (Proposition 7.4.15) are listed for each matroid (they are the complements of the modular-cut generators listed in Figure 7.4). The unique [and hence freest (see Proposition 7.5.3)] erection of M^* (besides the trivial one, Z) comes from the principal modular ideal consisting of subsets of *abcdef*, because, for example, this set is the 1-closure of the basis *abc* of M^*. (Equivalently, as in Proposition 7.5.4 for $F = E$, it is the union of the modular triple of nonspanning circuits *ad*, *be*, and *cf*.) Thus, the lattice of erections of M^* has two elements. The matroid $E_{ad}(M^*)$, the free principal erection (Proposition 7.5.4) of the flat *ad* (in M^*), is also identified. The only flats that have principal erections are *ad* (and its automorphs *be* and *cf*) and E. The reader can check these facts and, as an exercise, label each solid point in Figures 7.4 and 7.8 to verify orthogonality. The hollow points, added to the drawing to identify nontrivial planes, give examples of bracing (Proposition 7.5.5).

In analogy with iterated truncation (see the remarks after Proposition 7.4.9) we can define iterated free erection. In particular, let $\bar{E}(M)$, the *total free erection* of M, be the matroid M freely erected as much as possible. Note that, for example, $r\{\bar{E}[T(M)]\} \geq r(M)$, because $T(M)$ has some erection (*viz.*, M), and so it has a free erection. On the other hand, as we shall see, there exist matroids M such that $r\{\bar{E}[T^2(M)]\} < r(M)$ [i.e., $M' = T^2(M)$ has the obvious double erection, but its free erection is inerectible.

When a matroid is pictured in the plane, as in Figure 7.la, one.can (if told it is of rank $k > 3$) freely erect it in one's mind subject to various conventions that give certain constraints (e.g., that certain points are meant to be coplanar, as in Figure 7.10).

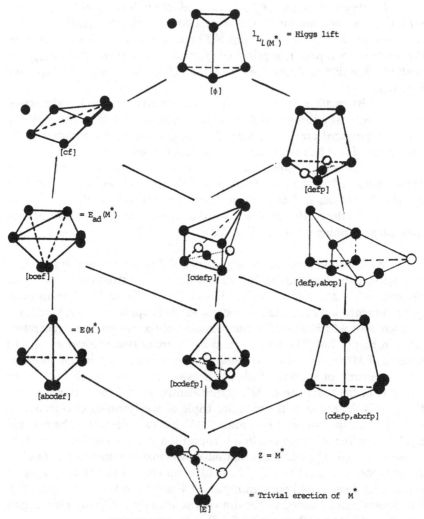

FIGURE 7.8. Some elementary lifts of M^*.

We present matroid bracing as an operation on matroids of rank $N \geq 3$ in order to formalize these conventions and, as a result, unambiguously represent any geometry in the plane where the only dependence is collinearity. In particular, for a matroid $M(E)$, we extend M to a matroid $M'(E \cup E')$ [with $M'(E) = M(E)$] such that $\bar{E}[T^{N-3}(M')] = M'$, where $T^{N-3}(M')$ is the truncation of M' to the plane. Thus, $M = \bar{E}[T^{N-3}(M')](E)$.

The idea is to rigidify a circuit C of M with a configuration of lines so that, in any truncation of M', it can be (freely) erected to rank $|C| - 1$, but no higher. To do this, let $I = (p_1, \ldots, p_n)$ be an (ordered) independent subset of M. Recursively, define the principal rank-2 extensions $\bar{p}_i$, where $\bar{p}_2$ is on the line $p_1 p_2$, $\bar{p}_3$ is on the line $p_3 \bar{p}_2$, and, in general, $\bar{p}_i$ is freely placed on the line $p_i \bar{p}_{i-1}$. Then it is easy to check that these extensions are compatible (see Exercise 7.13) and that $M'(E \cup \bar{p}_i) = M(E) +_F \bar{p}_i$, where $F = \mathrm{cl}_M(p_1, \ldots, p_i)$. If a subset $\hat{E}$ of E is isomorphic, as a submatroid, to the uniform matroid $U_{n,K}$ ($n \geq 3$, $K > n$), we choose a basis $(p_1, \ldots, p_n)$ for $\hat{E}$ and construct $\bar{p}_2^0, \bar{p}_3^0, \ldots, \bar{p}_n^0$ as before. For each other point, p_{n+i} ($i \leq K - n$), we construct another sequence $\bar{p}_2^i, \bar{p}_3^i, \ldots, \bar{p}_{n-1}^i$, where $\bar{p}_2^i$ is freely placed on the line $p_1 p_2$, each $\bar{p}_j^i$ ($2 < j < n - 1$) is freely placed on the line $p_j \bar{p}_{j-1}^i$ (distinct from all other $\bar{p}_j^{i'}$), and $\bar{p}_{n-1}^i$ is placed on the two lines $p_{n+i} \bar{p}_n^0$ and $\bar{p}_{n-2}^i p_{n-1}$ (where, when $n = 3$, $\bar{p}_1^i := p_1$). It is perhaps easier to extend, for fixed i, by the points $\bar{p}_j^i$ in reverse order starting with $\bar{p}_n^i = \bar{p}_n^0$ and then putting $\bar{p}_{n-1}^i$ (as freely as possible) on the line $\bar{p}_n^i p_{n+i}$ and on the flat $\mathrm{cl}_{M'}(p_1, \ldots, p_{n-1})$. Then each successive $\bar{p}_j^i$ ($j = n - 2, n - 3, \ldots, 2$) is placed on the line $\mathrm{cl}_{M'}(\bar{p}_{j+1}^i, p_{j+1})$ and on the j-flat $\mathrm{cl}_{M'}(p_1, \ldots, p_j)$.

This construction is made for each such dependent uniform subgeometry of M, and the reader can check that all such single-element extensions $\bar{p}_j^i$ are compatible [and that $M'(E \cup \bar{p}_j^i) = M +_F \bar{p}_j^i$ for some flat F of M].

We could do this for every circuit C ($|C| \geq 4$) of M, but, instead, we only brace C if (after previously added braces) C is still erectible in $T^{N-|C|+1}(M')$.

We call these simultaneous compatible extensions a *bracing* $B(M)$ of M. To illustrate, we (minimally) brace some familiar matroids in Figure 7.9. Here the points E of M are in black, the points added in the extension are in white, and V is the Vamos cube, where the collinearities in the bracing are shown with dashed lines for clarity.

The details in the following proposition outlining the properties of bracing are left to the reader.

7.5.5. Proposition. *Let $M' = B(M)$ be a bracing of $M(E)$ by the points* E'.

1. $M'(E) = M(E)$.
2. M' *has no erection.*
3. *If $r(M') > 3$, then $E[T(M')] = M'$ (i.e., free erection is the inverse of truncation).*
4. *If $r(M') > 3$, then there is a bracing B' of $T(M)$ such that B' has $T[B(M)]$ as a subgeometry.*
5. *If $r[T^i(M')] \geq 3$, then $\bar{E}[T^i(M')] = M'$.*
6. *If $\hat{M}$ is a (rank-preserving) weak-map image of M with the same lines, then there is a bracing of $\hat{M}$, $B(\hat{M})$, that contains the bracing $B(M)$ as a subgeometry (merely add braces to all dependent sets of $\hat{M}$ that are independent in M).*
7. *If M'' is a spanning subgeometry of M and $\bar{E}[T^{N-3}(M)] = M$, where $N = r(M)$, then $r\{\bar{E}[T^{N-3}(M')]\} \geq r(M')$.*

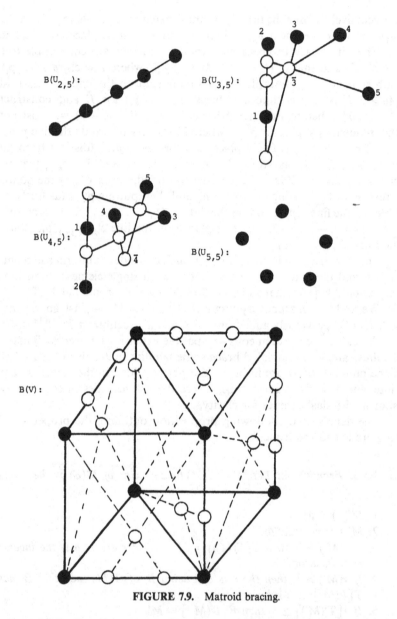

FIGURE 7.9. Matroid bracing.

7.5.6. Corollary. 1. *Let M be a rank-4 matroid, and let $\mathcal{R} = \{M'\colon M'$ is a weak-map image of M, and $T(M) = T(M')\}$. Then there exists a matroid $\bar{M}$ all of whose erections form a poset that is isomorphic to $\mathcal{R}$ under the weak-map order.*

2. *There exists a matroid that is doubly erectible but whose free erection is inerectible.*

3.   *Free erection is not, in general, a vector operation. However, when F is sufficiently large, bracing is vectorial, so that $B(M)$ is representable whenever M is.*

4.   *$B(C_4)$ is not graphical (although any added chord makes a circuit graphically inerectible). Further, $B(C_4')$ is not transversal, where C_4' is the transversal matroid consisting of a 4-circuit, three of whose points are doubled (M^* in Figure 7.2).*

Proof. 1.   Let $\bar{M} = B(M)$, and apply Proposition 7.5.5, parts 6 and 7.

2.   Consider $T[B(V)]$ (Figure 7.9). Its free erection is $B(V)$, which is inerectible. However, $T[B(V)]$ has another erection $\bar{V}'$ that is a partially braced V', and $\bar{V}'$ is in turn erectible. (See Proposition 7.5.5, part 7, and Exercise 7.33.)

3. $T[B(V)]$ of Figure 7.9 is clearly representable over the reals (it is drawn with straight lines). However, $B(V)$ is not representable, as V is not (see Exercise 7.6). On the other hand, the extensions made in any bracing $B(M)$ are always either principal or on the intersection of two flats L and F spanned by previously added points with $r(L) = 2, r(F) = m, r(L \cup F) = m + 1$. Further, the associated modular cut has no other generators. Thus, $B(M)$ can be coordinatized in a transcendental extension of F whenever F coordinatizes M.

$\square$

7.6. DIRECT SUM AND ITS GENERALIZATIONS

An operation as basic as deletion or contraction is that of *direct sum*: $M_1(E_1) \oplus M_2(E_2) := (M_1 \oplus M_2)(E_1 \cup E_2)$, which we can define as the matroid on the disjoint union of the groundsets E_1 and E_2, whose circuit family $\mathscr{C}_{M_1 \oplus M_2}$ is given by $\{C: C$ is a circuit of M_1, or C is a circuit of $M_2\}$. As a special case, we have, when $|E_2| = 1$, the addition of a loop or isthmus. We can characterize direct sum (easily) for any cryptomorphism:

7.6.1.   Proposition.   *The direct sum $M_1(E_1) \oplus M_2(E_2)$ can be described as follows (for all subsets $A_1 \subseteq E_1, A_2 \subseteq E_2$):*

> *Geometric lattice: The Cartesian product $L(M_1) \times L(M_2)$ on pairs $\{(x, y): x \in L(M_1), y \in L(M_2)\}$ [this is the lattice with $\vee, \wedge,$ and $\leq$ defined componentwise: $(x, y) \leq (x', y')$ if $x \leq x'$ and $y \leq y'$]. (See Chapter 3, Section 3.5.)*
> *Closure: $cl_{M_1 \oplus M_2}(A_1 \cup A_2) = cl_{M_1}(A_1) \cup cl_{M_2}(A_2)$.*
> *Rank: $r_{M_1 \oplus M_2}(A_1 \cup A_2) = r_{M_1}(A_1) + r_{M_2}(A_2)$.*
> *Closed sets: $\{K_1 \cup K_2: K_1$ is closed in M_1, and K_2 is closed in $M_2\}$.*
> *Hyperplanes: $\{H_1 \cup E_2: H_1$ is a hyperplane of $M_1\} \cup \{E_1 \cup H_2: H_2$ is a hyperplane of $M_2\}$.*

Bonds: $\{B: B$ is a bond of M_1 or B is a bond of $M_2\}$.
Spanning sets: $\{S_1 \cup S_2: S_1$ is a spanning set of M_1, and S_2 is a spanning set of $M_2\}$.
Bases: $\{B_1 \cup B_2: B_1$ is a basis of M_1, and B_2 is a basis of $M_2\}$.
Independent sets: $\{I_1 \cup I_2: I_1$ is independent in M_1, and I_2 is independent in $M_2\}$.
Circuits: $\{C: C$ is a circuit in M_1 or $M_2\}$.
Dependent sets: $\{D: D \cap E_1$ is dependent in M_1, or $D \cap E_2$ is dependent in $M_2\}$.

Direct sum is so called because it is a categorical direct sum (or coproduct) in the category of either weak maps or strong maps. The operation is self-dual and, in addition, commutes with several of our previous operations:

7.6.2 Proposition. 1. $[M_1(E_1) \oplus M_2(E_2)]^* = M_1^*(E_1) \oplus M_2^*(E_2)$.
 2. (a) If A_1 and I_1 are disjoint subsets of E_1: $[(M_1 \oplus M_2) - A_1]/I_1 = [(M_1 - A_1)/I_1] \oplus M_2$.
  (b) $[(M_1 \oplus M_2) - A]/I = \{[M_1 - (A \cap E_1)]/(I \cap E_1)\} \oplus \{[M_2 - (A \cap E_2)]/(I \cap E_2)\}$.
 3. $(M_1 +_{\mathscr{M}} p) \oplus M_2 = (M_1 \oplus M_2) +_{\mathscr{M}'} p$, where $\mathscr{M}' = \{F_1 \cup F_2: F_1 \in \mathscr{M}, F_2$ is a flat of $M_2\}$.
 4. $(M_1 \times_{\hat{\mathscr{M}}} p) \oplus M_2 = (M_1 + M_2) \times_{\hat{\mathscr{M}}'} p$, where $\hat{\mathscr{M}}' = \{A_1 \cup A_2: A_1 \in \hat{\mathscr{M}}, A_2 \subseteq E_2\}$.
 5. $M_1 \oplus M_2 = M_2 \oplus M_1$.
 6. $(M_1 \oplus M_2) \oplus M_3 = M_1 \oplus (M_2 \oplus M_3)$.
 7. $(M_1 \oplus M_2)(E_1) = (M_1 \oplus M_2)/E_2 = M_1$.

Other operations we have studied destroy the direct sum. For example, $T(M_1 \oplus M_2)$ is never a direct sum [unless $r(M_1)$ or $r(M_2) = 0$], and $(M_1 \oplus M_2) +_{\mathscr{M}} p$ and $(M_1 \oplus M_2) \times_{\hat{\mathscr{M}}} p$ are direct sums only if $\mathscr{M}$ and $\hat{\mathscr{M}}$ are as in Proposition 7.6.2, parts 3 and 4 (i.e., $\mathscr{M}$ has, as generators, flats from only one of the matroids; and $\hat{\mathscr{M}}$ contains all the subsets of one of the matroids).

Most of the interesting matroid classes are closed under direct sums. A hereditary class closed under direct sum is called a *variety*. In particular, our four basic classes have easy constructions for direct sum.

7.6.3. Proposition. 1. *If M_1 and M_2 are graphical (respectively planar graphical), with connected graphic representations G_1 and G_2, then $M(G_1) \oplus M(G_2)$ is a graphical (respectively planar graphical) matroid and has, as a connected representing graph, $G_1 \wedge G_2$ consisting of the graphs G_1 and G_2 wedged together at a vertex.*

2. *If M_1 and M_2 are vectorial over F with matrix representations N_1 and N_2, then $M_1 \oplus M_2$ is vectorial with representing matrix*

$$N_1 \oplus N_2 = \begin{bmatrix} N_1 & 0 \\ 0 & N_2 \end{bmatrix}.$$

(If N_1 and N_2 both contain an identity submatrix, so does $N_1 \oplus N_2$.)

3. *If M_1 and M_2 are both affine over F, so is $M_1 \oplus M_2$. It can be represented synthetically by taking two skew flats F_1 and F_2 of dimension $dim(M_1)$ and $dim(M_2)$, respectively, in $\mathrm{AG}[dim(M_1) + dim(M_2) + 1, F]$, and then representing M_1 on F_1 and M_2 on F_2.*

Analytically, if M_i has affine coordinates N_i ($i = 1, 2$), $M_1 \oplus M_2$ is represented affinely by the matrix

$$\begin{bmatrix} 1 & 0 \\ N_1 & 0 \\ 0 & N_2 \end{bmatrix},$$

where $1 = (1, \ldots, 1)$, a vector with $|E_1|$ 1's, and 0 is a vector with $|E_2|$ 0's.

4. *If the transversal matroid M_i has the presentation*

$$R_i \subseteq E_i \times T_i,$$

then $M_1 \oplus M_2$ is transversal and has the presentation $R_1 \cup R_2 \subseteq (E_1 \cup E_2) \times (T_1 \cup T_2)$, where $R_1 \cup R_2|_{E_i} = R_i$.

5. *Because M_1 is a submatroid of $M_1 \oplus M_2$ (see Proposition 7.6.2, part 7), if $M_1 \oplus M_2$ is graphical (respectively vectorial, affine, or transversal), so are M_1 and M_2.*

Inverse to direct sum is *direct-sum decomposition*. We say a matroid $M(E)$ is *connected* if M is not a (nontrivial) direct sum [i.e., $M = M_1(A) \oplus M_2(E - A)$ only if $A = \varnothing$ or $A = E$]. Otherwise, M is termed *separable*. The direct-sum decomposition of M is defined by the collection $M_i(E_i)$ ($i = 1, \ldots, k$), where, for each i, $M_i(E_i)$ is connected, and

$$M(E) = \bigoplus_{i=1}^{k} M_i(E_i).$$

It is a fundamental theorem in matroid theory that direct-sum decomposition is unique (so that the operation of direct sum on multisets of connected matroids and that of direct-sum decomposition on arbitrary matroids are inverses).

This result has already been proved (Corollary 3.5.3) and involves showing that M is connected if and only if each pair of points lies in a common circuit. To give cryptomorphic equivalents to this property of *circuit connectivity*, call a subset $E' \subseteq E$ a *separator* of M if $M(E) = M(E') \oplus M(E - E')$ [so that $M(E')$ is a union of direct-sum components]. Then we

have the following, where we note that a subset is a separator if and only if
its complement is, and a matroid is connected if and only if it has no proper
separator.

7.6.4. Proposition. *A subset* E' *of a matroid* $M(E)$ *is a separator if
and only if any of the following hold:*

Geometric lattice: E' *is a flat* x, *and in the lattice,*
  (a) x *has a unique complement.*
  (b) $x \wedge (y \vee z) = (x \wedge y) \vee (x \wedge z),$
        $x \vee (y \wedge z) = (x \vee y) \wedge (x \vee z),$
        $y \wedge (x \vee z) = (y \wedge x) \vee (y \wedge z),$ *or, equivalently,*
        $y \vee (x \wedge z) = (y \vee x) \wedge (y \vee z)$ *for all* $y, z \in L.$
  (c)   *If* p *is an atom and* y *is an element such that* $p \leq x \vee y$, *then*
        $p \leq x$, *or* $p \leq y.$
  (d)   *If* p *is an atom and* q *is a coatom with* $p \nleq q$, *then either* $p \leq x$
        or $x \leq q.$
  (In lattice theory, such elements x, *corresponding to separators, are
  variously termed distributive, neutral, or central, and the lattice of a
  connected matroid is called simple.)*
Closure: For all $A \subseteq E$, $\overline{E' \cup A} = E' \cup \bar{A}$. *Equivalently, for all* A,
$\overline{E' \cap A} = E' \cap \bar{A}.$
Rank: $r(E') + r(E - E') = r(E).$
Closed sets: $E' \cup K$ *is closed for all closed sets* K *of* M.
Hyperplanes: All hyperplanes contain E' *or* $E - E'$.
Bonds: No bond intersects E' *in a proper nonempty subset of the bond.*
Spanning sets: For all spanning sets S *of* M, $S \cap E'$ *spans* $M(E')$, *and*
$S - E'$ *spans* $M(E - E').$
Bases: For all bases B, $B \cap E'$ *is a basis of* $M(E')$. *Equivalently, if* B'
is a basis of $M(E')$, *and* B'' *of* $M(E - E')$, *then* $B' \cup B''$ *is a basis of*
M.
Independent sets: If I' *is an independent subset of* $M(E')$ *and* I'' *is an
independent subset of* $M(E - E')$, *then* $I' \cup I''$ *is independent in* M.
Circuits: No circuit intersects E' *in a proper nonempty subset of the
circuit. Equivalently, for any* $p \in E'$ *and* $q \in E - E'$, p *and* q *do not
lie in a common circuit.*
Dependent sets: For all dependent sets D, *either* $D \cap E'$ *or* $D - E'$ *is
dependent.*
Minors: $M - E' = M/E'$. *Equivalently,* $r(M - E') = r(M/E').$
Set-theoretical: E' *is a Boolean expression (using unions, intersections,
and complements) of separators.*

A direct-sum component (connected separator) can be recognized in
all our classes. In a graph, it is a block. In a matrix {either in the form $[IA]$
over a field or as a basic-circuit incidence matrix $A(M; B)$, where a basis B

labels the rows of A, and a basis complement labels its columns}, it is an indecomposable block. Further, in isthmus-free transversal presentations, it is (one of the vertex parts of) a connected component of the bipartite graph.

One of the classes *not* closed under direct sum is the hereditary class of coordinatizable (vectorial) matroids. An example (though not minimal) of a noncoordinatizable matroid is $PG(2,3) \oplus PG(2,2)$, because both components are minors, but the first is not coordinatizable over fields of characteristic 2, and the latter is coordinatizable only over such fields. [As an interesting historical footnote to this example, Rado (1957) recognized that a way to construct a noncoordinatizable matroid was to take the direct sum of $PG(2,2)$ with a matroid M not coordinatizable over any field of characteristic 2. However, he never produced such an M.] The foregoing example points out the following, in analogy with many branches of mathematics: Connected matroids are the building blocks of matroid theory. Many properties, when understood for connected matroids, are easily extended to direct sums. As an example, if it can be shown that the Whitney numbers (of the first or second kind) are logarithmically concave for connected geometries, then they are logarithmically concave for all geometries.

A hereditary class that is closed under direct sums is called a *variety*. We can scan the list of hereditary classes in Table 7.1 and identify those that are varieties. The reader can verify that those that are not closed under direct sums are $\mathscr{C}_i$ ($i = 5, 6, 7, 10, 11, 19, 20, 21$). An easy way to see this is in terms of excluded minors:

7.6.5. Proposition. *A hereditary class $\mathscr{H}$ is a variety if and only if all of its excluded minors are connected.*

Proof. If $\mathscr{E}(\mathscr{H})$ contains the matroid $M = M_1 \oplus M_2$, then each M_i as a proper minor of M is in $\mathscr{H}$, but $M_1 \oplus M_2 \notin \mathscr{H}$.

Conversely, assume that $M = M_1 \oplus M_2$ is not in $\mathscr{H}$, and let $M' = [(M_1 \oplus M_2) - A]/I$ be a smallest minor of M not in $\mathscr{H}$. By Proposition 7.6.2, part 2, $M' = M'_1 \oplus M'_2$, where each M'_i is a minor of M_i. Then we have one of the following: M'_1 is empty, and $M_2 \notin \mathscr{H}$; M'_2 is empty, and $M_1 \notin \mathscr{H}$; or $M'_1 \oplus M'_2$, a separable matroid, is in $\mathscr{E}(\mathscr{H})$. $\square$

For a hereditary class $\mathscr{H}$ that is not closed under direct sum, we can turn $\mathscr{H}$ into the variety $\mathscr{H}^\oplus = \{\oplus_i M_i : M_i \in \mathscr{H}\}$. Clearly, $\mathscr{E}(\mathscr{H}^\oplus)$ must contain any connected matroid M such that $M - p$ or M/p is a separable member of $\mathscr{E}(\mathscr{H})$. The determination of $\mathscr{E}(\mathscr{H}^\oplus)$ thus leads naturally to the study of connected extensions and coextensions of separable matroids. This is our first generalization of direct sum. Not only are such matroids "almost separable," but they also generalize some of the categorical properties of direct sum. A matroid whose contraction by a point increases its number of connected components is called a *parallel connection*. These matroids generalize the graph-theoretical operation explored in Chapter 6. As was shown there, the parallel connection of graphs G_1 and G_2 identifies each of two distinguished vertices (termed a *port* in electrical-circuit theory) in G_1 with two corresponding vertices of G_2. Because vertices do not exist in matroid theory, we add a (distinguished) edge in each G_i that joins these vertices and then identify the two distinguished edges in G_1 and G_2 to create $P(G_1, G_2)$ (where the identified distinguished edge "keeps track" of the port). This serves the purpose of being matroidal (an edge of a graph readily generalizes to a point of an arbitrary matroid) and also fits in well with the electrical-circuit idea that identifying vertices $v_1 \in G_1$ with $v_2 \in G_2$ and $w_1 \in G_1$ with $w_2 \in G_2$ has the same electrical properties as identifying $v_1 \in G_1$ with $w_2 \in G_2$ and $w_1 \in G_1$ with $v_2 \in G_2$. Thus, the parallel connection is a good example of the symbiotic relationship between matroid theory and graph theory (although, as we shall see, applications of the parallel connection can be found for other classes).

We summarize some of the properties of parallel connection. It is an instructive exercise for the reader to prove the following propositions by generalizing the graph-theoretical arguments of Chapter 6. Otherwise, one can consult Brylawski (1971). A *basepointed matroid* M is a pair $[M(E), p]$ with $p \in E$. Thus, (M, p) is merely a matroid with a special point. The dual $(M, p)^*$ of (M, p) equals (M^*, p). The *parallel connection* of two basepointed matroids $[M_1(E_1), p_1]$ and $[M_2(E_2), p_2]$ is the basepointed matroid $(P, \hat{p}) := P(M_1, M_2)$ on the groundset $(E_1 - p_1) \cup (E_2 - p_2) \cup \{\hat{p}\}$, where, if we make the identification $p_1 = p_2 = \hat{p}$, the closed sets of P are those K such that $K \cap E_1$ is closed in M_1, and $K \cap E_2$ is closed in M_2. If the basepoint is a separator of either M_1 or M_2, then $P(M_1, M_2)$ is an appropriate direct sum [if p_1 is a loop of M_1, then $P(M_1, M_2) \simeq M_1 \oplus (M_2/p_2)$, and if p_1 is an isthmus of M_1, then $P(M_1, M_2) \simeq (M_1 - p_1) \oplus M_2$]. We shall henceforth consider parallel connection only when the basepoint is not a separator (of either matroid).

For the following cryptomorphic descriptions it will be convenient to identify the three basepoints (as p) and, occasionally, to consider separately for a given family of subsets of P those members that contain p and those that do not.

7.6.6. Proposition. *Let* $[M_1(E_1), p]$ *and* $[M_2(E_2), p]$ *be two matroids neither of whose basepoints* p *is a separator. Then the parallel connection* $P(E_1 \cup E_2)$ *of* M_1 *and* M_2 *can be specified in any of the following ways for all* $A_1 \subseteq E_1$ *and* $A_2 \subseteq E_2$:

Closure:

$$\mathrm{cl}_P(A_1 \cup A_2) = \begin{cases} \mathrm{cl}_{M_1}(A_1 \cup p) \cup \mathrm{cl}_{M_2}(A_2 \cup p) & \text{if either} \\ \quad p \in \mathrm{cl}_{M_1}(A_1) \quad \text{or} \quad p \in \mathrm{cl}_{M_2}(A_2), \\ \mathrm{cl}_{M_1}(A_1) \cup \mathrm{cl}_{M_2}(A_2) & \text{otherwise.} \end{cases}$$

Rank:

$$r_P(A_1 \cup A_2) = \begin{cases} r_{M_1}(A_1 \cup p) + r_{M_2}(A_2 \cup p) - 1 & \text{if either} \\ \quad r_{M_1}(A_1 \cup p) = r_{M_1}(A_1) \quad \text{or} \\ \quad r_{M_2}(A_2 \cup p) = r_{M_2}(A_2), \\ r_{M_1}(A_1) + r_{M_2}(A_2) & \text{otherwise.} \end{cases}$$

[*In particular, for any closed set K of P,* $r_P(K) = r_{M_1}(K \cap E_1) + r_{M_2}(K \cap E_2) - r(K \cap \{p\})$.]

Closed sets: $\{K: K \cap E_1$ *is closed in* M_1, *and* $K \cap E_2$ *is closed in* $E_2\}$.

Hyperplanes: $\{H_1 \cup E_2: H_1$ *is a hyperplane of* M_1 *that contains* $p\} \cup \{E_1 \cup H_2: H_2$ *is a hyperplane of* M_2 *that contains* $p\} \cup \{H_1 \cup H_2: H_i$ *is a hyperplane of* M_i *that does not contain* $p\}$.

Bases: $\{B_1 \cup B_2 \cup p: B_1 \cup p$ *is a basis of* M_1, *and* $B_2 \cup p$ *is a basis of* $M_2\} \cup \{B_1 \cup B_2: p \notin B_1 \cup B_2, B_1$ *is a basis of* M_1, *and* $B_2 \cup p$ *is a basis of* M_2; *or* $B_1 \cup p$ *is a basis of* M_1, *and* B_2 *is a basis of* $M_2\}$.

Independent sets: $\{I_1 \cup I_2: I_1$ *is independent in* M_1, I_2 *is independent in* M_2, *and either* $I_1 \cup p$ *or* $I_2 \cup p$ *is independent*$\}$.

Circuits: $\{C: C$ *is a circuit of* M_1 *or of* $M_2\} \cup \{C_1 \cup C_2: C_1 \cup p$ *is a circuit of* M_1, *and* $C_2 \cup p$ *is a circuit of* $M_2\}$.

Binary cycles $(: = $ *unions of disjoint circuits*)*:* $\{C: C$ *is the symmetric difference of a cycle of* M_1 *and a cycle of* $M_2\}$. (*Note that neither M, nor* M_2 *need be binary in this description.*)

We now give some properties of parallel connection:

7.6.7$_P$. Proposition. *For all i, let* $M_i(E_i)$ *be a matroid with basepoint p (which is not a separator). The parallel connection P then has the following properties:*

1. $P(M_1, M_2) = P(M_2, M_1)$.
2. $P[P(M_1, M_2), M_3] = P[M_1, P(M_2, M_3)]: = P(M_1, M_2, M_3)$.
3. $P(M_1, M_2)$ *is connected if and only if both* M_1 *and* M_2 *are connected. Further,* $P(M_1, M_2 \oplus M) = P(M_1, M_2) \oplus M$.
4. *If* $A_1 \cup I_1 \subseteq E_1 - p$, *then* $P[(M_1 - A_1)/I_1, M_2] = [P(M_1, M_2) - A_1]/I_1$ (*even if* $p \in M_1$ *becomes a loop or isthmus in the minor*). *Similar results are obtained for extensions, coextensions, quotients, and lifts of one of the components.*
5. $P(M_1, M_2) - (E_2 - p) = M_1$, *so that the parallel connection contains each* M_i *as a submatroid.*
6. $P(M_1, M_2)/p = (M_1/p) \oplus (M_2/p)$. *Conversely, if M is connected and* M/p *is separable* $[M/p = M_1(E_1') \oplus M_2(E_2')]$, *then* $M = P(M - E_2', M - E_1')$.
7. *A connected basepointed matroid* (M, p) *has a parallel decomposition into parallel-irreducible connected matroids* $(M_i, p): M = P(M_1, \ldots, M_k)$ *that is unique up to a permutation of the components.*

8. *In the category of basepointed matroids whose morphisms are strong maps that preserve the basepoint, parallel connection is the direct sum. (Equivalently, in the strong-map category, parallel connection is a one-point pushout.)*

A special case of the parallel connection: $P(M_1, C_2)$, where C_2 is a two-point circuit, is called the *parallel extension* of the point p. (Its multiplicity is increased by 1. We have informally referred to this operation previously as "doubling" a point.)

The dual of the parallel connection is the *series connection*:

$$S(M_1, M_2) := P^*(M_1, M_2) := [P(M_1^*, M_2^*)]^*.$$

Cryptomorphisms for series connection are easily derived by duality arguments. We continue our proviso that the basepoint is not a separator.

7.6.8. Proposition. *The series connection $S(E) = S[M_1(E_1), M_2(E_2)]$ can be described in any of the following terms:*

Rank:

$$r_S(A) = \begin{cases} r_{M_1}(A \cap E_1) + r_{M_2}(A \cap E_2) \\ \quad \text{if } r_{M_1}[(A \cap E_1) - p] = r_{M_1}(A \cap E_1) \\ \quad \text{or if } r_{M_2}[(A \cap E_2) - p] = r_{M_2}(A \cap E_2), \\ r_{M_1}(A \cap E_1) + r_{M_2}(A \cap E_2) - 1 \quad \text{otherwise.} \end{cases}$$

Closed sets: $\{E - (K_1 \triangle K_2): E_i - K_i$ is closed in M_i, and $\triangle$ is symmetric difference$\} \cup \{K_1 \cup K_2: K_i$ is closed in $M_i, p \notin K_i\}$.
Bases: $\{B_1 \cup B_2: B_i$ is a basis of $M_i\}$.
Independent sets: $\{I_1 \cup I_2: I_1$ is independent in M_1, and I_2 is independent in $M_2\}$.
Circuits: $\{C: p \notin C$, and C is a circuit of M_1 or of M_2, or $C = C_1 \cup C_2$, where C_i is a circuit of M_i and $p \in C_1 \cap C_2\}$.

Properties of series connections are similar to those of parallel connections. In particular, replacing P by S, statements 1, 2, 3, 4, and 7 of Proposition 7.6.7 all hold; parts 5 and 6 must be dualized: $S(M_1, M_2)$ contains each M_i as a contraction; $S(M_1, M_2) - p = (M_1 - p) \oplus (M_2 - p)$; and if M is connected and $M - p$ is separable $[M - p = M_1(E_1) \oplus M_2(E_2)]$, then $M = S(M/E_2, M/E_1)$. In the following, we refer to the properties in Proposition $7.6.7_P$ for the series connection as the properties of Proposition $7.6.7_S$. We also have such dual concepts as the *series coextension* $S(M, C_2)$ of a point (corresponding to adding a vertex of degree 2 to an edge of a graph). Previous constructions guarantee that series coextension is also a vector, (transcendental) affine, and transversal construction.

An important relationship between series and parallel connections (when p is neither an isthmus nor a loop in M_i) is that $P(M_1, M_2) - p = S(M_1, M_2)/p$. This latter matroid, $S_2(M_1, M_2)$, is called the *2-sum of M_1 and*

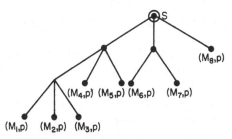

FIGURE 7.10. Series nonunary rooted tree.

M_2 by Seymour (1980). S_2 is a self-dual operator, and $S_2(M_1, M_2)$ has both M_1 and M_2 as minors.

Series and parallel connections can act in concert, where, for example, we evaluate the (*series*) *nonunary rooted tree* in Figure 7.10 as $M = S\{P[S(M_1, M_2, M_3), M_4, M_5], P(M_6, M_7), M_8\}$. (We make the convention that interior nodes an even distance from the root are series connections.) The same rooted tree labeled with P, whose end nodes are (M_i^*, p), gives the dual matroid M^*. When each $M_i = C_2$ (the two-point circuit), a *series-parallel network* (see Chapter 8) results, and conversely.

We now give some applications of series and parallel connections to minors.

7.6.9. Proposition. 1. (*Crapo 1967*) *If M is connected and $p \in M$, then either $M - p$ or M/p is connected.*

2. *Every connected pointed matroid has a unique series-parallel decomposition giving a tree, as in Figure 7.10, where each end node is series-parallel irreducible. Thus, the theory of connected pointed matroids is cryptomorphic to the theory of nonunary rooted trees T whose end nodes are labeled with pointed, connected, series-parallel irreducible matroids, and where, if $|T| > 1$, the root is labeled P or S.*

3. *If M' is a connected minor of a connected matroid M, then there exists a sequence of connected minors $M = M_1, M_2, \ldots, M_n = M'$ such that for each i, $M_{i+1} = M_i - p_i$ or $M_{i+1} = M_i/p_i$ for some $p_i \in M_i$.*

4. *If $\mathcal{H}$ is a hereditary class with $\mathcal{E}$ its class of excluded minors each of whose members has at most two direct-sum components, then $\mathcal{E}(\mathcal{H}^{\oplus})$, the excluded minors for the variety generated by $\mathcal{H}$, are the connected members in $\mathcal{E}$ along with the minor-minimal members of the following families:*

$\mathcal{E}_S = \{S(M_1 +_{\mathcal{M}_1} p, M_2 +_{\mathcal{M}_2} p): M_1 \oplus M_2 \in \mathcal{E}, \mathcal{M}_i$ is a proper modular cut of M_i, and the series connection is made at $p\}$.

$\mathcal{E}_p = \{P(M_1 \times_{\hat{\mathcal{M}}_1} p, M_2 \times_{\hat{\mathcal{M}}_2} p): M_1 \oplus M_2 \in \mathcal{E}, \hat{\mathcal{M}}_i$ is a proper modular ideal of $M_i\}$.

$\mathcal{E}_{SP} = \{S[P(M_1 \times_{\hat{\mathcal{M}}_1} p, C_2), M_2 +_{\mathcal{M}_2} p]: M_1 \oplus M_2 \in \mathcal{E}\}$.

Proof. 1. If M/p is separable, then M is a parallel connection. In this case, Proposition 7.6.5 guarantees that every pair of points of $M - p$ lies in a circuit. Thus, $M - p$ is connected.

2. If M is not series-parallel irreducible, then exactly one of M/p or $M - p$ is separable. The separation will then specify the label of the root and the matroid label on each second-level node. We can then further decompose [where, if (M, p) is a parallel connection, each basepointed component at the second level is either a series connection or irreducible].

3. This was shown in Brylawski (1972). Essentially, the proof proceeds by induction, where points are deleted or contracted until, say, at the *i*th stage, a deletion by p creates a separable matroid. This means that M_i is a series connection, with M' as a minor of one of its series components. By induction, each other component can be reduced to a loop, keeping a connected matroid at every stage.

4. It is elementary to observe that each member of $\mathscr{E}_p \cup \mathscr{E}_S \cup \mathscr{E}_{SP}$ is connected and has $M_1 \oplus M_2$ as a minor. If $M(E)$ is a minor-minimal connected matroid with $(M_1 \oplus M_2)(E')$ as a minor, then for some $p \in E - E'$, $M - p$ (respectively M/p) is separable and contains M_1 in one component and M_2 in another. The rest of the proof proceeds as in the foregoing part 3. {Note that $\mathscr{E}_{SP} = P[S(M_2 +_{\mathcal{M}_2} p, C_2), M_1 \times_{\hat{\mathcal{M}}_1} p].$} □

We now construct the series and parallel connections, where possible, for our four basic classes.

7.6.10. Proposition. *Let* $M_1(E_1 \cup p)$ *and* $M_2(E_2 \cup p)$ *be two matroids with basepoint* p *(which is not a loop or an isthmus).*

1. *Assume that* M_1 *and* M_2 *are vectorial over* F *by matrices* N_1 *and* N_2, *where the illustrated column vector represents* p:

$$N_1 = \left[N_1' \quad \begin{matrix} 0 \\ \vdots \\ 0 \\ 1 \end{matrix} \right]$$

and

$$N_2 = \left[\begin{matrix} 1 \\ 0 \\ \vdots \\ 0 \end{matrix} \quad N_2' \right].$$

Then $P(M_1, M_2)$ *and* $S(M_1, M_2)$ *are represented by matrices* P *and* S, *respectively:*

$$
P = \begin{bmatrix}
 & E_1 & & p & & E_2 & \\
 & & & 0 & & & \\
 & N'_1 & & \vdots & & 0 & \\
 & & & 0 & & & \\
\hline
 & & & 1 & & & \\
\hline
 & & & 0 & & & \\
 & 0 & & \vdots & & N'_2 & \\
 & & & 0 & & &
\end{bmatrix},
$$

$$
S = \begin{bmatrix}
 & E_1 & & p & & E_2 & \\
 & & & 0 & & & \\
 & N'_1 & & \vdots & & 0 & \\
 & & & 0 & & & \\
 & & & 1 & & & \\
\hline
 & & & 1 & & & \\
 & & & 0 & & & \\
 & 0 & & \vdots & & N'_2 & \\
 & & & 0 & & &
\end{bmatrix}.
$$

2. *If M_1 and M_2 are affine, then so is $P(M_1, M_2)$, where, in* $AG[r(M_1) + r(M_2) - 2, F]$, *complementary affine flats F_1 and F_2 are chosen such that* $\dim(F_1) = r(M_1) - 1$, $\dim(F_2) = r(M_2) - 1$ [*thus,* $\dim(\overline{F_1 \cup F_2}) = r(M_1) + r(M_2) - 2$], *and* $F_1 \cap F_2 = p$. *We then place M_1 on F_1 and M_2 on F_2 (translating each so that they intersect at p). Analytically, if N_i gives affine coordinates for M_i and (by a suitable translation) p is represented by the zero (column) vector* $\mathbf{0}$, *then the vectors of P have coordinates given in the matrix*

$$
\begin{bmatrix}
N_1 & 0 & \\
 & & 0 \\
0 & N_2 &
\end{bmatrix}.
$$

(*Because, for example, dimension is additive, for P, we see that for affine matroids, parallel connection is perhaps a more natural operation than even direct sum.*)

In general, for affine M_1 and M_2, $S(M_1, M_2)$ is affine only when F is sufficiently large (although $S - p$ is always affine). (Over the binary field, for example, S is never affine.) On the other hand, if neither M_1 nor M_2 is affine, neither is $S(M_1, M_2)$.

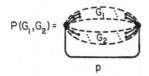

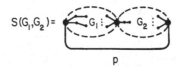

FIGURE 7.11. Parallel and series connections of graphs.

3. *As we saw in Chapter 6, if M_i is representable by the graph G_i as in Figure 7.11, then $S(G_1, G_2)$ and $P(G_1, G_2)$ are as shown in Figure 7.11. Both are planar if and only if both G_1 and G_2 are planar. Further, a graph G has no subgraph homeomorphic to K_4, the complete graph on four vertices, if and only if $M(G)$ is a series-parallel network (i.e., can be built up using only series and parallel connections from two-point circuits).*

4. *Let M_1 and M_2 be transversal matroids with presentations $R_1 \subseteq (E_1 \cup p) \times T_1$ and $R_2 \subseteq (E_2 \cup p) \times T_2$, respectively. Then $S(M_1, M_2)$ has the presentation $S(R_1, R_2) \subseteq (E_1 \cup E_2 \cup p) \times (T_1 \cup T_2)$, where $S|_{(E_i \cup p) \times T_i} = R_i$ (i.e, the respective bipartite graphs R_1 and R_2 are wedged together at p).*

$P(M_1, M_2)$ is transversal if and only if M_1 and M_2 have transversal presentations with $|R_i(p)| = 1$, in which case $P(M_1, M_2)$ has the presentation $P(R_1, R_2)$ given by identifying the edge $[p, R_1(p)]$ with the edge $[p, R_2(p)]$ in the bipartite graphs R_1 and R_2.

Proof. All the foregoing representations are immediate consequences of Proposition $7.6.7_P$, part 6, and its dual, Proposition $7.6.7_S$, part 6. For example, the analytical affine representation for P comes from the fact that if N represents M affinely, with p represented by the zero vector, then M/p has a vector representation given by the respective columns of N.

If M_1 and M_2 are binary affine matroids, and p is not an isthmus of either, then p is in a circuit C_i in each M_i, both of which must necessarily be even. Therefore, $C_1 \cup C_2$ is an odd circuit of $S(M_1, M_2)$ and cannot be represented in $AG(n, 2)$.

It is clear for the series transversal presentation given in Part 4, that the independent sets are those specified in Proposition 7.6.8. For the parallel connection (removing isthmuses, if necessary, from M_1 and M_2), if $P' \subseteq (E_1 \cup E_2 \cup p) \times T$ presents $P(M_1, M_2)$, then the number of vertices related to $E_1 \cup p$, $|P'(E_1 \cup p)| = r(M_1)$, $|P'(E_2 \cup p)| = r(M_2)$, and $|P'(E_1 \cup E_2 \cup p)| = |T| = r(M_1) + r(M_2) - 1$. Hence, $|P'(p)| \leq |P'(E_1 \cup p) \cap P'(E_2 \cup p)| = 1$, and each M_i has the presentation $R_i = P'_{E_i \cup p}$, where for each i, $|R_i(p)| = 1$. On the other hand, if the bipartite graphs presenting M_1 and M_2 are joined as indicated, presenting the matroid M, then $P|_{(E_1 \cup E_2) \times [T - P(p)]}$ presents the contraction M/p by Proposition 7.4.17, part 4, and (because it is a disconnected graph), M/p has separators E_1 and E_2. But $M - E_2$ is presented by R_1, and $M - E_1$ is presented by R_2, and so we are done, by Proposition 7.6.7$_P$, part 6.

The parallel connection, a "freest" way to join two matroids at a point, can in turn be generalized to a construction that gives the freest way to join two matroids over a common submatroid. This is not always possible. Given $M_1(E_1 \cup A)$ and $M_2(E_2 \cup A)$ with a common submatroid $M(A)$, there may be no matroid M at all on $E_1 \cup E_2 \cup A$ with $M(E_1 \cup A) \simeq M_1$ and $M(E_2 \cup A) \simeq M_2$ (see Exercise 7.13, part b, showing two incompatible extensions). On the other hand, a matroid may exist, but no freest one. For example, we can paste two triangular prisms together on a common lateral face, resulting in the Vamos cube V, which is the freest such common extension in rank 4. However, we can also paste the prisms together in rank 5, resulting in the erected cube $E(V')$ (see Exercise 7.33).

We can, however, always form such a freest compatible extension (which we now formulate for combinatorial geometries G and G') when the common subgeometry x is a modular flat of one of the geometries. If x is a modular flat of G but is not closed in G', we extend G by the points $cl_{G'}(x) - x$, creating a new geometry $\bar{G}$ with a modular flat $\bar{x}$ isomorphic to the flat $\bar{x}$ in G' (see Exercise 7.11, part b). We can then make the construction defined later via the common flat $\bar{x}$: $P_x(G, G') := P_{\bar{x}}(\bar{G}, G')$.

Assume now that x is a flat of both G and G' and is modular in G. The *generalized parallel connection* (or *strong join*) $P_x[G(E), G'(E')]$ is the geometry on the groundset $(E - x) \cup (E' - x) \cup x$ whose closed sets are given by

$$\{K: K \cap E \text{ is closed in } G, \text{ and } K \cap E' \text{ is closed in } G'\}.$$

Generalized parallel connections are pushouts from x in the category of injective strong maps and are explored elsewhere (Brylawski 1975b) (where other cryptomorphic descriptions are given along with applications to the characteristic polynomial). Further properties for the generalized parallel connection are found in Lindström (1968), where he shows that the foregoing closed-set description gives a geometry if and only if a pushout exists in the category of injective strong maps, and (when $G = G'$) it gives a geometry only if x is modular in G [in which case we have the symmetric connection $P_x[(G, G)]$. We mention only one other property of $P_x[G(E), G'(E')]$ that generalizes Proposition 7.6.7$_P$, part 6:

If $P_x(G, G')$ is a parallel connection, then $P_x(G, G')(E) = G$, $P_x(G, G')$ $(E') = G'$, and $\overline{P_x(G, G')}/x = \overline{G/x} \oplus \overline{G'/x}$ (always staying within the geometry category). Conversely, if P is a geometry with a modular flat x of the sub-geometry $P - A'$, and if P/x has separators A and A', then $P = P_x(P - A', P - A)$. (Here, $E = A \cup x$, and $E' = A' \cup x$.)

The most striking application of the generalized parallel connection is by Seymour (1980), where he defines the *3-sum* of G and H by $P_x(G, H) - x$ for binary geometries G and H with a common three-point line x (necessarily a modular flat of both). He shows that a regular geometry (one in the hereditary class $\mathscr{C}_{15} = \bar{\mathscr{R}}$ in Table 7.1) is the direct sum, 2-sum, or 3-sum of graphic geometries (those in $\mathscr{C}_{16} = \bar{\mathscr{G}}$), cographic geometries (those in $\mathscr{C}_{17} = \bar{\mathscr{G}}^*$), and a number of copies of the binary 10-point matroid R_{10} which is represented over F_2 by all distinct binary vectors of length 5 with exactly three 1's. Two related classes include $\mathscr{C}_{22} = \overline{\mathscr{P}\mathscr{G}}$ and $\mathscr{C}_{23} = \bar{\mathscr{R}}'$. The matroid $P_x(G, H) - x$ is called the *modular sum* of G and H and is denoted $S_x(G, H)$. There is not much difference between generalized parallel connection and modular sum. Modular sum is "more efficient," whereas generalized parallel connection guarantees that G and H are both subgeometries. Note that for matroids, if we parallel extend every point of x in H, creating $\bar{H}$, then $S_x(G, \bar{H}) = P_x(G, H)$.

Classes closed under generalized parallel connection and hence modular sum include $\mathscr{C}_{15} = \bar{\mathscr{R}}$, $\mathscr{C}_{12} = \bar{\mathscr{F}}_2$ (binary geometries), $\mathscr{C}_{13} = \bar{\mathscr{F}}_3$ (ternary geometries), and $\mathscr{C}_{16} = \bar{\mathscr{G}}$. The generalized parallel connection of two graphs essentially consists of taking two graphs and identifying a common clique.

The classes $\mathscr{C}_{18} = \overline{\mathscr{P}\mathscr{G}}$ of planar graphic geometries and $\mathscr{C}_{17} = \bar{\mathscr{G}}^*$ of cographic geometries are not closed, because if C_3 is a common triangle of three copies of K_4, then $S_{C_3}[P_{C_3}(K_4, K_4), K_4] = K_{3,3}$, and the latter geometry is neither planar graphic nor cographic. The class $\mathscr{C}_{14} = \overline{\mathscr{S}\mathscr{P}}$ of series-parallel networks is closed under generalized parallel connection, because for these geometries the operation practically reduces to (ordinary) parallel connection. (A modular subgraph of a series-parallel network can have only two vertices of attachment with the rest of the graph, and these vertices will give the resulting graph a two-vertex cut-set.)

As the previous examples (Proposition 7.6.10) suggest, transversal matroids are not closed under modular sum, and the generalized parallel connection of two affine matroids is affine *when it is vectorial*. The deter-mination of when the generalized parallel connection of two vector geometries is vectorial (even over extension fields) reduces to one of common representations.

7.6.11. Proposition. 1. Let $P_x(G_1, G_2)$ be the generalized parallel connection of $G_1(E_1 \cup x)$ and $G_2(E_2 \cup x)$. Then $P_x(G_1, G_2)$ is vectorial over F if and only if there exists a representation N_1 for G_1 and a representation N_2 for G_2 and there is a linear transformation that is nonsingular on N_1 taking the columns indexed by x in N_1 to those indexed by x in N_2.

In this case, x has a common representation M_x in representations for G_1 and G_2, respectively, and $P_x(G_1, G_2)$ is represented by

$$N = \begin{bmatrix} M_2 & 0 & 0 \\ M_1 & M_x & M_1' \\ 0 & 0 & M_2' \end{bmatrix},$$

and

$$\begin{bmatrix} M_2 & 0 \\ M_1 & M_1' \\ 0 & M_2' \end{bmatrix} \text{ represents the associated modular sum,}$$

where

$$N_1 = \begin{bmatrix} M_2 & 0 \\ M_1 & M_x \end{bmatrix} \text{ represents } G_1,$$

$$N_2 = \begin{bmatrix} M_x & M_1' \\ 0 & M_2' \end{bmatrix} \text{ represents } G_2.$$

Further, if N_i gives affine coordinates for G_i, then N gives affine coordinates for $P_x(G_1, G_2)$.

2. Let $G(E)$ and $G'(E')$ be the geometries represented over any field of characteristic not equal to 2 by the matrix $\bar{N}$:

$$
\begin{array}{c}
E: \quad a \quad b \quad c \quad d \quad x \quad y \quad z \quad w \\
\bar{N} = \begin{bmatrix} 1 & 1 & 1 & 1 & 0 & 0 & 0 & 0 \\ 0 & 1 & 0 & 1 & 1 & 0 & 1 & 1 \\ 0 & 0 & 1 & 1 & 0 & 1 & 1 & -1 \end{bmatrix} \\
E': \quad a' \quad b' \quad c' \quad d' \quad x' \quad z' \quad y' \quad w'
\end{array}
$$

Then $L = xyzw \simeq x'y'z'w' = L'$ is a modular four-point line in both G and G'. Further, $P_L(G, G') - L$ (where we identify a point in L with its primed counterpart in L') is an eight-point geometry (pictured by the solid points in Figure 7.12) that is representable only over fields of characteristic 3. Similar examples exist for fields of order 2^m and 3^m ($m > 1$) (where we paste two projective planes together across a common line after interchanging two of the points).

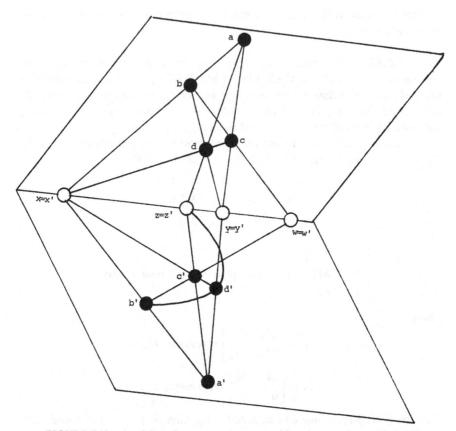

FIGURE 7.12. An eight-point geometry representable only over characteristic 3.

Thus, a modular sum of vector matroids (over F) is vectorial in general only when $F = F_2$ or F_3.

Proof. 1. $P_x(G_1, G_2)$ contains both G_1 and G_2 as subgeometries, and so, if represented, it must give a common representation of M_x in a representation for G_1 and a representation for G_2. Conversely, if P is the matroid represented by N, then clearly P/x has a direct-sum decomposition of the form $(P - E')/x \oplus (P - E)/x$ [and x is modular in $P(E_1 \cup x) = G_1(E_1 \cup x)$ by assumption]. Thus, $P = P_x(G_1, G_2)$.

2.   We easily check that L is modular in G (and L' is modular in G'). When $M = P_L(G, G') - L$ is represented in $PG(3, F)$, the plane spanned by the points $abcd$ intersects the plane spanned by $a'b'c'd'$ in a line intersected by the six lines ab, ac, ad, bc, bd, and cd (in four distinct points). Further, the eight points a, b, c, d, x, y, z, w must have coordinates projectively equivalent to those given in $\bar{N}$. (A geometric reason is that each of the points x, y, z, and

w is on at least three independent planes spanned by points in M). Equivalently, we can show algebraically that any matrix representation for M (if it exists) is projectively unique (see the chapter on coordinatizations in the next volume) and gives the specified coordinates. But the cross-ratio $(w, z; x, y)$ computed in $\bar{N}$ is $[(-1 - 0)(1 - \infty)]/[(1 - 0)(-1 - \infty)] = -1$ (see Hirschfeld 1979), whereas the cross-ratio of the same points in an embedding of G': $(w', z'; x', y') = (w, y; x, z)$ is $\frac{1}{2}$. Because $-1 = \frac{1}{2}$ only in characteristic 3, the cube M is representable only in that characteristic. $\quad\square$

Our final generalization of direct sum takes the independent-set description of Proposition 7.6.1 and removes the restriction that E_1 and E_2 be disjoint sets. The *union* $(M_1 \vee M_2)(E)$ of two matroids on the same groundset E is a matroid whose independent sets are given by

$$\{I_1 \cup I_2 : I_1 \text{ is independent in } M_1, \text{ and } I_2 \text{ is independent in } M_2\}.$$

This important operation, originated by Nash-Williams (1966), has surprisingly few easily stated cryptomorphic equivalents.

7.6.12. Proposition. *Let $M_1(E)$ and $M_2(E)$ be two matroids on the set E. Then the matroid union $M_1(E) \vee M_2(E) := (M_1 \vee M_2)(E)$ is a matroid on the same set that can be characterized in any of the following ways:*

Closure: $\mathrm{cl}_{M_1 \vee M_2}(A) = A \cup \{p: \text{there exist } A_1, A_2 \subseteq A, \text{ with } A_1 \cap A_2 = \varnothing, \text{ such that } p \in \mathrm{cl}_{M_i}(A_i), A_1 \subseteq \mathrm{cl}_{M_2}(A_2), \text{ and } A_2 \subseteq \mathrm{cl}_{M_1}(A_1)\}.$
Rank:

$$r_{M_1 \vee M_2}(A) = \min_{A' \subseteq A}\left[r_{M_1}(A') + r_{M_2}(A') + |A - A'|\right]$$
$$= \max_{A' \subseteq A}\left[r_{M_1}(A') + r_{M_2}(A - A')\right].$$

Bases: $\{B_1 \cup B_2 : B_i \text{ is a basis for } M_i, \text{ and for all } p \notin B_1 \cup B_2, q \in B_1 \cap B_2, \text{ neither } (B_1 \cup p) - q \text{ nor } (B_2 \cup p) - q \text{ is a basis}\}$
$= \text{maximal sets of the form } \{B_1 \cup B_2 : B_i \text{ is a basis of } M_i\}.$
Independent sets: $\{I_1 \cup I_2 : I_i \text{ is independent in } M_i\}.$
Circuits: Minimal sets of the form $\{C : A \cap C \text{ contains a circuit of } M_1, \text{ or } C - A \text{ contains a circuit of } M_2 \text{ for all } A \subseteq C\}.$

The reader can check that this defines a matroid. {One way to do this is based on the fact that $r_{M_1} + r_{M_2}$ is a nondecreasing, nonnegative, integer-valued submodular function. For such a function f, the associated function $\bar{f}$ where $\bar{f}(A) = \min_{A' \subseteq A}[f(A') + |A - A'|]$, always gives the rank function of a matroid (Edmonds and Rota 1966).} Matroid union has a good characterization in the sense of Edmonds, because if we know the structure of M_1 and M_2, we can ascertain (in polynomial time) that a subset A of E is independent in $M_1 \vee M_2$ using the max definition of the rank function and can ascertain that A is dependent from the min definition. Hence, it is not surprising that there is a polynomial algorithm that computes $r_{M_1 \vee M_2}(A)$ for any $A \subseteq E$ from the rank functions r_{M_1} and r_{M_2} (Greene and Magnanti 1975).

The dual of matroid union is termed *matroid intersection* by Welsh (1976): $M_1 \wedge M_2 := (M_1^* \vee M_2^*)^*$. (This is not to be confused with the "intersection" of Exercise 7.20, part b, because here spanning sets, not independent sets, are intersected.) Perhaps a more useful operation is the "half-dual union," $M_1 \vee M_2^*$. The reader will note that $r(M_1 \vee M_2^*) = |E| - r(M_2) + \max\{|I| : I$ is independent in both M_1 and $M_2\}$. Thus, we have a polynomial algorithm that computes the size of a largest common independent set for two matroids. As noted by Welsh (1976), the analogous computation for three matroids is NP-hard. Matroid union has several connections with our previous operations.

7.6.13. Proposition. 1. $(M_1 \vee M_2)(E') = M_1(E') \vee M_2(E')$ for all $E' \subseteq E$. (A similar result holds for extension.)

2. Let $M_1(E_1)$ and $M_2(E_2)$ be two matroids, and let E be any set that contains E_1 and E_2. We can then (trivially) extend the definition of matroid union to this case by letting $M_i' = M_i(E_i) \oplus (E - E_i)$, where $E - E_i$ is adjoined to M_i as a set of loops. Then $(M_1 \vee M_2)(E) := (M_1' \vee M_2')(E)$ has the independent-set description of Proposition 7.6.12. It reduces to ordinary matroid union when $E_1 = E_2 = E$, and to direct sum when E_1 and E_2 partition E. Finally, it is the series connection $S(M_1, M_2)$ when $E = E_1 \cup E_2$ and $E_1 \cap E_2 = p$ (a nonseparator).

3. $M_1 \vee M_2 = M_2 \vee M_1$, and $(M_1 \vee M_2) \vee M_3 = M_1 \vee (M_2 \vee M_3) := M_1 \vee M_2 \vee M_3$, where $r_{M_1 \vee M_2 \vee M_3}(A) = \min_{A' \subseteq A}[r_{M_1}(A') + r_{M_2}(A') + r_{M_3}(A') + |A - A'|]$.

4. If $r(M_2) = 1$, so that M_2 is determined by its subset A of loops, then $M_1 \vee M_2$ is the principal lift $L_{E-A}(M_1)$.

The reverse of matroid union (i.e., "union decomposition") is a difficult concept. [For example, any matroid M is a nontrivial "union component" of a Boolean algebra: $B_{|E|}(E) = M \vee M^*$.] Characterizing those matroids that are *reducible*, i.e. the union of two matroids, each of positive rank, remains an unsolved problem. For graphical matroids we have the result of Lovász and Recski (1973), which states that M is reducible if and only if it is separable or is a single-element extension of a separable matroid (i.e., a series connection).

Matroid union has the following representations:

7.6.14. Proposition. 1. Let $M_1(E)$ and $M_2(E)$ be vector matroids with matrix representations N_1 and N_2, respectively. Then $M_1 \vee M_2$ need not be

*vectorial unless the field is sufficiently large. It has the transcendental repre-
sentation*

$$\begin{bmatrix} N_1 \\ N_2^\lambda \end{bmatrix},$$

*where column c_i of N_2^λ is the ith column of N_2 multiplied by the transcendental
λ_i.*

2. (a) *If M_1 and M_2 are affine, $M_1 \vee M_2$ need not be affine, even
if it is vectorial. It will be affine if F is sufficiently large. If L_i is the set
of loops of M_i, and $M_i - L_i$ is affine (where $L_1 \cap L_2 = \varnothing$), then $M_1 \vee M_2$
is affine in a transcendental extension. Affine coordinates in $AG[r(M_1) +
r(M_2) - 1, F(\{\lambda_i\})]$ are given for all $p_i \in E - (L_1 \cup L_2)$ by the column vectors*

$$\begin{bmatrix} \lambda_i \\ \lambda_i \mathbf{v}_i \\ \mathbf{w}_i \end{bmatrix},$$

*where $\mathbf{v}_i$ gives affine coordinates for p_i in M_1, and $\mathbf{w}_i$ gives affine coordinates
for p_i in M_2. If $p_i \in L_1$, it is given coordinates*

$$\begin{bmatrix} \mathbf{0} \\ \mathbf{w}_i \end{bmatrix},$$

and if $p_i \in L_2$, it is given the vector

$$\begin{bmatrix} \mathbf{0} \\ \mathbf{v}_i \\ \mathbf{0} \end{bmatrix}.$$

(b) *We can represent matroid union synthetically as in Figure 7.13 by
placing a set of points E_i corresponding to the matroid $M_i - L_i$ on the generally
placed flat F_i [in an ambient affine space of dimension at least $r(M_1 \vee M_2) -
1$]. Then, for each $p \in E - (L_1 \cup L_2)$, we replace the two points $p' \in E_1 - L_1$
and $p'' \in E_2 - L_2$ by the single point $\bar{p}$ freely placed on the line joining p'
to p''. When $p \in L_1$ (or $p \in L_2$), we represent p by $\bar{p} = p'' \in F_2$ ($\bar{p} = p' \in F_1$,
respectively).*

3. (a) *If M_1 and M_2 are transversal with presentation $R_1 \subseteq E_1 \times T_1$
and $R_2 \subseteq E_2 \times T_2$, then the matroid union $(M_1 \vee M_2)(E)$, $E \supseteq E_i$, is transversal with presentation $R \subseteq E \times (T_1 \cup T_2)$, where $R^{-1}(T_i) = R_i^{-1}$ (i.e., the respective bipartite graphs are wedged together at the vertex-set $E_1 \cap E_2$).*

(b) *M is transversal if and only if it is the union of rank-1 matroids.*

Proof. 1.   Let A be a square submatrix of

$$N = \begin{bmatrix} N_1 \\ N_2^\lambda \end{bmatrix}.$$

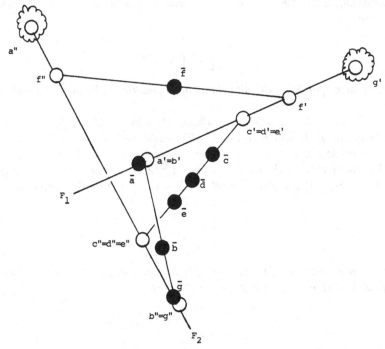

FIGURE 7.13. Matroid union.

We calculate its determinant by the Laplace expansion relative to its rows in N_1:

$$\det(A) = \sum_{i_1 < \ldots < i_m} \pm \det(A_1) \det(A_2) \lambda_{i_1} \ldots \lambda_{i_m},$$

where A_i is a submatrix of N_i, and each summand is multiplied by a different subset of λ_i's. Thus, $\det(A) \neq 0$ if and only if its columns can be partitioned into two subsets C_1 and C_2 such that C_1 indexes a nonsingular determinant of N_1, and C_2 indexes a nonsingular determinant of N_2. Thus, the independent column subsets of N obey Proposition 7.6.12.

    2. (a) Consider the binary matrix M:

$$M = \begin{bmatrix} 1 & 0 & 0 & 0 & 0 & 1 & 0 & 0 & 1 & 1 & 1 \\ 0 & 1 & 0 & 0 & 0 & 1 & 1 & 0 & 0 & 1 & 1 \\ 0 & 0 & 1 & 0 & 0 & 1 & 1 & 1 & 0 & 0 & 1 \\ 0 & 0 & 0 & 1 & 0 & 0 & 1 & 1 & 1 & 0 & 1 \\ 0 & 0 & 0 & 0 & 1 & 0 & 0 & 1 & 1 & 1 & 1 \end{bmatrix}.$$

(The first 10 columns of this matrix represent the regular matroid R_{10} that we discussed in the remarks preceding Proposition 7.6.11.) The reader can verify that M is affine (all of its circuits are even) and that any 10-element subset can be partitioned into two bases. Therefore, $M \vee M = C_{11}$, an 11-point circuit that is binary but not affine.

When transcendentals are added to the field, we get the affine representation of $M_1 \vee M_2$ in part 2 by first adding a row of 1's to the affine coordinates of M_1 and M_2, respectively, obtaining vector representations N_1 and N_2. We then multiply each column of N_1 by a transcendental and use the vector representation of part 1. Finally, we delete the (superfluous) row of 1's from N_2.

(b) The synthetic construction comes from the fact that the analytical representation puts M_1 and M_2 on skew flats F_1 and F_2 and the fact that the analytical construction in (a) gives p_i, the coordinates of a generic point on the line from $\mathbf{v}_i \in F_1$ to $\mathbf{w}_i \in F_2$.

3.   (a) A subset $I \subseteq E$ can be matched by R into $T_1 \cup T_2$ if and only if it can be partitioned into two subsets I_1 and I_2, where I_i can be matched into T_i by R_i. Thus, Proposition 7.6.12 is satisfied. [Note that, in general, this will create isthmuses, and $r(M_1 \vee M_2)$ will be less than $|T_1 \cup T_2|$.]

(b) If $T = \{t\}$, then $R \subseteq E \times \{t\}$ represents a rank-1 matroid whose loops are $E - R^{-1}(t)$. We can thus apply part 3(a) or use Proposition 7.6.13, part 4, in conjunction with Exercise 7.26, part a, showing the equivalence of union with a rank-1 matroid and principal lift.   □

7.7. LOWER TRUNCATIONS

The (upper) truncation of a geometry G (of rank n) has the easy lattice-theoretical interpretation of removing the copoints H from $L(G)$. When a similar operation is attempted for the set A of atoms, the submodular law fails (if and only if $n > 3$), and new flats must be introduced to restore the lattice property of being geometric. The resulting geometry will then give a matroid interpretation to the lines of G and the flats that contain them.

We describe two ways to make this construction. Both ideas are essentially due to Dilworth (Dilworth 1944; Crawley and Dilworth 1973). The first gives a rank-n construction in which $L' = L(G) - A$ is a sublattice, and the second has rank $n - 1$ and contains L' as an infimum subsemilattice. Both have suggestive affine constructions. Our first construction has a more general context: embedding an arbitrary lattice as a sublattice of a geometric lattice $L(G)$. We describe the construction of $L(G)$ in the following lemma. The reader desiring a proof is referred to the book by Crawley and Dilworth (1973), or, for a more general setting, to Chapter 10.

7.7.1. Lemma. 1. *Given a finite lattice L', we define recursively from the top element, $1_{L'}$, the rank function:*

(a) $\rho(1_{L'}) = N$ *(an arbitrary constant).*

(b) $\rho(q) = \rho(q') - 1$ if q is an infimum-irreducible element of L' uniquely covered by q'.

(c) Otherwise, $\rho(q) = min[\{\rho(x) + \rho(y) - \rho(x \vee y): x \wedge y = q\}]$. Then the function $r'(x):= \rho(x) - \rho(0_{L'})$ is intergervalued, submodular, normalized [so that $r'(0_{L'}) = 0$], and strictly increasing, in that if $x < y$, then $r'(x) < r'(y)$.

2.   For each supremum-irreducible $q' \in L'$, let $E_{q'}$ be an $[r'(q') - r'(q)]$-element set, where q' uniquely covers q. Now define the groundset

$$E := \cup\{E_{q'}: q' \text{ is a supremum-irreducible element of } L'\},$$

and for all $x \in L'$, let

$$A_x := \bigcup_{q' \leq x} E_{q'} \qquad (A_{0_L} = \varnothing).$$

Then the subsets $\{A_x: x \in L\}$ form a closure system $\mathscr{C}$ with a partial rank function, $\bar{r}$, where $\bar{r}(A_x):= r'(x)$. This function has the following properties:

(a) $\bar{r}(\varnothing) = 0$.
(b) $\bar{r}(A_x) - \bar{r}(A_y) \leq |A_x - A_y|$ for all $A_x \supseteq A_y$ (i.e., for all $x \geq y$).
(c) $\bar{r}(A_x \vee A_y) + \bar{r}(A_x \cap A_y) \leq \bar{r}(A_x), + \bar{r}(A_y)$, where $A_x \cap A_y = A_{x \wedge y}$, and $A_x \vee A_y = \cap\{A_z: A_z \supseteq A_x \cup A_y\} = A_{x \vee y}$.

3.   If we define the function r on E by

$$r(A):= \min_{C \in \mathscr{C}} [\bar{r}(C) + |A - C|],$$

then r is the rank function of a geometry $G(L')$ that agrees with $\bar{r}$ on subsets in $\mathscr{C}$.

4.   The elements of $\mathscr{C}$ are closed in $G(L')$ and form a sublattice of its geometric lattice $L[G(L')]$. Thus, L' is embedded in $L[G(L')]$ as a sublattice.

We apply the foregoing construction, called the *Dilworth completion* of L', to the case in which the atoms A are removed from a geometric lattice $L = L(G)$ [i.e., the lattice L' on the elements $L - A$, where $x \vee_{L'} y = x \vee_L y$, and $x \wedge_{L'} y = x \wedge_L y$ if $r(x \wedge_L y) > 1$, but $x \wedge_{L'} y = 0_{L'}$ otherwise].

We call the resulting geometry $G(L')$ constructed in Lemma 7.7.1, part 3, the *line completion* of G, and it is denoted G^c. We can readily check that the function r' of Lemma 7.7.1, part 1, when applied to $L(G) - A$, agrees with the rank function of G on all closed sets of G of rank greater than 1. If $r(G) \leq 3$, then $L - A$ is modular and is the geometric lattice of a line (or point). If $r(G) > 3$, then G must contain two skew lines, so that $r'(l) = 2$ for all lines l, and these elements (which are atoms of $L - A$) are the only

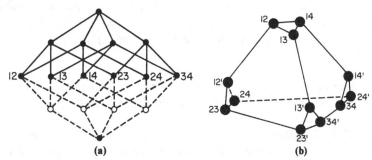

FIGURE 7.14. Line completion and point-pair completion.

supremum-irreducibles. We thus have the following:

7.7.2. Proposition. *Let G be a geometry with $r(G) \geq 4$. Then $L(G) - A$ is a sublattice of the geometric lattice of G^c, where G^c is a geometry on the set*

$$E = \cup\{p_l, p'_l : l \text{ is a line of } G\}$$

with rank function

$$r(E') = \min_{x \in L(G)} \left[r_G(x) + |E' - \cup\{p_l, p'_l : l \leq x\}| \right].$$

G^c can be constructed by making two principal extensions on each line of G and then deleting the (original) points of G.

A related construction (which is somewhat easier to describe) makes $|l| \cdot (|l| - 1)$ principal extensions on each line before deleting the points of G. In the resulting geometry, $G^{\bar{c}}$, each p_{qs} on l corresponds to an ordered pair of points (q, s) of G, with $l = \{\overline{q,s}\}$. Clearly, G^c is a subgeometry of $G^{\bar{c}}$, where all but two of the points on each line are deleted. $G^{\bar{c}}$ is called the *point-pair completion* of G. We picture $L(B_4 - A)$ and $B_4^c = B_4^{\bar{c}}$ in Figure 7.14.

Note that the construction can be realized in Euclidean space E^n ($n \geq 3$) for any neighborly polytope P, where P^c is the vertex truncation of P, with each point then pulled (along its line).

From the rank-function description of G^c (or $G^{\bar{c}}$) of Proposition 7.7.2 we have the following, where Proposition 7.7.3, part 2, follows from part 1 using Rado's generalization of the Hall marriage theorem (Rado 1942).

7.7.3. Proposition. 1.   *A subset $I = \{p_{l_1}, \ldots, p_{l_m}\}$ of G^c is independent (where the lines l_i need not be distinct) if and only if for all subsets $\{p_{l_{i_1}}, \ldots, p_{l_{i_j}}\}$ of I, $r(l_{i_1} \cup \cdots \cup l_{i_j}) \geq j$.*
2.   *A subset I is independent if and only if there exist points $p_i \in l_i$ ($i = 1, 2, \ldots, m = |I|$) such that $\{p_1, \ldots, p_m\}$ is independent in G.*

3.   Similarly, in $G^{\bar{c}}$, $A = \{p_{q_1 s_1}, \ldots, p_{q_m s_m}\}$ is independent if and only if for all subsets $\{p_{q_{i_1} s_{i_1}}, \ldots, p_{q_{i_j} s_{i_j}}\}$, $r_G(\{q_{i_1}, \ldots, q_{i_j}, s_{i_1}, \ldots, s_{i_j}\}) \geq j$.

4.   Equivalently, A is independent if and only if $\{p_1, \ldots, p_m\}$ is independent where for each i, $p_i \in \{q_i, s_i\}$.

Because the structure of G^c is so elementary when G is a line or is planar, assume in the following proposition that $r(G) \geq 4$.

7.7.4.   Proposition.   1. G^c is never binary (or graphic). However, a different (and much more complicated) construction embeds any given lattice as a sublattice of the geometric lattice of a complete graph K_N (where, in general, $N \gg n$).

2.   If G is vectorial over F, where p_i is coordinatized by the vector $\mathbf{v}_i$, then $G^{\bar{c}}$ (and hence G^c) is vectorial over $F(\{\lambda_{ij} : \{p_i, p_j\} \in G \times G, i \neq j\})$, where p_{ij} is represented by the vector $\mathbf{v}_{ij} = \mathbf{v}_i + \lambda_{ij}\mathbf{v}_j$.

Conversely, if $G^{\bar{c}}$ (or G^c) is vectorial over F, then so is G.

3.   If F is sufficiently large and G is affine, then G^c has the affine description of placing two points freely on each line of G. [Similarly, in $G^{\bar{c}}$, we place two points freely on the line joining any pair of (distinct) points of G.]

4.   If G has a transversal presentation $R \subseteq E \times T$, then $G^{\bar{c}}$ is transversal with presentation

$$R^c \subseteq [(E \times E) - diag] \times T, \quad where \ R(p_{qs}) = R(q) \cup R(s).$$

(G^c is also transversal and has a presentation consisting of a suitable restriction of R^c.)

Proof.   1.   As can be seen from Figure 7.15, G^c contains $U_{3,6}$ as a subgeometry and is not binary. The construction that embeds any lattice into a finite (but extremely large) partition lattice is due to Pudlák and Tůma (1977).

2.   If $r(G) = n$, then $\{\mathbf{v}_{i_1 j_1}, \ldots, \mathbf{v}_{i_n j_n}\}$ is an independent set of vectors if and only if its determinant D is nonzero. Expanding D using column multilinearity, we get a sum of different products of transcendentals multiplied by determinants of the form $\det[\mathbf{w}_1, \ldots, \mathbf{w}_n]$, where for each k, $\mathbf{w}_k = \mathbf{v}_{i_k}$ or $\mathbf{v}_{j_k}$. But one of these determinants is nozero if and only if $\{p_{i_1 j_1}, \ldots, p_{i_n j_n}\}$ is a basis of $G^{\bar{c}}$ by Proposition 7.7.3, part 4.

Conversely, if $[G(E)]^c$ is represented over $\bar{F}$ by the vectors $\{\mathbf{v}_{ij}\}$, we coordinatize $G(E)$ as follows: Let p_i, p_j, and p_k be independent in G. Then $\{p_{ij}, p_{ji}, p_{ik}, p_{ki}\}$ has rank 3 and spans a plane π in $PG(n-1, \bar{F})$. In this plane, $cl_p(p_{ij}, p_{ji})$ and $cl_p(p_{ik}, p_{ki})$ intersect. We coordinatize p_i by this intersection. For any point p_l that is independent of p_i, p_j, and p_k, the line $cl_p(p_{il}, p_{li})$ intersects π in a point, and this point must be p_i, so that p_i, p_{il}, and p_{li} are collinear. Finally, if p_m depends on $\{p_i, p_j, p_k\}$, we replace p_j (or p_k) by some p_l that is independent of the other three points and argue as before to show that p_i, p_{im}, and p_{mi} are collinear.

3. This is the affine restatement of part 2.

4. Again, this follows easily from the independent-set description of Proposition 7.7.3, part 4. $\square$

A related construction, called the *Dilworth truncation*, denoted G^d or $T^d(G)$, intersects the lines of G with a generally placed hyperplane. This construction, like line completion, is coordinatizable over an extension field of any field that coordinatizes G [and over no other characteristic for $r(G) > 4$] and has the added advantage of reducing the rank by 1 (so that, for example, many coordinatization problems reduce to the plane). G^d is a geometry on the set $E = \{p_l: l \text{ is a line of } G\}$ and is defined in any of the following ways. A fuller discussion and all missing proofs appear elsewhere (Brylawski 1985).

7.7.5. Proposition. *The Dilworth truncation G^d is characterized by the following cryptomorphisms:*

> *Geometric lattice: Although $L(G^d)$ is difficult to axiomatize completely, we note that it contains $L(G) - A$ as an infimum subsemilattice, and no smaller geometric lattice has this property.*
> *Rank: $r_{G^d}(\varnothing) = 0$, and, for all $A \neq \varnothing$,*
>
> $$r(A) = \min_{\pi \in \Pi(A)} [r_\pi(A)],$$
>
> *where*
>
> $$r_\pi(A) = \sum_{i=1}^{k} r_G(\cup\{l: p_l \in A_i\}) - k.$$
>
> *Here, π partitions A, written $\pi \in \Pi(A)$, if $\pi = \{A_1, \ldots, A_k\}$, $A = A_1 \cup \cdots \cup A_k$, $A_i \cap A_j = \varnothing$, and $A_i \neq \varnothing$ for all $i \neq j$.*
> *Bases: See Proposition 7.7.6.*
> *Independent sets: $\{I: r_G(l_{i_1} \cup \cdots \cup l_{i_m}) \geq m + 1 \text{ for all nonempty subsets } \{p_{l_{i_1}}, \ldots, p_{l_{i_m}}\} \text{ of } I\}$. (See also Proposition 7.7.6.)*
> *Circuits: Minimal nonempty members of the family $\{C: r(\cup\{l: p_l \in C\}) = |C|\}$.*

Proof. The equivalence of the characterizations in terms of rank, independent sets, and circuits, and the proof that they give a matroid structure (in fact, a combinatorial geometry), can be deduced from more general results of Edmonds (1970). In that work he defines a matroid from a "β_0-function" that is nonnegative (on nonempty subsets of E), nondecreasing, and submodular. The function

$$\beta_0(A) = r_G(\cup\{l: p_l \in A\}) - 1$$

is clearly such a function, and we can apply Propositions 8 and 15 from Edmonds (1970). $\square$

As before, we can give a point-pair analogue of Dilworth truncation that, in general, is a matroid whose simplification is G^d. The *matroidal Dilworth truncation* $G^{\bar{d}}$ is a matroid on the set of unordered pairs of points of $G(E)$,

$$E_{G^{\bar{d}}} = \{p_{ij}: \{p_i, p_j\} \subseteq E, i \neq j\},$$

where, for all lines l of G, the atom p_l of G^d corresponds to the multiple point $\{p_{ij}: \{p_i, p_j\} \subseteq l\}$ of $G^{\bar{d}}$.

7.7.6. Proposition. *A set of points* $\{p_{i_1 j_1}, \ldots, p_{i_m j_m}\}$ *is independent in* $G^{\bar{d}}$ *if and only if there is an ordering on the points and choice of indices such that for all* $k = 1, 2, \ldots, m$, $\{p_{i_1}, p_{i_2}, \ldots, p_{i_k}, p_{j_k}\}$ *is independent in* G. *(It is a basis if and only if, in addition,* $m = n - 1$, *so that, in particular,* $\{p_{i_1}, p_{i_2}, \ldots, p_{i_{n-1}}, p_{j_{n-1}}\}$ *is a basis of* G.)

Proof. The sufficiency is clear, because the condition will obviously guarantee that the independent-set characterization of Proposition 7.7.5 is satisfied.

Conversely, if $r_G(l_{i_1} \cup \cdots \cup l_{i_k}) \geq k + 1$ for all nonempty k-subsets of lines where $l_{i_s} = \mathrm{cl}_G(p_{i_s}, p_{j_s})$, then the G-rank of any m-set certainly exceeds its cardinality, and by the Rado matching theorem there will be an independent subset of points $I' = \{p'_1, \ldots, p'_m\}$, where each p'_k is from a different pair $\{p_{i_k}, p_{j_k}\}$. Because the rank of $\{p_{i_1}, \ldots, p_{i_m}, p_{j_1}, \ldots, p_{j_m}\}$ is at least $m + 1$, the independent set I' can be extended to an independent set I of size $m + 1$. Assume that $I = I' \cup \{p_{j_m}\}$. Then we repeat the process on the $(m - 1)$-element independent set $\{p_{i_1 j_1}, \ldots, p_{i_{m-1} j_{m-1}}\}$. $\qquad\square$

We illustrate the construction of $G^{\bar{d}}$ (or G^d) when $G = C_3 \oplus p_1 \oplus p_3$ in Figure 7.15.

We now list some properties of G^d and $G^{\bar{d}}$.

7.7.7. Proposition. 1. *As a labeled matroid,* $G^{\bar{d}}$ *determines* G *up to isomorphism, because* $\{p_1, \ldots, p_n\}$ *is a basis of* G *if and only if* $\{p_{12}, p_{13}, \ldots, p_{1n}\}$ *is a basis of* $G^{\bar{d}}$. *[Considering the matroids* $\mathrm{PG}(n, q)$ *and* $\mathrm{PG}(n, q) - p$, *we see that the isomorphism class of* $G^{\bar{d}}$ *does not characterize* G *up to isomorphism (Mazzocca 1983).]*

2. *If* $p_i \in E$, *then the submatroid* $G^{\bar{d}}(\{p_{ij}: p_j \in E - p_i\})$ *is isomorphic to the contraction* $G(E)/p_i$, *and the submatroid* $G^{\bar{d}}(\{p_{jk}: j \neq i, k \neq i\})$ *is isomorphic to* $(G - p_i)^{\bar{d}}$. *Thus,* $G^{\bar{d}}$ *can be thought of as a way to piece together all the one-point contractions of* G *into a single matroid.*

3. *Using part 2, in the case in which* p_i *is an isthmus, we see that* $(G \oplus p)^{\bar{d}}$ *combines* $G^{\bar{d}}$ *and* G *into a single matroid that, in addition, shows the construction of the Dilworth truncation of* G^d *from* G. *[This is illustrated by the isthmus* p_1 *of* $G = C_3 \oplus p_1 \oplus p_3$ *of Figure 7.15, and, in fact, Figure 7.15 illustrates* $(C_3 \oplus p_1 \oplus p_3 \oplus p)^d$. *The idea of illustrating a construction* $G \to O(G)$ *as a single matroid encompassing both* G *and* $O(G)$ *is a*

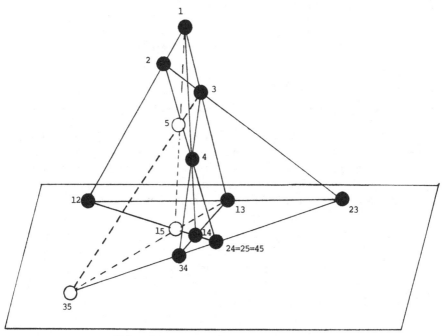

FIGURE 7.15. Matroidal Dilworth truncation.

fundamental idea of Mason (1977), and the reader should view our Figures 7.2b, 7.3, 7.6, 7.7, 7.13, 7.14, and 7.15 in that context.]

4. $B_5^d = B_5^{\bar{d}}$ is the (three-dimensional) Desargues configuration (the black points of Figure 7.15). Similarly, $U_{4,5}^d$ is the two-dimensional Desargues configuration, and, in general, when $r(G) > 3$, $T(G^d) = [T(G)]^d$.

We now illustrate the Dilworth truncation for our classes. Because $G^d = L_n$ (an n-point line with $n = |\{l: l \text{ is a line of } G\}|$) when $r(G) = 3$, and is a point when $r(G) = 2$, we assume that $r(G) > 3$ in the following.

7.7.8. Proposition. *Let $r(G) = n > 3$.*

1. *G^d is never transversal.*

2. *G^d is graphical if and only if G is the Boolean algebra B_n. In this case, $G^d = G^{\bar{d}} = B_n^d$ is the geometry of the complete graph K_n. Thus, the geometry of a graph G can be visualized by putting its vertices freely in space as isthmuses and intersecting its edges with a generally placed hyperplane (which intersects each edge in the point corresponding to that edge). Note also that because $B_n^d = (B_{n-1} \oplus p)^d$, we can construct B_n^d recursively from the construction that gave B_{n-1}^d by Proposition 7.7.7, part 3. Here, K_n is partitioned into K_{n-1} and a vertex bond corresponding to B_{n-1}.*

3. *If G is vectorial over F by the matrix $M = [\mathbf{v}_i: p_i \in E]$ [with $r(G) = n$], then $G^{\bar{d}}$ is vectorial over the transcendental extension field*

$F(\lambda_1, \ldots, \lambda_n)$ by the matrix

$$M^{\bar{d}} = \left\{ \mathbf{v}_{ij} := (\mathbf{v}_j \cdot \lambda)\mathbf{v}_i - (\mathbf{v}_i \cdot \lambda)\mathbf{v}_j \colon p_{ij} \in \binom{E}{2} \right\}.$$

Here, $\lambda = (\lambda_1, \ldots, \lambda_n)$, so that

$$\mathbf{v}_{ij}(k) = \mathbf{v}_i(k)[\mathbf{v}_j(1)\lambda_1 + \mathbf{v}_j(2)\lambda_2 + \cdots + \mathbf{v}_j(n)\lambda_n] - \mathbf{v}_j(k)[\mathbf{v}_i(1)\lambda_1 + \cdots + \mathbf{v}_i(n)\lambda_n].$$

Although this represents $G^{\bar{d}}$ in an $n \times \binom{|E|}{2}$ matrix, the rank of the matrix is $n - 1$, as each vector is contained in the hyperplane $\{\mathbf{v}: \mathbf{v} \cdot \lambda = 0\}$.

4.   Conversely, if $G^{\bar{d}}$ is vectorial over F, then so is G. In particular, if $\{p_1, \ldots, p_n\}$ is a basis B for G, the submatroid $G^{\bar{d}}(\{p_{ij}: 1 \leq i < j \leq n\})$ is isomorphic to $M(K_n)$ and, via a projective transformation, can be represented over F by the vectors $\mathbf{v}_{ij}$ ($i < j$), where

$$\mathbf{v}_{1j}(j) = 1,$$

$$\mathbf{v}_{1j}(k) = 0 \quad \text{for } k \neq j,$$

and, for $i \neq 1$,

$$\mathbf{v}_{ij}(i) = 1,$$

$$\mathbf{v}_{ij}(j) = -1,$$

$$\mathbf{v}_{ij}(k) = 0 \quad \text{for } k \neq i, j.$$

Then we represent p_1 by the n-vector $\mathbf{v}_1$, where

$$\mathbf{v}_1(1) = 1,$$

$$\mathbf{v}_1(k) = 0 \qquad (k \neq 1),$$

and the point p_m ($1 < m \leq n$) by the n-vector $\mathbf{v}_m$, where

$$\mathbf{v}_m(1) = \mathbf{v}_m(m) = 1,$$

$$\mathbf{v}_m(k) = 0 \qquad (k \neq 1, m).$$

For a point p_m with $m > n$, there are two points of B, p_i and p_j, such that $\{p_m, p_i, p_j, p_k\}$ is independent for all $p_k \in B - \{p_i, p_j\}$. We then represent p_m by the vector $\mathbf{v}_m$, where

$$\mathbf{v}_m = \overline{\mathbf{v}_i \mathbf{v}_{im}} \cap \overline{\mathbf{v}_j \mathbf{v}_{jm}}.$$

5.   If G is affine in a sufficiently large affine space $A = \mathrm{AG}(n - 1, F)$, $G^{\bar{d}}$ is represented by the intersection of each line spanned by G with a hyperplane in general position. (Here, "general position" means that intersections of lines of G with H have the sufficiently large dimension specified in Proposition 7.7.5.) Conversely, G is affine over any field that represents $G^{\bar{d}}$.

Proof. 1. Because G contains B_4, G^d contains $M(K_4)$, which is not transversal.

2. If $G \neq B_n$, then $|G^d| > \binom{n}{2}$ and is larger than $M(K_n)$, the largest graphical geometry of rank $n - 1$. On the other hand, if $G = B_n$, and C is a circuit of G^d, then, by Proposition 7.7.5, C is a minimal set such that $|\{p_{i_k}, p_{j_k}: p_{i_k j_k} \in C\}| = r_G(\{p_{i_k}, p_{j_k}: p_{i_k j_k} \in C\}) = |C| = m$. Thus, the points $\{p_{i_k j_k}\}$ can be ordered so that $i_1 = j_m$, and for all $k < m$, $i_{k+1} = j_k$. But these are precisely the circuits of K_n. For an arbitrary graph, we apply the submatroid construction of Proposition 7.7.7, part 2.

3 and 4. See Brylawski (1985).

5. Although it is intuitively clear that this can be done in a sufficiently large affine space, the rigorous proof follows from part 3. Conversely, part 4 constructs G disjoint from the hyperplane "at infinity" representing G^d. □

To illustrate part 4 of the preceding proposition, we coordinatize the geometry $G \cup G^d$ of Figure 7.15:

$$
\begin{array}{ccccccccc}
12 & 13 & 14 & 23 & \begin{matrix}24\\(=25=45)\end{matrix} & 34 & 15 & 35 & \begin{matrix}1 \ \ 2 \ \ 3 \ \ 4 \ \ \ \ 5\end{matrix}
\end{array}
$$

$$
\left[
\begin{array}{cccccccc:ccccc}
0 & 0 & 0 & 0 & 0 & 0 & 0 & 0 & 1 & 1 & 1 & 1 & 1 \\
1 & 0 & 0 & 1 & 1 & 0 & 1 & 1 & 0 & 1 & 0 & 0 & \dfrac{1}{a+1} \\
0 & 1 & 0 & -1 & 0 & 1 & 0 & -1-a & 0 & 0 & 1 & 0 & 0 \\
0 & 0 & 1 & 0 & -1 & -1 & a & a & 0 & 0 & 0 & 1 & \dfrac{a}{a+1}
\end{array}
\right]
$$

$$(a \neq 0, -1)$$

[Here the coordinates for 5 were obtained by intersecting the lines $\overline{1, 15}$ and $\overline{3, 35}$, obtaining the equation

$$
\begin{bmatrix} 1 \\ 0 \\ 0 \\ 0 \end{bmatrix} + b \begin{bmatrix} 0 \\ 1 \\ 0 \\ a \end{bmatrix} = \begin{bmatrix} 1 \\ 0 \\ 1 \\ 0 \end{bmatrix} + c \begin{bmatrix} 0 \\ 1 \\ -1-a \\ a \end{bmatrix},
$$

so that $b = c = 1/(a + 1)$.]

7.8. INDEX OF CONSTRUCTIONS

In Table 7.3 we give the appropriate theorem, exercise, or figure where a specific construction is defined or exemplified. For example, $7.4.7^+$ refers to remarks following Proposition 7.4.7, and $7.4.7^-$ means that the remarks

TABLE 7.3. Index of Constructions

Construction	Notation	Definition	Cryptomorphisms	Properties	Affine const.	Vector const.	Graphic const.	Transversal const.
Adjoint, complete	$G(L^*)$	Ex. 7.59						
Bracing	$B(M)$	$7.5.5^-$		7.5.5, 7.5.6 Fig. 7.9 Ex. 7.37, 7.38	$7.5.6(3)^\lambda$ Fig. 7.9	$7.5.6(3)^\lambda$	7.5.6(4)	7.5.6(4)
Coextension	$M = M'/I \leftrightarrow M'$	$7.4.14^-$		7.4.14, 7.6.2(3), 7.6.2(4), $7.6.5^+$, 7.6.7(4) Ex. 7.53	DNE*	DNE*	DNE*	DNE*
free	$M \times p$	$7.4.15^+$	$7.4.16^+$	$7.4.15^+$, 7.4.16	$7.4.17(3)^\lambda$	$7.4.17(1)^\lambda$	DNE 7.4.17(2)	7.4.17(4)
principal	$M \times_A p$	$7.4.15^+$	$7.4.16^+$	$7.4.15^+$, 7.4.16	$7.4.17(3)^\lambda$	$7.4.17(1)^\lambda$	DNE 7.4.17(2)	7.4.17(4)
series	$S(M, C_2)$	$7.6.7^+$			$7.6.10(2)^\lambda$	7.6.10(1)	$7.6.7^+$	7.6.10(4)
single-element (via the modular ideal $\hat{\mathcal{M}}$)	$M \times_{\hat{\mathcal{M}}} p$	$7.4.14^+$, 7.4.15		7.4.15	DNE (Ex. 7.28) Fig. 7.7	DNE (Ex. 7.28)	DNE*	DNE 7.3.7
trivial								
Comap		Ex. 7.50						
Complete union	$(G \oplus H)^f(E \times E)$	Ex. 7.54						
Completion (see line completion, Dilworth completion)		Ex. 7.51			$Ex. 7.51^\lambda$	$Ex. 7.51^\lambda$	DNE Ex. 7.51	Ex. 7.51
Contraction (by I)	M/I	$7.4.3^-$	7.4.3	7.6.2(7), 7.6.7(4), 7.6.7(5)$_s$, 7.6.11	DNE*	7.4.4	7.4.5	DNE*

	Notation							
geometric single-element	$\overline{M/I}$ or $[x,1]$ M/p	$7.4.3^+$ $7.4.3^-$	7.4.3	$7.6.7(6)_F$, 7.7.7(2), Ex. 7.28, 7.29, 7.49	$7.4.4^\Delta$ DNE 7.4.6	$7.4.4, 7.4.3^+$ 7.4.4	$7.4.5, 7.4.3^+$ 7.4.5	DNE* DNE 7.4.5
Deletion of a loop or isthmus	$M - E'$ $M - p$	(see Submatroid) (see Loop or Isthmus, adjoining a)						
single-element	$M - p$	$7.3.1^-$	7.3.1	$7.6.7(6)_S$, 7.7.7(2)	$7.3.1^-$ Fig. 7.1	$7.3.1^-$ Fig. 7.1	$7.3.1^-$ Fig. 7.1	$7.3.1^-$ Fig. 7.1
Dilworth completion	$G(L)$	7.7.1		Ex. 7.38(c), 7.59 7.7				
Dilworth truncation	G^d or $T^d(G)$	7.7.5	7.7.5, 7.7.6	Ex. 7.54, 7.56, 7.57, 7.58, 7.60	$7.7.8(5)^\Delta$ Fig. 7.15 Ex. 7.57, 7.58	$7.7.8(3)^\Delta$; see also 7.7.8(4)	DNE 7.7.8(2)	DNE 7.7.8(1)
complete	G^D	Ex. 7.56			Ex. 7.56^Δ	Ex. 7.56^Δ	DNE*	DNE*
kth	$G^{d(k)}$	Ex. 7.55			Ex. 7.55^Δ	Ex. 7.55^Δ	DNE*	DNE*
matroidal	G^d	$7.7.6^-$	7.7.6	7.7.7	7.7.8(5) Fig. 7.15	$7.7.8(3)^\Delta$; see also 7.7.8(4)	DNE 7.7.8(2)	DNE 7.7.8(1)
Direct sum (and direct-sum decomposition)	$M_1 \oplus M_2$	$7.6.1^-$	7.6.1	$7.6.1^+$, 7.6.2, $7.6.2^+$, 7.6.4, $7.6.4^+$, 7.6.7(3), 7.6.7(6), 7.6.9 Ex. 7.40, 7.42, 7.49, 7.50	7.6.3(3)	$7.6.3(2), 7.6.4^+$	$7.6.3(1)$, $7.6.4^+$	$7.6.3(4)$, $7.6.4^+$
Drop (see Quotient) Dual operation	$O^*(M)$	$7.4.1^-$		$7.4.14^-$, $7.4.17^+$, $7.6.8^-$, $7.6.13^-$ Ex. 7.18				

(Continued)

TABLE 7.3. (*Continued*)

Construction	Notation	Definition	Cryptomorphisms	Properties	Affine const.	Vector const.	Graphic const.	Transversal const.
Dual, orthogonal (see Orthogonal dual)								
Erection	$M = T(M') \leftarrow M'$ or $E_{\alpha}(M)$	$7.5.2^-$	7.5.2, Ex. 7.35	7.5.3, Ex. 7.33, 7.35, 7.36, 7.38	DNE 7.5.6(3)	DNE 7.5.6(3)	Ex. 7.35(c)	DNE
$\mathscr{C}$-erection		Ex. 7.34			Ex. 7.34			
free	$E(M)$	$7.5.3^-$		7.5.5, 7.5.6, Ex. 7.32	DNE 7.5.6(3)	DNE 7.5.6(3)	Ex. 7.35(c)	
principal	$E_F(M)$	$7.5.3^+$	7.5.4	7.5.4, Ex. 7.51	DNE*	DNE*		
total free	$\bar{E}(M)$	$7.5.5^-$		7.5.5	DNE 7.5.6(3)	DNE 7.5.6(3)		
Extension	$M = M' - A \leftarrow M'$	$7.3.2^-$		$7.3.4, 7.6.5^+, 7.6.7(4), 7.6.13(1)$	DNE*	DNE* (but see 7.3.4)	DNE*	DNE*
compatible		Ex. 7.13		$7.25, 7.610^+,$ Ex. 7.13, 7.47				
free	$M + p$	$7.3.2^+$	7.3.3		$7.3.5^+$	$7.3.5^4, 7.3.5^+$	DNE 7.3.6	7.3.7
modular	$M +_{\mathscr{A}_F} p$	Ex. 7.11		$7.6.10^+,$ Ex. 7.11	DNE*	DNE*	DNE*	DNE*
parallel	$P(M_1, C_2)$ or $M +_q p$	$7.6.7^+$		Ex. 7.9, 7.11, 7.12, 7.15, 7.16	7.3.5 or 7.6.10(2)	7.3.5 or 7.6.10(1)	7.3.6 or 7.6.10(3)	DNE 7.3.7
principal (on F)	$M +_F p$	$7.3.2^+$	7.3.3		$7.3.5^4,^+$	$7.3.5^4, 7.3.5^+$	DNE 7.3.6	DNE 7.3.7

Operation	Symbol							
single-element (via the modular cut $\mathscr{M}$) trivial	$M +_{\mathscr{M}} p$	$7.3.2^-$, $7.3.2^+$	Ex. 7.8	$7.3.2^-$, $7.3.2$ Ex. 7.6, 7.7				
Hereditary class constructions	$\bar{\mathscr{C}}, \cup \mathscr{C}_\lambda, \oplus \mathscr{C}_\lambda, \mathscr{C}^\oplus$, etc.	Ex. 7.50 Table 7.1$^+$		Table 7.1, 7.4.14, 7.6.5, 7.6.9(4) Ex. 7.24, 7.50				
Induction, matroid		Ex. 7.48					DNE	DNE Ex. 7.49(d)
Intersection, matroid (see also Union, dual)		Ex. 7.20(b)		Ex. 7.20(b)				
Intertwining Isthmus, adjoining an	$I(M, M')$ $G \oplus p$	Table 7.1$^+$ $7.3.1^+$	7.2.2	$7.2.2^-$, $7.3.2^-$, 7.5.4.3, $7.6.1^-$, $7.6.6^-$, 7.7.7(3) Ex. 7.2	7.6.7(4)	$7.2.2^-$	$7.2.2^-$	Ex. 7.3
Lift	$L_{\mathscr{M}}(M)$	7.4.17$^+$			DNE* Fig. 7.6	DNE*	DNE*	DNE*
$\mathscr{C}$-lift		7.4.17$^+$						
elementary		7.4.17$^+$		Fig. 7.8	DNE* Fig. 7.8	DNE*	DNE*	DNE*
Higgs	$L(M)$	7.4.17$^+$		Ex. 7.20		*	7.4.17$^+$	Ex. 7.26(a)
ith Higgs	$M_i(E)$	Ex. 7.20						
principal	$L_A(M)$	7.4.17$^+$	Ex. 7.26(c)	7.6.13(4) Ex. 7.26	Ex. 7.26(b)$^\lambda$	Ex. 7.26(b)$^\lambda$	DNE*	Ex. 7.26(a)
Line completion	G^c or $G(L - A)$	7.7.2	7.7.2, 7.7.3	7.7.2$^+$	7.7.4(3)$^\lambda$ Fig. 7.14	7.7.4(2)$^\lambda$	DNE 7.7.4(1)	7.7.4(4)

(Continued)

TABLE 7.3. (*Continued*)

Construction	Notation	Definition	Cryptomorphisms	Properties	Affine const.	Vector const.	Graphic const.	Transversal const.
Loop, adjoining a	$G \oplus p$	$7.3.1^+$	7.2.1	$7.2.1^-$, $7.3.2^-$, 7.4.13(6), 7.4.14, $7.6.1^-$, $7.6.6^-$ Ex. 7.1, 7.24	DNE $7.2.1^-$	$7.2.1^-$	$7.2.1^-$	$7.2.1^-$
Minor	$(M/I)(E')$ or $(M/I) - A$ or $[M'] \le [M]$	$7.4.11^-$	7.4.12	$7.4.11, 7.4.12^+$, 7.4.13, $7.6.9^-$, 7.6.9 Ex. 7.25, 7.30, 7.50	DNE 7.4.6	$7.4.13^-$	$7.4.5, 7.3.1^-$	DNE 7.7.5
excluded		$7.4.13^-$		$7.4.11^-$ Ex. 7.22, 7.23, 7.24		Table 7.1	Table 7.1	
geometric visible	$\overline{(G/A)} - B$	$7.4.13^-$		7.4.11(2)	$7.4.13^{\lambda-}$	$7.4.13^-$, $7.4.3^+$ $7.4.13^-$	$7.4.5, 7.4.3^+$	DNE
Modular sum	$S_x(G,H)$ or $P_x(G,H) - x$	$7.6.11^-$		Ex. 7.46	DNE 7.6.11(2) Fig. 7.13	Only for F_2, F_3: 7.6.11(1)	$7.6.11^-$	DNE 7.6.10(4)
Orthogonal dual	M^*	7.4.1		$7.2.1^+$, 7.4.13(1), 7.4.15, 7.6.2(1), $7.6.5^+$, $7.6.8^-$ Ex. 7.18	DNE 7.4.2(2) Fig. 7.2a	7.4.2(1), $7.4.2^+$	DNE [except for planar graphs: 7.4.2(4), Fig. 7.2b]	DNE 7.4.2(3) Fig. 7.2c
Parallel connection (or decomposition)	$P(M_1, M_2)$	$7.6.5^+$	7.6.6	$7.6.5^+$, $7.6.7_p$	7.6.10(2)	7.6.10(1)	$7.6.5^+$, 7.6.10(3)	DNE 7.6.10(4)
generalized	$P_x(G_1, G_2)$	$7.6.11^-$		$7.6.11^-$ Ex. 7.46	DNE 7.6.11(2) Fig. 7.14	Only for F_2, F_3: 7.6.11(1)	$7.6.11^-$	DNE 7.6.10(4)

Operation	Definition							
Perspectivity	$M \diagdown M' \diagup M''$				Fig. 7.9	Ex. 7.31(a)		
Point-pair completion	G^{ξ}	7.7.2, 7.7.3	7.7.2⁺		7.7.4(3)$^\lambda$; Fig. 7.14	7.7.4(2)$^\lambda$; Ex. 7.31(a)	DNE 7.7.4(1)	DNE 7.7.4(1); 7.7.4(4)
Projectivity	$M_1 \diagdown M_2 \diagup M_3$	7.5.1	7.5.1⁻		7.5.1	Ex. 7.31(a)$^\lambda$	Ex. 7.31(a)$^\lambda$	DNE*
Quotient	$M = N - 1 \to N/I$	7.4.7(1)	7.4.7⁻	7.6.2(2), 7.6.7(4); Ex. 7.21(b)	DNE*	DNE*	DNE*	DNE*
$\mathcal{C}$-quotient	$M = N - 1 \to N/I,\ N \in \mathcal{C}$		7.4.8⁻		DNE*; 7.4.8⁻; Fig. 7.3	DNE*; 7.4.8(2); Ex. 7.17, 7.39	DNE*; 7.4.8(1); Ex. 7.17, 7.39	DNE*
elementary	$(M +_{\mathcal{M}} p)/p$ or $T_{\mathcal{M}}(M)$		7.4.7⁻	Ex. 7.21(a)	DNE 7.4.8⁻	DNE; Ex. 7.6	DNE; Ex. 7.6	DNE Ex. 7.6
nullity-k	$M = N - 1 \to N/I$		7.4.7⁻; Ex. 7.51(b)	Ex. 7.20	DNE*; Ex. 7.51(b)$^\lambda$	DNE*	DNE*	DNE*; DNE*
Rank-k completion	$G^{c(k)}$				Ex. 7.51(b)$^\lambda$			7.7.4(4); Ex. 7.51(b)
Restriction (see Submatroid)								
Reverse operation	Rev(O)	Table 7.2; 7.6.8; Ex. 7.18		Table 7.2				
Self-dual operation								
Series connection (or decomposition)	$S(M_1, M_2)$	7.6.8	7.6.8⁻	7.6.7$_s$, 7.6.8⁺; Ex. 7.39; Fig. 7.12	7.6.10(2)$^\lambda$; Ex. 7.39	7.6.10(1); Ex. 7.39	7.6.10(3); Ex. 7.39	7.6.10(4)
Series-parallel decomposition			7.6.9			Ex. 7.43	7.6.9; Ex. 7.43	
Simplification	$\bar{M}$	7.4.3⁺; 7.3.5	7.4.3⁺	7.4.13(3)	7.4.3⁺	7.4.3⁺	7.4.3⁺	7.4.3⁺
Subgeometry (see Submatroid)								
Submatroid	$M(E') = M(E) - (E - E')$	7.3.1	7.3.1⁻	7.3.1⁻, 7.6.2(7), 7.6.7(4), 7.6.7(5), 7.6.13(1), Ex. 7.10, Ex. 7.5	7.3.1⁻; Fig. 7.1	7.3.1⁻	7.3.1⁻; Fig. 7.1	7.3.1⁻; Fig. 7.1
spanning			7.3.1⁻					

(Continued)

TABLE 7.3. (*Continued*)

Construction	Notation	Definition	Cryptomorphisms	Properties	Affine const.	Vector const.	Graphic const.	Transversal const.
3-sum	$S_{L_3}(G,H)$	$7.6.11^-$						
Truncated lift	$TL(M)$	$7.5.1^-$	$7.5.1^-$		Fig. 7.7			DNE Fig. 7.2c
Truncation	$T(M)$	$7.4.9^-$	7.4.9	7.5.5, 7.5.6	Fig. 7.1(a)$^{\lambda}$	$7.4.10^{\lambda}$	DNE $7.4.9^+$	DNE*
principal	$T_F(M)$	$7.4.9^-$	7.4.9	$7.4.9^+$	$7.4.10^{\lambda}$	$7.4.10^{\lambda}$	DNE*	DNE*
complete principal	$\bar{T}(M)$	$7.4.9^+$			$7.4.10^{\lambda}$	$7.4.10^{\lambda}$	DNE*	DNE*
in special position (see Dilworth truncation, Line lower completion)	G^{dil}	Ex. 7.54(c)				Ex. 7.54		
2-sum	$S_2(M_1,M_2)$	$7.6.8^-$		Ex. 7.51				
Union, half-dual	$M_1 \vee M_2^*$	$7.6.13^-$		7.6.13	$7.6.14(2)^{\lambda}$	$7.6.14(1)^{\lambda}$	DNE 7.6.14	7.6.14(3)
matroid	$M \vee M'$	$7.6.12^-$	7.6.12	Ex. 7.51	Fig. 7.13		$7.6.14^-$	7.6.14(3)
decomposition, matroid		$7.6.14^-$						
dual	$M_1 \wedge M_2$	$7.6.13^-$	Ex. 7.47	Ex. 7.47	DNE Ex. 7.47(d)	Ex. 7.47(d)$^{\lambda}$	DNE Ex. 7.47(d)	DNE Ex. 7.47(d)
generalized matroid	$M_1(E_1) \vee M_2(E_2)$	7.6.13(2)		Ex. 7.47	$7.6.14(2)^{\lambda}$	$7.6.14(1)^{\lambda}$	DNE*	7.6.14(3)
Weak-map image	$M \geq M'$	Ex. 7.12	7.4.7(2)	Ex. 7.12	DNE*	DNE*	DNE*	DNE*
rank-preserving	$M \geq M'$	Ex. 7.12	7.4.7(2)		DNE $7.4.7^+$	DNE $7.4.7^+$	$7.4.7^+$	DNE $7.4.7^+$

precede Proposition 7.4.7. DNE means that the appropriate construction does not exist in general, and the cited reference either proves this or explains under what circumstances it does exist. When a vector or affine construction needs a (transcendental) extension field, a lambda (λ) is appended.

Finally, an asterisk means that a construction exists because a more general one, also listed, exists, and DNE* means that a more specific construction, also listed, fails to exist.

EXERCISES

7.1. Prove that $p_j \notin B$ is a loop if and only if $\mathbf{c}_j$ is the zero column in $A(M; B)$.

7.2. Show that $p_i \in B$ is an isthmus if and only if $\mathbf{r}_i = \mathbf{0}$ in the basic-circuit incidence matrix $A(M; B)$.

7.3. Prove that in a transversal matroid presented by $R \subseteq E \times T, (|T| = r(M))$, the set of isthmuses is the maximal set $E' \subseteq E$ such that E' is matched into a subset $T' \subseteq T$, and $R(E - E') = T - T'$.

7.4. Characterize, for each cryptomorphism, when a matroid is a Boolean algebra.

7.5. Show that if M contains a basis B disjoint from E', then the basic-circuit incidence matrix $A'[M(E' \cup B), B]$ is the basic-circuit incidence matrix $A(M, B)$ restricted to those columns indexed by E'.

7.6. (a) Prove, using Proposition 7.3.2, that the triangular prism $M - p$ in Figure 7.1a cannot be extended by placing the point p on only two of the dotted lines (i.e., show that there is no modular cut of a triangular prism that contains only two of the vertical lines, but that there are modular cuts containing none, one, or three of them).

(b) Use the theory of modular cuts to show that the "Vamos cube" V is not representable (i.e., cannot be a subset of a projective space). In particular, let V be the rank-4 matroid on the points $\{a, b, c, d, a', b', c', d'\}$ with four-point circuits (planes) $\{aba'b', bcb'c', cdc'd', ada'd', aca'c'\}$. Then the only modular cut that contains the two lines aa' and cc' is $\mathscr{K}$, the set of all flats, even though aa' and cc' do not form a modular pair in V.

7.7. Show that G is a geometry if and only if $G - p$ is a geometry and the modular cut $\mathscr{M}$ that extends $G - p$ contains no atom.

7.8. Describe single-element extensions in terms of other cryptomorphisms by reinterpreting Proposition 7.3.2, where $M = N - p$. For example, in terms of the rank function, subsets A such that $r_N(A \cup p) = r_N(A)$ are those whose closures form a modular cut. The subset of hyperplanes of M on which p depends, $\mathscr{H}_L = \{H: H \cup p \text{ is a hyperplane of } N\}$ forms a *linear subclass*: If H and $H' \in \mathscr{H}_L$, with $r_M(H \cap H') = r_M(H) - 1$, then $H'' \in \mathscr{H}_L$ for all $H'' \supseteq H \cap H'$. In terms of bases (White 1974), show that, characteristically, for all bases B, B' of M, if $(B \cup p) - q$ is a basis of N, there is a point $q' \in B'$ such that $(B' \cup p) - q'$ and $(B \cup q') - q$ are bases of N.

*7.9. Which other classes are closed under principal extensions? For example, show that the class of *oriented matroids* (Bland 1978) and the class of *gammoids* (Ingleton and Piff 1973) are closed under this operation.

7.10. Prove that any submatroid $M(E')$ with $r(E) - r(E') = k$ can be constructed by a sequence of deletions by points $p_1, \ldots, p_m, p_{m+1}, \ldots, p_{m+k}$, where $\{p_{m+1}, \ldots, p_{m+k}\}$ are isthmuses of the spanning submatroid $M - (\{p_1, \ldots, p_m\})$.

7.11. (a) Prove that principal extension on F and deletion of q commute: $(M +_F p) - q = (M - q) +_{F-q} p$ if and only if $r(F - q) = r(F)$.

   (b) Prove that if F is a modular flat of M and we make the extension $F' = M(F) +_{\mathscr{M}} p$, then there is a unique extension of M, $M' = M +_{\mathscr{M}'} p$, such that $M'(F \cup p) = F'$. Prove, further, that F' is modular in M', and $\mathscr{M}' = \{x: x \geq y$ for some $y \in \mathscr{M}\}$. [Recall that a flat F is modular if $r(F) + r(F') = r(F \cup F') + r(F \cap F')$ for all flats F' of M.]

7.12. (a) Prove that the intersection of two modular cuts $\mathscr{M}_1$ and $\mathscr{M}_2$ is a modular cut.

   (b) Show that $\mathscr{M}_1 \subseteq \mathscr{M}_2$ if and only if $M_1 = M +_{\mathscr{M}_1} p \geq M +_{\mathscr{M}_2} p = M_2$ in the *weak* ($-map$) *order* (i.e., where every independent set of M_2 is independent in M_1). M_1 is termed *freer* than M_2.

   (c) Deduce from (a) and (b) that the weak order is a lattice ordering $\bar{L}$ on all single-element extensions of M, with $M + p$ as the top element.

   (d) Show that $M +_F p$ is the freest single-element extension of M by p, with $r(F \cup p) = r(F)$. Further, show that $\bar{L}$ has a supremum subsemilattice isomorphic to the geometric lattice L of M (which is a sublattice if and only if $\bar{L} = L$, or, equivalently, if and only if M is modular).

   (e) Let M_1 and M_2 be two single-element extensions of M. Show, in general, that in $\bar{L}$ neither $M_1 \wedge M_2$ nor $M_1 \vee M_2$ is representable in any of our classes even when M_1 and M_2 both are.

   (f) Assume that two extensions M_1 and M_2 have *compatible representations* in that each is represented in such a way that $M_1 - p$ and $M_2 - p$ represent M with the same graph, matrix, affine point set, or transversal presentation. Show that $M_1 \wedge M_2$ is not, in general, representable in any of our classes, and $M_1 \vee M_2$ is not, in general, graphical or vectorial over the same field.

   Show, however, that $\bar{M} = M_1 \vee M_2$ has a transversal representation given by $R = R_1 \cup R_2$ [i.e., $R(p) = R_1(p) \cup R_2(p)$ and $R_1(q) = R_2(q) = R(q)$ for all $q \neq p$]. Further, over sufficiently large fields, $\bar{M}$ is vectorial where p is represented by the vector $\lambda_1 \mathbf{v}_1 + \lambda_2 \mathbf{v}_2$, where λ_i is a transcendental and $\mathbf{v}_i$ represents p in M_i. $\bar{M}$ has the synthetic affine construction of putting a point $\bar{p}$ freely on the line joining p_1 and p_2, where the point p_i represents p in the affine representation M_i.

*7.13. (Cheung 1974a). An interesting problem for extensions is to determine when two extensions are *compatible*, so that given extensions $M_1 = M +_{\mathcal{M}_1} p_1$ and $M_2 = M +_{\mathcal{M}_2} p_2$ there exists a matroid M' on the set $E \cup \{p_1, p_2\}$ with $M' - p_1 = M_2$ and $M' - p_2 = M_1$.

(a) Define the modular cut $\mathcal{M}_{12}$ as the smallest cut of $M_1 = M +_{\mathcal{M}_1} p_1$ that contains all closed sets of the form $\{\mathrm{cl}_{M_1}(A): A \in \mathcal{M}_2\}$. Define $\mathcal{M}_{21}$ similarly, and call $\mathcal{M}_1$ and $\mathcal{M}_2$ *compatible* if $M_{12} = (M +_{\mathcal{M}_1} p_1) +_{\mathcal{M}_{12}} p_2 = (M +_{\mathcal{M}_2} p_2) +_{\mathcal{M}_{21}} p_1$.

Prove that there exists a matroid M', with $M' - p_i = M_i$, if and only if $\mathcal{M}_1$ and $\mathcal{M}_2$ are compatible, in which case M_{12} is the freest such matroid.

(b) Show that if $E = \{a, b, c, d, e, f\}$ and $M(E)$ is the uniform matroid $U_{3,6}$ whose bases consist of all three-element subsets of E, then $\mathcal{M}_1 = \{ab, cd, E\}$ and $\mathcal{M}_2 = \{ab, cd, ef, E\}$ are not compatible.

(c) Prove that $\mathcal{M}_1$ and $\mathcal{M}_2$ are compatible if and only if $\mathcal{M}_1 \cap \mathcal{M}_2$ is compatible with both $\mathcal{M}_1$ and $\mathcal{M}_2$.

(d) Prove that any two principal modular cuts are compatible.

7.14. Prove as a corollary to Proposition 7.3.6 that G is a binary coaffine geometry if and only if it is a Boolean algebra.

7.15. (a) Show that the transversal matroid M in Figure 7.1c can be built up from a Boolean algebra by a sequence of principal extensions, but that the transversal matroid $M - p$ has no point that is added to its deletion by principal extension.

*(b) Compare the class $\mathcal{T} \cap \mathcal{T}^*$ of transversal matroids whose duals are transversal and the class $\mathcal{T} \cap \mathcal{P}$, where $\mathcal{P}$ is the class of matroids that can be built up by principal extensions starting with a Boolean algebra.

7.16. Prove that $M = M' +_F p_j$ if and only if both $M - p_j = M'$, and whenever B is a basis for M' with $r(F \cap B) = r(F)$, then in the basic-circuit incidence matrix $A(M; B)$, $a_{ij} = 1$ if and only if $p_i \in F \cap B$.

7.17. Prove Proposition 7.4.8, and show that M' is a graphical quotient of M, where M and M' are represented by the graphs in Figure 7.16. Note that the apparent contradiction with Proposition 7.4.8, part 1, comes from the fact that graphical representation of matroids is unique only up to 2-isomorphism, not isomorphism (see Chapter 6). Construct a similar example for vector quotients. Does such a problem result for 3-connected graphs or for matroids vectorial over F_2 or F_3?

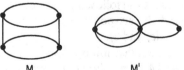

M   M'   FIGURE 7.16.

**7.18. Give examples of self-dual operations and, more generally, of families of operations $\mathcal{O}$ normalized by the dual operator: If $O \in \mathcal{O}$, then $O^* \in \mathcal{O}$.

7.19. If M is a matroid that is vectorial over F, then the row-space of its representing matrix is a (linear) code over F. Interpret duality, deletion, and contraction in the language of coding theory.

7.20. (a) (Higgs 1968; Kennedy 1975) If $M'(E)$ is a nullity-k quotient of $M(E)$, then the ith Higgs lift $M^i(E)$ $(0 \le i \le k)$ of $M \mapsto M'$ is defined in terms of its independent sets:

$$\mathscr{I}(M^i) = \{I: I \text{ is independent in } M, \text{ and the nullity} \\ \text{of } I \text{ in } M' \text{ is at most } i\}.$$

Prove that $M^k = M$, $M^0 = M'$, M^i is an elementary quotient of M^{i+1}, and M^i is the freest matroid of rank $r(M') + i$ that is both a quotient of M and a lift of M'.

(b) Let $M'(E)$ be a matroid of nullity k on the set E. Show that there is matroid $M_i'(E)$ whose independent sets are those subsets of nullity at most i in M'. Show, in fact, that M_i' is the Higgs lift of M' iterated i times $(i \le k)$. Thus, for a quotient M of M', the ith Higgs lift $M^i(E)$ of $M \mapsto M'$ is the "intersection" of (the independent-set families of) M_i' and M.

** 7.21. (a) Explore the lattice structure of the elementary-quotient lattice $L_g(M)$ (see Figure 7.4). For example, what is $l(L)$, the size of its largest chain? What is the size of its largest antichain? How do elementary operations on M affect $L_g(M)$? For example, show that $L_g(M/p)$ is isomorphic to a supremum subsemilattice of $L_g(M)$ but that $L_g(M - p)$ may be bigger than $L_g(M)$. For example, $4 = l\{L_g[AG(2,3)]\} > l\{L_g[PG(2,3)] \simeq L[PG(2,3)] = 3$.

(b) For a hereditary class $\mathscr{C}$ (such as matroids coordinatizable over F) and a matroid $M \in \mathscr{C}$, what is the structure of $\mathscr{Q}_{\mathscr{C}}(M) = \{Q: Q$ is an elementary $\mathscr{C}$-quotient of $M\}$?

7.22. Prove the excluded-minor characterizations for the first 10 hereditary classes of Table 7.1. For example, to show for $\mathscr{C}_{10}$ that $\mathscr{E}(\mathscr{L})$ is $\{C_4, C_3 \oplus C_3\}$, argue on the maximal size of the union of two three-element circuits in any geometry that contains no four-element circuit. Prove the characterizations for $\mathscr{E}(\mathscr{C}_{19})$, $\mathscr{E}(\mathscr{C}_{20})$, and using $\mathscr{E}(\mathscr{C}_{16})$, for $\mathscr{E}(\mathscr{C}_{17})$ and $\mathscr{E}(\mathscr{C}_{18})$.

7.23. Origami matroids $\bar{\mathcal{O}}$ (Kahn and Kung 1982) are transversal matroids with a presentation $R \subseteq E \times T$ such that for all points $p \in E$, $R(p) = t_i$ or $\{t_i, t_{i+1}\}$ for some i.

(a) Prove that $\mathscr{E}(\bar{\mathcal{O}})$ includes the following geometries: $M - p$ of Figure 7.1, $\mathscr{U}_{3,5}$, $M(K_4)$, and the six-point rank-3 geometry whose three-point lines are abc, cde, and efa.

*(b) Find the class of all excluded minors for $\bar{\mathcal{O}}$.

*7.24. Generalize Proposition 7.4.14 by showing how to find the excluded-minor class $\bar{\mathscr{E}}$ of $\bar{\mathscr{C}}$, where $\bar{\mathscr{C}}$ is the geometric hereditary class derived from the hereditary class $\mathscr{C}$.

7.25. (a) A perfect matroid design (PMD) is a matroid all of whose flats of rank r have the same size k_r. Prove that if A is a closed set of a PMD M, the minor $M(A)/B$ is also a PMD.

   (b) Show that in a PMD, the number of flats of rank r that contain a given flat x of rank s depends only on the parameters r and s (as well as the sequence $\{k_i\}$). We call matroids that obey this property *homogeneous*. Show that $M(K_n)$ is homogeneous but not a PMD for $n \geq 4$. Use the scum theorem to prove that if a matroid M is a submatroid of a homogeneous matroid, then it is a spanning submatroid of a homogeneous matroid.

**(c) Can Proposition 7.3.4 be generalized to show that every matroid is a submatroid of a homogeneous matroid?

7.26. (a) Show directly that transversal matroids form the smallest class containing loops and closed under principal lifts.

   (b) Let $M(E)$ be represented by the matrix $N = [I_n A]$ over F, where I_n is indexed by the basis B, and the first column of A is indexed by p_{n+1}. For any subset C such that $p_{n+1} \in C \subseteq E - B$, show that the principal lift $L_C(M)$ is represented over $F(\{\lambda_i : p_i \in C - p_{n+1}\})$ by the matrix $N' = [I_{n+1} A']$, where

$$
a'_{ij} = \begin{cases} a_{ij} & \text{if } p_j \notin C, i \neq n+1, \\ 0 & \text{if } p_j \notin C, i = n+1, \\ a_{ij} - a_{i,n+1} \cdot \lambda_j & \text{if } p_j \in C, i \neq n+1, \\ \lambda_j & \text{if } p_j \in C, i = n+1. \end{cases}
$$

   (c) Show that the principal lift $L_C(M)$ (where C is the complement of a spanning set) is characterized by its bases

$$
\mathscr{B} = \{B \cup p : B \text{ is a basis of } M, p \in C - B\}.
$$

   For any basis $B \cup p$ with $p \in C$ and $B \cap C = \varnothing$, show that for the basic-circuit incidence matrix $A' = A[L_C(M); B \cup p]$ and all $s \in E - (C \cup B)$, $A'_{ps} = 0$ and $A'_{qs} = A_{qs}$, where $A = A(M; B)$. For a point $s \in C - p$, both $A'_{ps} = 1$ and $A'_{qs} = 1$ if and only if either $A_{qs} = 1$ or $A_{ps} = 1$.

*7.27. Describe the operations of Section 7.4 in other matroid classes. For example, represent the principal lift for oriented matroids and contraction for gammoids.

7.28. A matroid is a *cube* if it is isomorphic to a geometry of rank 4 on the eight points $\{a, b, c, d, a', b', c', d'\}$ that includes, among its four-point planes $\mathscr{P}$, $aba'b'$, $bcb'c'$, $cdc'd'$, and $ada'd'$. Examples of cubes include the *real cube* (with 8 additional four-point planes), $AG(3, 2)$ (with 10 additional four-point planes), and the Vamos cube (see Exercise 7.6).

*(a) Prove that any dependent proper flat of a cube must be a four-point plane that intersects each plane of $\mathscr{P}$ in two points. Show that the dual of a cube is a cube, and count the number of cube isomorphism classes. How many are isomorphic to their duals? Which are coordinatizable over the reals?

(b) Show that a planar matroid is a one-point contraction of some cube if and only if it is isomorphic to one of the first 10 isomorphism classes of Figure 7.17. In each case, construct the modular ideal for

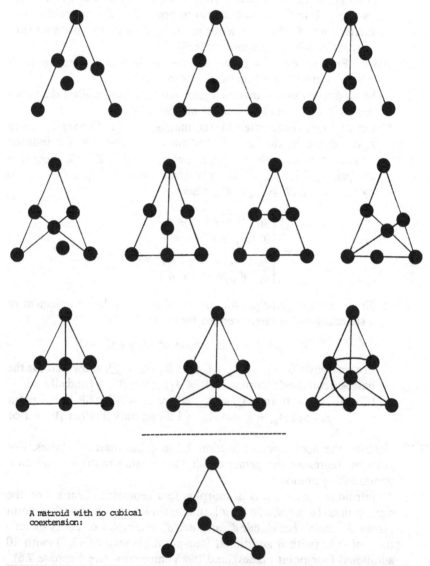

A matroid with no cubical coextension:

FIGURE 7.17. Matroids with cubical coextensions, and a matroid with no cubical coextension.

the coextension and prove via modular ideals that no cubic coextension exists for any other planar matroid, in particular for the final example.

**7.29. A non-Desarguesian projective plane such as $P' = PG'(2, 9)$ (with its Desarguesian counterpart P) gives an example of a pair of matroids that are not isomorphic but whose one-point contractions are isomorphic in pairs. Similarly, $P'* \neq P*$, but there is a bijection $f: P'* \to P*$ such that $P'* - p \simeq P* - f(p)$ for all p.

Find an example of two nonisomorphic matroids $G(E)$ and $G'(E)$ such that $G - p \simeq G' - p$ and $G/p \simeq G'/p$ for all $p \in E$.

**7.30. Explore hereditary and hereditarily recognizable properties. A property $\mathscr{P}$ is *hereditary* if the matroids that have this property form a hereditary class (i.e., if M has $\mathscr{P}$, then so do $M - p$ and M/p). $\mathscr{P}$ is *hereditarily recognizable* if, whenever M has $\mathscr{P}$, $M - p \simeq M' - p'$, and $M/p \simeq M'/p'$; then M' has $\mathscr{P}$.

7.31. (a) Show that if $|F|$ is sufficiently greater than $|M|$, and M and M' are equicardinal spanning subsets of $PG(n, F)$, then M is projectively equivalent to M' where each perspectivity is linear.

(b) Let M_1 and M_2 be two matroids on the set $E = abcd$, where, in M_1, a is a loop, bc is a double point, and d is an isthmus, whereas in M_2, ab and cd are double points. Show that for any $A \subseteq B \subseteq E$, $|r_{M_1}(B) - r_{M_1}(A) - r_{M_2}(A) + r_{M_2}(B)| \leq 1$, whereas $d(M_1, M_2) = \bar{r}(M_1, M_2) = 2$.

*(c) Prove that $\bar{r}(M_1, M_2) = d(M_1, M_2)$ for all matroids, or give a counterexample and find an invariant that better approximates $d(M_1, M_2)$.

7.32. Let $M' = M - p$ be the triangular prism of Figure 7.1.

(a) Construct the principal erection $E_F(M' +_F q)$, where F is, respectively, $\varnothing$, a, ad, $abde$, and $abcdef$.

(b) Show that M' has no principal erection with respect to any flat.

7.33. Which cubes are erectible? In particular, show that the Vamos cube V (Exercise 7.6) is not erectible, but when the complementary diagonal plane $bdb'd'$ is added as a sixth plane, the resulting cube V', a weak-map image of V, is erectible.

*7.34. Explore the class of $\mathscr{C}$-erectible matroids: matroids M with erections M' such that M, M', and $M' + p$ are all in $\mathscr{C}$. (When $\mathscr{C}$ is the class of matroids representable over F, the $\mathscr{C}$-erectible matroids are those with coaffine erections.)

7.35. (a) (Crapo 1971) Prove that M is erectible if and only if M^* has a proper modular cut that includes all dependent hyperplanes.

(b) For a linear subclass $\mathscr{H}$ (see Exercise 7.8), define a *connected component* $\mathscr{K}$ of $\mathscr{H}$ to be a class in the equivalence relation among hyperplanes in $\mathscr{H}$ generated by the relation $H \sim H'$ if $r(H \cap H') = r(H) - 1$. Show that M is erectible if and only if there is a linear subclass $\mathscr{H}$ of M^* that contains all dependent hyperplanes such that for each connected component $\mathscr{K}$ of $\mathscr{H}$, $\cap\{H: H \in \mathscr{K}\} \neq \varnothing$.

(c) Show that the only erectible graphical geometry is a circuit, C_n, and hence for any graphical matroid M, any erection is (trivially) graphical.

7.36. (a) (Las Vergnas 1975) Let N and N' be two erections of M. If N covers N' in the weak order, then there exists a basis B of N such that the independent sets of N' are those independent sets I of N that are not bases contained in $c(B)$, where $c(B)$ is the $(|B| - 1)$-closure of B.

(b) Show that the converse of part (a) is false. Let $M = U_{3,7}(\{a, b, c, d, e, f, g\})$, and let N_1, N_2, and N_3 be three erections of M, where N_1 has the dependent plane $abcdef$, N_2 has dependent planes $abcd$, $cdef$, and $abef$, and N_3 has dependent planes $abcd$ and $abef$. Then, clearly, N_3 does not cover N_1 in the lattice $\mathscr{E}(M)$, but N_1 can be obtained from N_3 by the 3-closure $abcdef$ (in N_3) of the basis $abce$.

7.37. Show that if we brace the Vamos cube by putting a point, for each of the five planes $xyx'y'$, on the intersection of lines xy' and yx', then in the resulting matroid, V', four of the five added points are coplanar. In any case, show that $\bar{E}[T(V')] = V'$, and V' has V as a subgeometry.

**7.38. (a) Complete the details of Proposition 7.5.4, especially part 1, and find ways, for a given matroid M, to construct a *minimal* bracing $B(M)$ that will nevertheless guarantee that the bracing is the total free erection of its truncation to the plane.

(b) Minimally brace the matroids in Figure 7.8 (only partial bracings are given).

**(c) For a matroid $M(E)$, let $\mathscr{PE}(M)$ be the poset of essential flats of M (ordered by containment). (Recall, an essential flat is one which is erectible as a subgeometry.) What posets are possible? Show, in particular, that any finite lattice is possible. [This latter result is essentially due to Dilworth, who showed that any lattice L can be realized within a geometric lattice (see Lemma 7.7.1). When a principal extension is made on each of the flats corresponding to L in the Dilworth completion, each becomes essential, and we can check that the other flats constructed in Lemma 7.7.1 are never erectible.]

**(d) Find a cryptomorphic description of matroids for the family of pairs $\{(E_i, r_i): E_i \subseteq E$ is an essential flat of M of rank $r_i\}$.

Special exercise (Kennedy, Crapo) Make up your own "dirty theorem" in matroid construction theory. You may use such terms as "(free) erection," "rigid" (for inerectible), "scum theorem," "excluded minors," and so forth.

*7.39. A useful feature of matroid theory is that it suggests analogies between different areas of combinatorics. Thus, a construction theorem in graph theory, when suitably interpreted via matroids for linear geometries, gives an analogous linear theorem. We illustrate this idea with a theorem of F. Jaeger based on a graph-theoretical result of Hajós. The reader should check the details.

(a) (Hajós 1961) Every loopless graph whose chromatic number is at least q contains a subgraph that can be obtained from K_q by repeated application of the following two operations:

(1) vertex identification (i.e., the graphical quotient of Proposition 7.4.8, part 1)

(2) Hajós union [i.e., series connection of graphs (Proposition 7.6.10, part 3)].

Conversely, K_q has chromatic number q, and both operations (1) and (2) stay within the class of non-$(q - 1)$-colorable graphs.

(b) (Jaeger 1981) Recall that the critical exponent of a vectorial loopless matroid M is the degree of the smallest extension field of F over which M is affine.

Every loopless linear matroid with critical exponent at least k contains a submatroid that can be obtained from $PG(k - 1, F)$ by repeated application of the following two operations:

(1) Linear quotient (Proposition 7.4.8, part 2)

(2) Series connection (Proposition 7.6.10, part 1).

Conversely, the critical exponent of $PG(k - 1, F)$ is k, and both operations (1) and (2) preserve the class of vector matroids with critical exponent at least k.

7.40. Prove (by induction) that a matroid M on at least three points is connected if and only if every triple $\{p_1, p_2, p_3\} \subseteq E$ is contained in a bond or a circuit of M.

7.41. Describe the operations of Section 6 in terms of basic-circuit incidence matrices.

*7.42. (Seymour 1980) Prove that every connected binary matroid with no points in series has a connected hyperplane.

*7.43. (Edmonds 1970; De Sousa and Welsh 1972; Brylawski 1975) Let M be a transversal matroid. Show that the following are equivalent:

(a) M is binary.

(b) M is a series-parallel network. (Conversely, every series-parallel network is a minor of a transversal matroid.)

(c) M has a presentation that, as a bipartite graph, is a forest.

*7.44. Use Polya theory and Proposition 7.6.8, part 2, to count isomorphism classes of series-parallel networks.

*7.45. Develop a theory for the generalized series connection, and find applications.

7.46. (a) Show that the generalized parallel connection $P_{C_3}[F_7, P(C_3, C_3)]$ is representable only over fields of characteristic 2, but that the modular sum $S_{C_3}[F_7, P(C_3, C_3)] \simeq P(C_3, C_3, C_3)$ is representable over every field.

(b) (J. Kahn) Find a matroid M such that $M - p = M(K_4) \oplus M(K_4)$, but M has the Fano plane F_7 as a minor as well as the "non-Fano plane" F_7' (a contraction of the real cube). Conclude that there is a nonvectorial matroid M with a one-point deletion which is graphical.

7.47. (a) Prove that the spanning sets of $M_1 \wedge M_2$ are given by $\{S_1 \cap S_2 : S_i$ spans $M_i\}$. Develop other cryptomorphisms for dual union and for the half-dual union.

(b) Given $M_1(E_1)$ and $M_2(E_2)$, with $E_1 \cap E_2 = B$, a common basis, define

$$(M_1 \cup M_2)(E_1 \cup E_2) := [(M_1^* \vee M_2^*)(E_1 \cup E_2)]^*$$

(Thus, $M_1 \cup M_2$ is the dual of the generalized matroid union defined in Proposition 7.6.13, part 2.)

Prove that $M_1 \cup M_2$ gives a freest compatible extension of B to M_1 and M_2.

(c) Show that if M_1 and M_2 are vectorial and are represented by $[I_n N_1]$ and $[I_n N_2]$, respectively, with I_n indexed by the common basis B, then $[I_n N_1^{\lambda} N_2]$ represents $M_1 \cup M_2$, where $^{\lambda} N_2$ is the submatrix N_2 with row r_i multiplied by the transcendental λ_i.

(d) Show that matroid dual union can create loops and hence is not an affine operation. Similarly, the dual union of two circuits is not graphic. It has, however, a transcendental linear construction, because we can combine Proposition 7.6.14, part 2, with Proposition 7.4.2, part 1. Finally, show that $M_1 = C_6$ and $M_2 = C_2 \oplus C_2 \oplus C_2$ are both transversal, but that $M_1 \wedge M_2 = L_3^{(2)}$ is not.

7.48. (a) Show that the following generalization of transversal matroids gives a matroid for all M and R. $M(E)$ is *induced* from $M'(E')$ via the bipartite graph $R \subseteq E \times E'$ if the independent sets of M are those that are matched by R into independent subsets of M'. The class of *induced matroids* $\mathscr{I}$ is the smallest class that contains Boolean algebras and is closed under inducing.

(b) Show that $\mathscr{I}$ is closed under all operations discussed earlier that are valid for the class $\mathscr{T}$ of transversal matroids by imitating transversal construction. In particular, show that $\mathscr{I}$ is closed under deletion and matroid union.

(c) Show that $\mathscr{I}$ is a subclass of the class of gammoids (see Proposition 7.4.2, part 3).

*(d) Show that if $M(E)$ is transversal, then $M/p \in \mathscr{I}$ for all $p \in E$. However, $\mathscr{I}$ is not, in general, closed under contraction. Find a gammoid not in $\mathscr{I}$.

*7.49. (Tutte 1966) Call a matroid M 2-separable if $|M| > 3$ and M is either separable or a nontrivial 2-sum [i.e., $M = M_1 \oplus M_2$, or $M = S_2(M_1, M_2)$, where $|M_i| < |M|$]. The *wheel* W_n ($n \geq 2$) is the matroid represented by a graph with $n + 1$ vertices that consists of an n-circuit C, each of whose vertices is connected by an edge to a central vertex v. The *whirl* $\bar{W}_n$ is a (nonbinary) matroid on the same set and with the same independent sets, except that the n-element set corresponding to the edges of C is now independent. ($W_3 = K_4$, $\bar{W}_2 = U_{2,4}$, etc.)

Prove that if M is a matroid that is not 2-separable, then M is a wheel or a whirl, or there is a point $p \in M$ such that $M - p$ or M/p is not 2-separable.

*7.50. (Seymour 1980) Define a *splitter* for hereditary class $\mathcal{H}$ to be a matroid $M \in \mathcal{H}$ such that if M is a minor of $M' \in \mathcal{H}$, then $M = M'$, or M' is 2-separable.

Prove that if M is neither a wheel nor a whirl, then it is a splitter if M and M^* are both geometries, and if the only single-element extension or coextension of M in $\mathcal{H}$ is trivial (by a loop, isthmus, parallel extension, or series extension).

*7.51. (a) Let G and H be two geometries on the same point set E. Now restrict $(G \oplus H)^c$ to the subset $E_G \times E_H$ by restricting to only one point for each line spanned by a point p in G and a point q in H. Show that the resulting matroid contains the union $(G \vee H)(E)$ as a subgeometry ("on the diagonal"). Describe the *complete union* $(G \oplus H)^c(E \times E)$ further. Show that it has transcendental linear and affine constructions by combining direct sum and Proposition 7.7.4. Further, if M_i is presented by R_i, then complete union is transversal and is presented by $R \subseteq (E \times E) \times (T_1 \cup T_2)$, where $R[(p,q)] = R_1(p) \cup R_2(q)$.

(b) When $L(G)$ is lower-truncated at level k and Dilworth-completed, the resulting geometry $G^{c(k)}$, the *rank-k completion of G*, results. Show that when $r(G) \geq 2k + 2$, $G^{c(k)}$ consists of G with each flat of rank k principally extended by k points, and then with the original points of G deleted. Give other cryptomorphic descriptions of $G^{c(k)}$ and of $G = G^{c(2)}$ in particular. Show that $G^{c(k)}$ is a subgeometry of $G^{c^k} := (\ldots (G^c)^c \ldots)^c$ and as such is linear, affine, and transversal.

*7.52. (a) Prove that closure and 2-closure coincide in G^d if and only if G is a direct sum of rank-3 matroids.

(b) Find cryptomorphic descriptions of G^d other than the ones in Proposition 7.7.5.

*7.53. Prove the two-dimensional Desargues theorem for $PG(2, F)$ by matroid theory. In particular, embed $PG(2, F)$ in $PG(3, F)$ and show that if only 9 of the 10 collinearities in Desargues's theorem hold, an appropriate coextension fails.

7.54. (a) Let G be a geometry vectorial over F that contains the six vectors

$$
\begin{array}{cccccc}
1 & 2 & 3 & 4 & 5 & 6
\end{array}
$$
$$
\begin{bmatrix}
 & & & & 1 & 1 \\
 & I & & & 1 & 0 \\
 & & & & 1 & 1 \\
 & & & & 1 & 0
\end{bmatrix}
$$

Let x be algebraic over F of degree at least 4, and consider the hyperplane $H = (1, x, x^2, x^3)^{\perp}$ in $P = PG[3, F(x)]$. Show that two distinct lines of $PG(3, F) \subseteq PG[3, F(x)]$ intersect H in distinct points. However, show that if $v_{ij} = [cl_p(v_i, v_j)] \cap H$, then v_{12}, v_{34}, and v_{56} are represented by

$$\begin{bmatrix} 12 & 34 & 56 \\ -x & 0 & -x \\ 1 & 0 & 1 \\ 0 & -x & -x \\ 0 & 1 & 1 \end{bmatrix},$$

whereas the corresponding points are independent in G^d. Conclude that H is not in sufficiently general position to represent G^d and that, in fact, P^d cannot be represented over an extension field of degree 4.

**(b) For a given geometry G coordinatizable over F [and, in particular, for $PG(n, F)$], determine $c^d(G)$, the *lower-truncation critical exponent* of G, the minimal degree of an extension field in which G^d can be constructed. In particular, show that $c(n, F) := c^d[PG(n, F)] \leq 2n^2$. Does $c(n, F)$ depend on F?

**(c) For a geometry G, let $G' = G(E) + H$ be an extension of G by a modular hyperplane H [i.e., G' is on the groundset $E \cup H$, $G'(H)$ is a hyperplane H of G', $G'(E) = G(E)$, and H intersects all lines spanned by points in G]. Define $G^{d(H)} := G'(H)$, the *truncation of G in special position*. The construction $G \mapsto G^{d(H)}$ is an example of a *comap* and has been studied by Crapo (1968). Describe the structure of the isomorphism classes of $\{G^{d(H)}\}$ for fixed G and variable H.

*7.55. Find other characterizations for the kth *Dilworth truncation* $G^{d(k)}$ that is a matroid on the set $\{p_x : x \text{ is a flat of } G \text{ of rank } k\}$ whose independent sets $\{p_{x_1}, \ldots, p_{x_m}\}$ are those such that any nonempty s-element union of the respective flats $\{x_i\}$ has G-rank at least $s + k - 1$. Imitate Proposition 7.7.8, parts 3 and 4, to show that $G^{d(k)}$ has transcendental linear and affine constructions.

*7.56. (Mason 1977) Show that we can construct a geometry that has rank $r(G) - 1$ and contains, as subgeometries, all Dilworth truncations $\{G^{d(k)} : k \geq 1\}$. In particular, G^D has as points $E = \{p_x : x \text{ is a nonempty flat of } G\}$, and $A = \{p_{x_1}, \ldots, p_{x_m}\}$ is independent if, whenever $J \subseteq [1, m]$ with $|J| = s \neq 0$,

$$r_G(\cup\{x_j : j \in J\}) \geq s - 1 + \min[\{r_G(x_j) : j \in J\}].$$

**7.57. Let $A(K_n)$ be the affine representation of K_n. Thus, $A(K_5)$ is the (three-dimensional) Desargues configuration, and Exercise 7.53 shows that if a planar geometry can be embedded in a higher-dimensional space, it satisfies D_2: the two-dimensional Desargues theorem. Conversely, if there is no embedding, the Desargues axiom (guaranteeing the 10th collinearity in D_2) is independent of other axioms in planar affine geometry.

Similarly, we use $A(K_3)$ to supply a proof of the following: There is a one-dimensional model of ordered geometry (the interval $[0, 1]$) in which a point can be interpolated between any two distinct points, but, in general, points cannot be extrapolated from two points. Extrapolation follows from interpolation as a theorem, however, if the one-dimensional geometry can be embedded in a planar model.

Use $A(K_{n+3})$ to find higher-dimensional analogues of the foregoing. In other words, find a statement $\mathscr{A}_n$ such that $\mathscr{A}_n$ is independent of the axioms of n-dimensional (ordered) neutral geometry. However, $\mathscr{A}_n$ is a theorem if the geometry is embeddable in an $(n + 1)$-dimensional model.

*7.58. (a) Show that a subset E' of points of $(B_n)^d$ can be separated from its complement $E - E'$ by a (Euclidean) hyperplane H if and only if E' intersects any line of G^d in a ray of points. [Thus, as in the closure description of $(B_n)^d$ (see Exercise 7.52, part a), hyperplane separation in $(B_n)^d$ can be determined by hyperplane separation on lines.] Show that when G is a circuit C_n ($n \geq 5$), some subset of G^d consists entirely of rays but is not hyperplane-separable.

(b) Use this property of B_n to count the number of such hyperplane separations of $A(K_n)$. Show that this is equal to the number of acyclic orientations of K_n by showing that an orientation of the edges of K_n is acyclic if and only if all triangles are oriented acyclically, and then setting up an acyclic orientation of K_n so that any acyclic reorientation gives rise to a unique hyperplane separation of $A(K_n)$. (Here, two orientations are equivalent if one just reverses all the arrows of the other.)

(c) Use subgeometries of $M(K_n)$ to give a synthetic combinatorial proof of the theorem of Greene and Zaslavsky that the number of acyclic orientations of a graph G is the number of hyperplane separations in one of its Euclidean representations.

**7.59. Explore other Dilworth completions. For example, if $L^*(G)$ is the order dual of the geometric lattice of G (i.e., the lattice "turned upside down"), the Dilworth completion of $L^*(G)$ is called the *complete adjoint* of G. A related (but noncanonical) adjoint was studied by Crapo and Cheung (1974b).

7.60. Find a non-Desargues configuration in V^d, the Dilworth truncation of the Vamos cube, to illustrate Proposition 7.7.8, parts 3 and 4.

REFERENCES

Atiyah, M. F., and MacDonald, I. G. (1969). *Introduction to Commutative Algebra.* Addison-Wesley, Reading, Mass.

Blackburn, J. E., Crapo, H. H., and Higgs, D. A. (1973). A catalogue of combinatorial geometries. *Math. Comp.* **27**: 155–66, with loose microfiche supplement A12–G12.

Bland, R. G., and Las Vergnas, M. (1978). Orientability of matroids. *J. Combin. Theory Ser. B* **24**: 94–123.

Brown, T. J. (1974). Transversal theory and F-products. *J. Combin. Theory Ser. A* **17**: 290–8.

Brylawski, T. (1971). A combinatorial model for series-parallel networks. *Trans. Amer. Math. Soc.* **154**: 1–22.

   (1972). A decomposition for combinatorial geometries. *Trans. Amer. Math. Soc.* **171**: 235–82.

   (1975a). An affine representation for transversal geometrics. *Studies Appl. Math.* **54**: 143–60.

   (1975b). Modular constructions for combinatorial geometries. *Trans. Amer. Math. Soc.* **203**: 1–44.

   (1977). The broken circuit complex. *Trans. Amer. Math. Soc.* **234**: 417–33.

   (1985). Coordinatizing the Dilworth truncation, in *Proceedings of a Conference on Matroid Theory,* Szeged, Hungary.

Cheung, A. L. C. (1974a). Compatible extensions of a combinatorial geometry. *Thesis,* Department of Pure Mathematics, University of Waterloo, Canada.

   (1974b). Adjoints of a geometry. *Canad. Math. Bull.* **17**: 363–5, 623.

Crapo, H. H. (1965). Single-element extensions of matroids. *J. Res. Natl. Bureau Standards B* **69B**: 55–65.

   (1967). A higher invariant for matroids. *J. Combin. Theory* **2**: 406–17.

   (1968). The joining of exchange geometries. *J. Math. Mech.* **17**: 837–52.

   (1970). Erecting geometries, in *Proceedings of Second Chapel Hill Conference on Combinatorial Mathematics,* pp. 74–99.

   (1971). Constructions in combinatorial geometry. Notes for combinatorial theory advanced science seminar, Bowdoin College, Brunswick, Me.

Crapo, H. H., and Rota G. C. (1970). *Combinatorial Geometries.* M.I.T. Press, Cambridge, Mass.

Cravetz, A. (1978). Essentials for matroid erection. Master's project, University of North Carolina, Chapel Hill.

Crawley, P., and Dilworth, R. P. (1973). *Algebraic Theory of Lattices.* Prentice-Hall, Englewood Cliffs, N.J.

De Sousa, J., and Welsh, D. J. A. (with internal reference to J. Edmonds) (1972). A characterisation of binary transversal matroids. *J. Math. Anal. Appl.* **40**: 55–9.

Dilworth, R. P. (1944). Dependence relations in a semimodular lattice. *Duke Math J.* **11**: 575–87.

Dowling, T. A., and Kelly, D. G. (1974). Elementary strong maps and transversal geometries. *Discrete Math.* **7**: 209–24.

Edmonds, J. (1970). Submodular functions, matroids, and certain polyhedra, in *Combinatorial Structures and Their Applications, Proceedings of an International Conference on Combinatorics,* pp. 69–87. Gordon & Breach, New York.

Edmonds, J., and Rota, G.-C. (1966). Submodular set functions, in *Abstracts of Waterloo Combinatorics Conference.*

Greene, C., and Magnanti, T. L. (1975). Some abstract pivot algorithms. *SIAM J. Appl. Math.* **29**: 530–9.

Hajós, G. (1961). Ueber eine Konstruktion nicht *n*-farbbarer Graphen. *Wiss. Z. Martin Luther Univ. Halle-Wittenberg Math.-Natur. Reihe* **10**: 116–17.

Hall, D. W. (1943). A note on primitive skew curves. *Bull. Amer. Math. Soc.* **49**: 935–7.

Higgs, D. (1968). Strong maps of geometries. *J. Combin. Theory* **5**: 185–91.

Hirschfeld, J. W. P. (1979). *Projective Geometries over Finite Fields.* Clarendon Press, Oxford.

Ingleton, A. W., and Piff, M. J. (1973). Gammoids and transversal matroids. *J. Combin. Theory Ser. B* **15**: 51–68.

Jaeger, F. (1981). A constructive approach to the critical problem. *Eur. J. Combinatorics* **2**: 137–44.

Kahn, J., and Kung, J. P. S. (1982). Varieties of combinatorial geometries. *Trans. Amer. Math. Soc.* **271**: 485–99.

Kennedy, D. (1975). Majors of geometric strong maps. *Discrete Math.* **12**: 309–40.

Knuth, D. (1975). Random matroids. *Discrete Math.* **12**: 341–58.

Las Vergnas, M. (1975). On certain constructions for matroids, in *Proceedings of the Fifth British Combinatorial Conference*, pp. 395–404.

Lindström, B. (1968). On strong joins and pushouts of combinatorial geometries. *J. Combin. Theory Ser. A* **25**: 77–9.

Lovász, L., and Recski, A. (1973). On the sum of matroids. *Acta Math. Acad. Sci. Hung.* **24**: 329–33.

Lucas, D. (1975). Weak maps of combinatorial geometries. *Trans. Amer. Math. Soc.* **206**: 247–79.

Mason, J. (1977) Matroids as the study of geometrical configurations, in *Higher Combinatorics*, edited by M. Aigner, pp. 133–76. D. Reidel, Dordrecht.

Mazzocca, F. (1983). On a characterization of Dilworth truncations of combinatorial geometries. *J. Geometry* **20**: 63–73.

Nash-Williams, C. St. J. N. (1966). An application of matroids to graph theory, in *Theory of Graphs International Symposium*, pp. 263–5. Dunod, (Paris).

Nguyen, H. Q. (1979). Weak cuts of combinatorial geometries. *Trans. Amer. Math. Soc.* **250**: 247–62.

Oxley, J. G., Prendergast, K., and Row, D. H. (1982). Matroids whose ground sets are domains of functions. *J. Austral. Math. Soc. (A)* **32**: 380–7.

Pudlák, P., and Tůma, J. (1977). Every finite lattice can be embedded in the lattice of all equivalences over a finite set (preliminary announcement). *Comment. Math. Univ. Carolinae* **18**: 409–14.

    (1980). Every finite lattice can be embedded in a finite partition lattice. *Algebra Universalis* **10**: 74–95.

Rado, R. (1942). A theorem on independence relations. *Quart. J. Math.* **13**: 83–9.

    (1957). Note on independence functions. *Proc. London Math. Soc. (3)* **7**: 300–20.

Robertson, N., and Seymour, P. (in press). Graph minors. V. Excluding a planar graph. *J. Combin. Theory Ser. B*.

Seymour, P. (1980). Decomposition of regular matroids. *J. Combin. Theory Ser. B* **28**: 305–59.

Tutte, W. T. (1966). Connectivity in matroids. *Canad. J. Math.* **18**: 1301–24.

Ungar, P. (1978). Dissections and intertwinings of graphs. *Amer. Math. Monthly* **85**: 664–6.

Welsh, D. J. A. (1976). *Matroid Theory.* Academic Press, New York.

White, N. L. (1974). A basis extension property. *J. London Math. Soc. (2)* **7**: 662–4.

CHAPTER 8

Strong Maps

Joseph P. S. Kung

8.1. MINORS AND STRONG MAPS

What are the maps, or morphisms, in the category of matroids? As is usual for objects arising in combinatorial analysis and universal algebra, there is more than one reasonable answer. In this chapter and the next, two approaches and their relationship will be discussed.

The notion of morphism is dependent on the notion of subobject. A reasonable notion of subobject for the category of matroids is that of a minor. To define a minor, we first recall from Chapter 7 the operations of contraction and deletion. Let M be a matroid on the set S, and let U and V be subsets of S. The *restriction* of the matroid M to the set U is the matroid $M(U)$ on the set U for which the rank of a subset A in U is simply its rank in M as a subset of S. We also say that the restricted matroid $M(U)$ is obtained from M by *deleting* the elements in $S - U$ from S, and we shall sometimes denote $M(U)$ by $M - (S - U)$. The *contraction* of M by the subset V is the matroid M/V on the set $S - V$ whose rank function is as follows: For $A \subseteq S - V, r(A) = r_M(A \cup V) - r_M(V)$. Contractions and deletions commute in the sense that for any pair of disjoint subsets U and V, the matroids $(M - U)/V$ and $(M/V) - U$ are the same matroid on the set $S - (U \cup V)$. A *minor* of the matroid M on the set S is a matroid on a subset of S obtained from M by a sequence of deletions and contractions.

Let us take minors as the subobjects in our category. What then are the morphisms? There is an obvious extension of the notion of a contraction; namely, we are allowed to contract by elements *not* in the set S. To be more precise, suppose that M^+ is a matroid on the set $S \cup E$ (where E is a set of new elements disjoint from S) such that the restriction $M^+(S)$ is the original matroid M on S. Such a matroid is called an *extension* of M (see Chapter 7). A matroid N can now be defined on the set S by contracting the elements E in the extension M^+. Writing this out, we have

$$M \to M^+ \to N,$$

where the first arrow is an embedding (or injection) and the second is a contraction.

The matroid N can be thought of as the projected image of the matroid M onto a subspace of appropriate dimension. It has the following important property:

For any set $A \subseteq S$, the fact that A is closed in N implies that A is closed in M. (8.1)

It turns out that the converse is also true: If M and N are matroids on S satisfying (8.1), then there exists an extension M^+ such that N is the contraction of M^+ by the added elements. Moreover, as will be seen in Section 8.4, (8.1) is just the right property to ensure that N is the "image" of M under a combinatorial analogue of a linear transformation.

Just as a linear transformation can map a nonzero vector to the zero vector, a strong map can map a point to the closure of the empty set. To make sure that this can be done – there may be no elements on $\bar\varnothing$ – we adjoint a *zero* to every matroid in the following manner: Let M be a matroid on S, and let o be an element not in S. The matroid M_o is the direct sum $M \oplus \{o\}$ on the set $S \cup o$, where $\{o\}$ is the matroid of rank zero on the single-element set $\{o\}$. We use the same symbol o for the zero of any matroid M_o.

We are now ready to give two equivalent definitions of a strong map. The first emphasizes the action of the strong map on the elements of the matroid.

8.1.1. Definition. *Let $M(S)$ and $N(T)$ be matroids. A strong map σ from $M(S)$ to $N(T)$ is a function $\sigma: S \cup o \to T \cup o$, mapping o to o, and satisfying the following condition:*

The inverse image of any closed set of N_o is also a closed set of M_o.

We abbreviate this as

$$\sigma: M(S) \to N(T).$$

The second definition, to be shown equivalent to the first in Proposition 8.1.3, is more lattice-theoretic in spirit. Given matroids $M(S)$ and $N(T)$ and a function $\sigma: S \cup o \to T \cup o$, we can define a map $\sigma^*: L(M) \to L(N)$ between their lattice of flats as follows:

If x is a flat with underlying set X, then $\sigma^*(x)$ is the flat in $L(N)$ with underlying set $\overline{\sigma(X)}$.

8.1.2. Definition. *A function $\sigma: S \cup o \to T \cup o$ mapping o to o is a strong map between the matroids $M(S)$ and $N(T)$ if*

(1) *the map $\sigma^*: L(M) \to L(N)$ is supremum-preserving, that is, for every pair of flats x and y,*

$$\sigma^*(x \vee y) = \sigma^*(x) \vee \sigma^*(y),$$

(2) *the map σ^* maps points to points or $\overline{\varnothing}$.*

It is evident from either definition that the composition of two strong maps is a strong map. In addition, a simple manipulation shows that * preserves composition [i.e., $(\tau \circ \sigma)^* = \tau^* \circ \sigma^*$].

From the operation of contraction we obtain a particularly important kind of strong map. Let $\sigma: M(S) \to N(T)$ be a surjective strong map. Then σ is said to be a *contraction map* if there exists a subset U of S such that U is mapped to o and the lattice of flats $L(N)$ is isomorphic to the interval $[\bar{U}, S]$ in $L(M)$. In particular, if $U \subseteq S$, the map $\sigma: S \cup o \to (S - U) \cup o$ sending $S - U$ to $S - U$ identically and $U \cup o$ to o is a contraction map from M to its contraction M/U by U. As will be shown later, every contraction map can be put into this form. Using this, it is immediate that contraction maps are strong maps.

Another kind of strong map arises from submatroids. Let $M(S)$ be a matroid, and let $T \subseteq S$. The injection $i: T \to S$ is a strong map from the restriction $M(T)$ to M. Such a strong map is called an *embedding*. As we shall see in the next section, contractions and embeddings "generate" all strong maps.

Our most immediate task is to show that our two definitions are indeed equivalent.

8.1.3. Proposition. *Let $M(S)$ and $N(T)$ be matroids, and let $\sigma: S \cup o \to T \cup o$ be a function mapping o to o. The following are equivalent:*

(a) *for any subset $A \subseteq S$, $\sigma(\bar{A}) \subseteq \overline{\sigma(A)}$.*
(b) *σ is strong according to Definition 8.1.1.*
(c) *σ is strong according to Definition 8.1.2.*

Proof. We first show that (a) implies (b). Let X be a closed set in $N(T)$. Because

$$\sigma[\overline{\sigma^{-1}(X)}] \subseteq \overline{\sigma\sigma^{-1}(X)} = \bar{X} = X,$$

we have

$$\overline{\sigma^{-1}(X)} \subseteq \sigma^{-1}\sigma[\overline{\sigma^{-1}(X)}] \subseteq \sigma^{-1}(X).$$

Hence, $\overline{\sigma^{-1}(X)} = \sigma^{-1}(X)$, and $\sigma^{-1}(X)$ is closed.

Next we show that (b) implies (c). Observe first that σ^* is order-preserving. For if u and v are flats with underlying sets U and V,

$$u \leq v \Rightarrow U \subseteq V \Rightarrow \overline{\sigma(U)} \subseteq \overline{\sigma(V)} \Rightarrow \sigma^*(u) \leq \sigma^*(v).$$

In particular, $\sigma^*(x \vee y)$ is greater than both $\sigma^*(x)$ and $\sigma^*(y)$. That is,

$$\sigma^*(x \vee y) \geq \sigma^*(x) \vee \sigma^*(y).$$

To prove the reverse inequality, let Z be the underlying subset of the flat $\sigma^*(x) \vee \sigma^*(y)$. Because Z is closed, $\sigma^{-1}(Z)$ is closed and contains X and Y. Hence, if U is the set underlying the flat $x \vee y$, then Z contains U. Thus,

$$\overline{\sigma(U)} \subseteq \overline{\sigma[\sigma^{-1}(Z)]} \subseteq \bar{Z} = Z.$$

This proves the reverse inequality. Now suppose that there exists a point p with underlying set P in $L(M)$ such that $\sigma^*(p)$ is of rank strictly greater than 1. Let q be a point (with underlying set Q) in $L(N)$ contained in $\sigma^*(p)$. Then

$$\bar{\varnothing} \subset \sigma^{-1}(Q) \subset P,$$

with the containments being strict. Thus, $\sigma^{-1}(Q)$ cannot be closed, and we have a contradiction.

We close the cycle by proving that (c) implies (a). Suppose that $A \subseteq S$ and that x is the flat with underlying set $\bar{A}$. Let $p, q, \ldots, r$ be all the points contained in A. Then

$$\sigma^*(x) = \sigma^*(p \vee q \vee \ldots \vee r) = \sigma^*(p) \vee \sigma^*(q) \vee \ldots \vee \sigma^*(r).$$

Reading this in terms of the underlying sets, we have

$$\sigma(\bar{A}) \subseteq \overline{\sigma(A)}. \qquad \square$$

Our next lemma is not particularly difficult, but very useful.

8.1.4. Lemma. *Let $\sigma: M(S) \to N(T)$ be a strong map. Then σ can be factored into a surjective strong map followed by an injective strong map.*

Proof. Let $T' \cup o$ be the image of σ in $T \cup o$. Then

$$M(S) \xrightarrow{\hat{\sigma}} N(T') \xrightarrow{i} N(T),$$

where $\hat{\sigma}$ is the map from $S \cup o$ to $T' \cup o$ identical with σ on S and i is the injection of $T' \cup o$ to $T \cup o$, is the natural choice for such a factorization.
$\square$

Injective strong maps are essentially embeddings of a submatroid into a matroid (simply relabel the elements). Thus, we can concentrate our attention on surjective strong maps.

Among surjective strong maps, we can further restrict our attention to quotient maps.

8.1.5. Definition. *Let M and N be matroids on the same set S such that the identity map id on $S \cup o$ is a strong map from M to N. Then N is said to be a quotient of M and id a quotient map from M to N. We abbreviate this situation by*

$$M \to N.$$

By relabeling elements, every surjective strong map can be considered a quotient map. Indeed, let $\sigma: M(S) \to N(T)$ be a surjective strong map, and let $\tilde{N}$ be the matroid on S with rank function

$$r_{\tilde{N}}(A) = r_{N_o}[\sigma(A)].$$

The matroid N can be obtained from $\tilde{N}$ by identifying parallel elements, deleting loops, and relabeling. [Two elements a and b are *parallel* if $r(\{a, b\}) = 1$; an element a is a *loop* if $r(\{a\}) = 0$.] From this, we see that the lattices of flats $L(N)$ and $L(\tilde{N})$ are isomorphic. Hence, any surjective strong map σ can be factored into

$$M(S) \overset{id}{\to} \tilde{N}(S) \overset{\tau}{\to} N(T),$$

where *id* is a quotient map and τ is a strong map such that $\tau^*: L(\tilde{N}) \to L(N)$ is an isomorphism.

Useful examples of quotient maps are canonical maps of matroids. Let M be a matroid on the set S, and B the free matroid on S. Then the identity map is a quotient map from B to M and is called the *canonical map* of the matroid M. Such examples show that "almost any property" of matroids is not preserved under strong maps.

Let us now write down several equivalent conditions for a matroid N to be the quotient of another matroid M.

8.1.6. Proposition. *Let M and N be two matroids on the set S. The following are equivalent:*

(a) *N is a quotient of M.*
(b) *If X is closed in N, then it is also closed in M.*
(c) *For any subset $A \subseteq S$, $\bar{A}^M \subseteq \bar{A}^N$.*
(d) *The induced map $id^*: L(M) \to L(N)$ is supremum-preserving, and it maps points to points or $\bar{\varnothing}$.*
(e) *Any bond of N is the union of a collection of bonds of M.*
(f) *M^* is a quotient of N^*. (Here, M^* is the orthogonal dual of M.)*
(g) *Any circuit of M is the union of a collection of circuits of N.*
(h) *For any pair of subsets A and B in S, with $A \subseteq B$,*

$$r_M(B) - r_M(A) \geq r_N(B) - r_N(A).$$

In other words, r_M grows faster than r_N.

Proof. The equivalence of (a), (b), (c), and (d) follows from the definition and Proposition 8.1.3. Because any closed set is the set intersection of the copoints containing it, (e) is equivalent to (b).

We next show that (c) implies (f). Recall from Chapter 5, Proposition 5.2.4, that the closure $A \mapsto \bar{A}^*$ in the orthogonal dual is given by

$$\bar{A}^* = A \cup \{a: a \notin \overline{(A \cup a)^c}\}.$$

Here, $(A \cup a)^c$ is the complement of $A \cup a$ in S. But, using (c), we have

$$a \notin \overline{(A \cup a)^c}^N \Rightarrow a \notin \overline{(A \cup a)^c}^M.$$

From this, we conclude that for any subset $A \subseteq S$,

$$\bar{A}^{N*} \subseteq \bar{A}^{M*},$$

that is, M^* is a quotient of N^*. The converse follows from the fact that $M^{**} = M$. In a similar vein, the fact that the circuits of M^* are precisely the bonds of M allows us to deduce the equivalence of (g) from the equivalence of (e) and (f).

To finish the proof, we first show that (d) implies (h). Because $id^{\#}$ preserves suprema and sends points to points or $\bar{\varnothing}$, $id^{\#}$ also preserves the relation of covers or equals. (This follows from: A flat y covers a flat x if and only if there exists a point p such that $y = x \vee p$.) Thus, any saturated chain stretched between the M-flats $\bar{A}^M$ and $\bar{B}^M$ in $L(M)$ is mapped by $id^{\#}$ onto a saturated chain (with possibly repeated elements) between the N-flats $\bar{A}^N$ and $\bar{B}^N$ in $L(N)$. But $r_M(B) - r_M(A)$ is the length of a saturated chain stretched between $\bar{A}^M$ and $\bar{B}^M$. Hence, (d) implies (h). Finally, we show that (h) implies (c). Let $A \subseteq S$, and suppose that $a \in \bar{A}^M$. This implies that $r_M(A \cup a) - r_M(A) = 0$. Using the inequality, we obtain $r_N(A \cup a) - r_N(A) = 0$, or $a \in \bar{A}^N$. $\square$

We end this section with the definition of a weak map. A function $\tau: S \cup o \to T \cup o$ is a *weak map* from the matroid $M(S)$ to $N(T)$ if for every subset $A \subseteq S$,

$$r_{N_o}[\tau(A)] \leq r_M(A).$$

8.1.7. Lemma. *A strong map $\sigma: M(S) \to N(T)$ is also a weak map.*

Proof. Use the fact that $\sigma^{\#}$ preserves suprema, and send points to points or $\bar{\varnothing}$. $\square$

The notion of a weak map is the other candidate for the notion of morphism in the category of matroids. It is the subject of the next chapter.

8.2. THE FACTORIZATION THEOREM

Undoubtedly the most fundamental result in the structure theory of strong maps is the factorization theorem. This theorem says that any strong map can be factored into an embedding followed by a contraction. Thus, the notion of strong map yields the "smallest" category – in the sense that there are as few morphisms as possible – in which the subobjects are minors.

There are two complementary approaches to the factorization theorem. The first emphasizes the action of the strong map on the lattice of flats. This is historically the first approach and is based on the lift construction. The second approach focuses on how the strong map changes the dependence relations between the elements of a matroid. This approach is more concise but less intuitive. We shall present both approaches in this section, with the remark that the two subsections can be read independent of each other.

8.2.A. The Lift Construction

Throughout this subsection we shall consider a surjective strong map $\sigma: M(S) \rightarrow N(T)$. As well as inducing the map $\sigma: L(M) \rightarrow L(N)$, σ also induces a map $\sigma^\dagger: L(N) \rightarrow L(M)$ defined in the following manner:

$$\sigma^\dagger(y) = \sup\{x \in L(M): \sigma(x) = y\}.$$

The maps σ^* and $\sigma^\dagger$ form a *Galois coconnection* between the lattices $L(M)$ and $L(N)$. That is to say, the maps satisfy the following:

(1) σ^* and $\sigma^\dagger$ are order-preserving maps.
(2) For all x in $L(M)$, $\sigma^\dagger\sigma^*(x) \geq x$, and for all y in $L(N)$, $\sigma^*\sigma^\dagger(y) \leq y$.

From the general theory of Galois connections (e.g., Ore 1962, Chapter 11), we have that (a) the function $\sigma^\dagger\sigma^*$ is a closure operator on $L(M)$, and the function $\sigma^*\sigma^\dagger$ is a coclosure operator on $L(N)$, (b) the collection of $\sigma^\dagger\sigma^*$-closed sets is closed under taking intersections and forms a lattice [and similarly for the $\sigma^*\sigma^\dagger$-coclosed sets in $L(N)$], and (c) the lattice of closed sets in $L(M)$ is isomorphic to the lattice of coclosed sets in $L(N)$. The $\sigma^\dagger\sigma^*$-closed sets in $L(M)$ are called simply σ-closed sets. Because $\sigma^*\sigma^\dagger(y) = y$, every flat in $L(N)$ is coclosed; hence, the lattice of σ-closed sets is isomorphic to $L(N)$. Summarizing, we have the following:

8.2.1. Lemma. *The collection of σ-closed sets is closed under taking intersections. As a lattice, it is isomorphic to $L(N)$.*

A closed set X in M is said to be σ-*independent* if its rank is preserved under σ, that is, if $r_N[\sigma(X)] = r_M(X)$. Of course, such a flat need not be σ-closed. The lift construction consists of "interpolating" a layer of σ-independent sets that are not σ-closed between the σ-closed σ-independent sets and the σ-closed but not σ-independent sets. To be more precise, consider the collection $\mathscr{L}$ of subsets of S consisting of the union of the σ-closed sets and the σ-independent sets.

8.2.2. Proposition. *The collection $\mathscr{L}$ is the collection of closed sets of a quotient of M.*

The matroid defined by the collection $\mathscr{L}$ of closed sets is called the (*first*) *Higgs lift of $N(T)$ toward $M(S)$* and is denoted Lift(N). When $M(S)$ is not specified, it is conventional to take it to be the free matroid on S, as is implicit in Chapter 7, Section 7.4.

To prove the proposition, we need the following characterization of the collection of closed sets of a quotient in terms of a "separation" property. Let $\mathscr{N}$ be a collection of subsets of S containing S and closed under taking intersections. Then $\mathscr{N}$ defines a closure operator (called $\mathscr{N}$-*closure*) on S as follows:

$$A \to \bigcap \{B: B \in \mathscr{N}, \text{ and } A \subseteq B\}.$$

The closed sets of this closure operator are precisely the subsets in $\mathscr{N}$.

8.2.3. Lemma. *A collection $\mathscr{N}$ of closed sets of a matroid $M(S)$ is the collection of closed sets of a quotient of $M(S)$ if and only if $\mathscr{N}$ contains S, $\mathscr{N}$ is closed under taking intersections, and*

$$\begin{aligned} &\textit{if } X \textit{ and } Y \textit{ are in } \mathscr{N}, X \subsetneqq Y, \textit{ and } a \notin Y, \textit{ then} \\ &Y \nsubseteq \overline{X \cup a}. \end{aligned} \tag{8.2}$$

Here, $\overline{X \cup a}$ is the $\mathscr{N}$-closure of $X \cup a$.

Proof. Suppose that $\mathscr{N}$ is the collection of closed sets of a quotient N of M. Let X, Y, and a be as described earlier. Because $\overline{X \cup a}$ covers or equals X, $Y \subseteq \overline{X \cup a}$ implies that $Y = \overline{X \cup a}$, or $a \in Y$. This is a contradiction.

Now let $\mathscr{N}$ be an intersection-closed collection of M-closed sets satisfying (8.2). We first prove that $\mathscr{N}$ defines a matroid N on S. Let X be in $\mathscr{N}$, and $a, b \notin X$. Suppose that $a \in \overline{X \cup b}$ but $b \notin \overline{X \cup a}$. Then the σ-closed sets X and $\overline{X \cup a}$ and the element b form a counterexample to (8.2). This proves the exchange property for closure [axiom (cl4), Chapter 2, Section 2.3].

We next observe that the function $id^{\#}$ sending a subset to its $\mathscr{N}$-closure is a function on $L(M)$ preserving suprema (because $A \vee B$ in the lattice of $\mathscr{N}$-closed sets is defined to be the $\mathscr{N}$-closure of $A \cup B$). It remains to check that $id^{\#}$ maps points to points or $\bar{\emptyset}$. Let p be a point of $L(M)$ with underlying set P for which this is false. Then there exists a $\mathscr{N}$-closed set Q such that

$$\bar{\emptyset} \subsetneqq Q \subsetneqq \bar{P}.$$

In particular, there exists an element a in P not in Q. Because p is a point in $L(M), \bar{a}^M = P$, and hence $\bar{a} = \bar{P}$. The $\mathcal{N}$-closed sets $\bar{\varnothing}$ and Q and the element a form a counterexample to (8.2). Thus, the identity map is a strong map between M and N. $\square$

Proof of Proposition 8.2.2. We begin with the observation that if X and Y are closed sets in $M, X \subseteq Y$, and Y is σ-independent, then X is also σ-independent. This follows from the fact that $r_M(Y)$ is the length of a saturated chain stretched between $\bar{\varnothing}$ and Y, and we can choose that chain to pass through X.

Let X and Y be subsets in $\mathcal{L}$. If both of them are σ-closed, then $X \cap Y$ is also σ-closed. If at least one of them is σ-independent, then $X \cap Y$ is σ-independent. Hence, $\mathcal{L}$ is closed under taking intersection.

We denote the $\mathcal{L}$-closure of a set A by $\bar{A}$, the σ-closure by $\bar{A}^\sigma$, and the M-closure by $\bar{A}^M$.

Our next task is to check that $\mathcal{L}$ satisfies (8.2). Let X, Y, and a be as described in (8.2). We need to prove that $Y \nsubseteq \overline{X \cup a}$. There are two cases.

Case 1: Y is σ-closed but not σ-independent. In the lattice of σ-closed sets, $\overline{Y \cup a}^\sigma$ covers Y, and $\overline{X \cup a}^\sigma$ covers $\bar{X}^\sigma$. If $Y \nsubseteq \overline{X \cup a}^\sigma$, then $Y \nsubseteq \overline{X \cup a}$. Hence, we can assume that $Y \subseteq \overline{X \cup a}^\sigma$. Because $a \notin Y$, $Y \neq \overline{X \cup a}^\sigma$, and

$$X \subseteq \bar{X}^\sigma \subseteq Y \subsetneq \overline{X \cup a}^\sigma.$$

Because $\overline{X \cup a}^\sigma$ covers $\bar{X}$ in the lattice of σ-closed sets, $\bar{X}^\sigma = Y$. We conclude that X is not σ-closed; because X is in $\mathcal{L}$, X must be σ-independent. Further, $\overline{X \cup a}^M$ is also σ-independent, because $r_N[\sigma(\overline{X \cup a}^M)] = r_N[\sigma(\overline{X \cup a}^\sigma)] = r_N[\sigma(X)] + 1 = r_M(X) + 1 = r_M(\overline{X \cup a}^M)$. Thus, $\overline{X \cup a} = \overline{X \cup a}^M$; because Y is not σ-independent, $Y \nsubseteq \overline{X \cup a}$.

Case 2: Y is σ-independent. Then X is also σ-independent. If $\bar{X}^\sigma \subsetneq \overline{X \cup a}^\sigma =$ the σ-closure of $\overline{X \cup a}^M$, then $\overline{X \cup a}^M$ is σ-independent and $Y \nsubseteq \overline{X \cup a}^M = \overline{X \cup a}$. On the other hand, if $\bar{X}^\sigma = \overline{X \cup a}^\sigma$, then neither $\overline{X \cup a}^M$ nor any closed set containing it is σ-independent. Because Y is a σ-independent set, $Y \nsubseteq \overline{X \cup a} = \overline{X \cup a}^\sigma$. $\square$

8.2.4. Lemma. *Suppose that the rank of N is strictly less than the rank of M. Then $r[Lift(N)] = r(N) + 1$.*

Proof. Let X be a minimal closed set of M whose σ-closure is the entire set S. Such a set X is σ-independent and has rank $r(N)$. {Proof: Consider a saturated chain stretched between $\bar{\varnothing}$ and X. If $r(N) \neq r_M(X)$, then there must be a "link" $Y \subseteq Z$ in that chain such that $\bar{Y}^\sigma = \bar{Z}^\sigma$. Let T be a relative complement of Z in the interval $[Y, X]$. Then $Y \vee T \neq X$, but $\overline{Y \cup T}^\sigma = \overline{Z \cup T}^\sigma = \bar{X}^\sigma = S$, contradicting the minimality of X.} Because $r(N) < r(M)$, X is not equal to S. Now let Y be a closed set of M such that $X \subsetneq Y \subsetneq S$. Because $\bar{Y}^\sigma = S$, Y is not σ-closed; because $r_M(Y) > r_M(X) = r(N)$, Y is not σ-

independent. In other words, $Y \notin \mathscr{L}$. Hence, X is covered by S and is a copoint in the matroid $L(N)$. Because X has rank $r(N)$, $r[\mathrm{Lift}(N)] = r(N) + 1$.

$\square$

Observing that the closed sets of N (which are the σ-closed sets) are also closed in $\mathrm{Lift}(N)$, we have a factorization of the strong map σ into

$$M \to \mathrm{Lift}(N) \overset{\sigma_1}{\to} N,$$

where $\mathrm{Lift}(N)$ is a quotient of M and σ_1 is an *elementary* strong map; that is, σ_1 is a strong map that decreases rank by at most 1. We can iterate this construction. Let

$$\mathrm{Lift}^0(N) = N \quad \text{and} \quad \mathrm{Lift}^{i+1}(N) = \mathrm{Lift}[\mathrm{Lift}^i(N)].$$

The matroid $\mathrm{Lift}^i(N)$ on the set S is called the ith *Higgs lift of N toward M*. The Higgs lifts form an *elementary factorization* of σ, in the sense that the strong maps in the sequence

$$M = \mathrm{Lift}^n(N) \to \mathrm{Lift}^{n-1}(N) \to \cdots \to \mathrm{Lift}^1(N) \to \mathrm{Lift}^0(N) = N$$

are all elementary strong maps. Here, n is the difference $r(M) - r(N)$. We have proved the following:

8.2.5. Proposition. *Every strong map can be factored into a sequence of elementary strong maps.*

Another way to look at the lift construction is through the nullity function. Let $A \subseteq S$. Its *σ-nullity* is defined by

$$n_\sigma(A) = r_M(A) - r_N[\sigma(A)].$$

The nullity of the strong map σ is simply $n_\sigma(S)$. It is easy to prove that n_σ is an order-preserving map and somewhat harder to prove that n_σ is an increasing set function. (*Hint:* Rephrase Proposition 8.1.6, part h.) Further, a closed set X in M is σ-independent if and only if its σ-nullity is zero. Thus, the closed sets of the first Higgs lift are the σ-closed sets together with the closed sets of σ-nullity strictly less than 1. Iterating this, we obtain the following:

8.2.6. Lemma. *The closed sets of the ith Higgs lift are the σ-closed sets together with the closed sets in M of σ-nullity strictly less than i.*

We are finally ready to prove the factorization theorem.

8.2.7. The factorization theorem. *Every strong map can be factored into an embedding followed by a contraction.*

Proof. We first consider a surjective strong map $\sigma: M(S) \to N(T)$. Let $R(E)$ be any matroid on a set E disjoint from S of rank n, where n is the nullity of σ. Consider the direct sum $M \oplus R$ on $S \cup E$. The function $(\sigma \oplus 0)$: $S \cup E \cup o \to T \cup o$, which is identical with σ on S and maps $E \cup o$ to o, is a strong map from $M \oplus R$ to $N \oplus Z$, where Z is the matroid on E of rank zero. Let M^+ be the nth lift of $N \oplus Z$ toward $M \oplus R$. We shall prove that

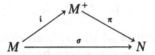

is a factorization of σ into an embedding i and a contraction π.

(a) To show that i is an embedding, it suffices to show that the restriction $M^+(S)$ is the matroid M. We observe first that if X is a closed set of M, then its $(\sigma \oplus 0)$-nullity is the same as its σ-nullity, which is less than or equal to n. If its $(\sigma \oplus 0)$-nullity is strictly less than n, then, by the previous lemma, it is a closed set in M^+. If its $(\sigma \oplus 0)$-nullity is exactly n, then the set $X \cup E$ must be closed in $N \oplus Z$ (for otherwise, $\bar{X}^\sigma$ would have σ-nullity greater than n); hence, $X \cup E$ is a $(\sigma \oplus 0)$-closed set and is closed in M^+. In particular, the closure of X in M^+ is contained in $X \cup E$. We conclude that the map $i^\#: L(M) \to L(M^+)$ is one-to-one and that i is an embedding.

(b) To show that π is a contraction, it suffices to show that the interval $[\bar{E}, \hat{1}]$ in M^+ is isomorphic to $L(N)$. To do so, observe that the $(\sigma \oplus 0)$-nullity of the closed set E in $M \oplus R$ is n. Hence, because nullity is an increasing set function, every closed set containing E in $M \oplus R$ has $(\sigma \oplus 0)$-nullity at least n; in particular, among the sets containing E, only the $(\sigma \oplus 0)$-closed sets are closed in the nth Higgs lift. Thus, the interval $[\bar{E}, \hat{1}]$ remains unchanged during the lift constructions. But $[\bar{E}, \hat{1}]$ is isomorphic to $L(N)$ in $L(N \oplus Z)$ (see Exercise 8.6).

To finish the proof, we tackle in a rather artificial manner the case of a nonsurjective strong map $\sigma: M(S) \to N(T)$. Consider the function $(\sigma \oplus 0)$: $S \cup T \cup o \to T \cup o$, which is identical with σ on S and sends $T \cup o$ to o. This is a surjective strong map from $M \oplus N$ to N and hence has a factorization into an embedding i and a contraction π. Composing this with j, the embedding of $M(S)$ into $(M \oplus N)(S \cup T)$, we obtain the factorization

$$M(S) \xrightarrow{j} (M \oplus N)(S \cup T) \xrightarrow{i} (M \oplus N \oplus R)(S \cup T \cup E) \xrightarrow{\pi} N(T). \qquad \square$$

Usually the matroid $R(E)$ in the proof is taken to be the free matroid on a set E of size n. When this is the case, the matroid M^+ is called the *Higgs major* of the strong map σ.

8.2.B. The Major Construction

In this subsection it is technically simpler to deal only with quotient maps. As was indicated in Section 8.1, this is not an essential restriction.

Our goal is to prove the following version of the factorization theorem:

8.2.8. The factorization theorem, alternative version. *Let M and N be matroids on the finite set S. Then N is a quotient of M if and only if there exists an extension M^+ of M by a set of elements E such that N equals the contraction M^+/E.*

To prove this, we introduce the concept of nullity. The *nullity* of a set A in S *relative to the quotient map* $M \to N$ is defined by

$$n_{M \to N}(A) = r_M(A) - r_N(A).$$

When there is no ambiguity, we abbreviate $n_{M \to N}$ to n. The *nullity* of the quotient map $M \to N$ is defined to be $n_{M \to N}(S)$. By Proposition 8.1.6, part h, nullity is an increasing set function. It is also a unit-increase function; that is, for $A \subseteq S$,

$$n(A) \leq n(A \cup a) \leq n(A) + 1.$$

We now construct the matroid M^+ directly. Let n be the nullity of $M \to N$, and let E be the set $\{1, 2, \ldots, n\}$, which we can assume to be disjoint from S. We specify a matroid M^+ on the set $S \cup E$ by describing its dependent sets. A subset of $S \cup E$ is a dependent set of M^+ if and only if it contains a subset of the following kind:

(a) A: A is contained in S and is dependent in M, or
(b) $A \cup J$: A is contained in S, J is contained in E, A is an independent set of M, and $|J| = n + 1 - n(A)$.

If M^+ is indeed a matroid, then it is easy to check that it is an extension of M by the set E such that $M^+/E = N$. (*Hint:* Use the rank cryptomorphism.)

There are many ways to show that M^+ is a matroid. One should try one's favorite cryptomorphism. Our choice is the following:

8.2.9. Cryptomorphism. *Let $\mathscr{I}$ be a collection of subsets of S. Then $\mathscr{I}$ is the collection of independent sets of a matroid on S if and only if it satisfies*

If $I \in \mathscr{I}$, $a \in S$, and $I \cup a \notin \mathscr{I}$, then $I \cup a$ contains a unique minimal subset not in $\mathscr{I}$. (8.3)

Proof. It is easy to check that the independent sets of a matroid satisfy (8.3). To prove the converse, let us call a subset that is not in $\mathscr{I}$ a dependent set. Consider the minimal dependent subsets or *circuits*. Being minimal, the circuits are pairwise incomparable. Moreover, if C_1 and C_2 are two distinct circuits and $a \in C_1 \cap C_2$, then $C_1 \cup C_2 - a$ cannot be in $\mathscr{I}$, because $C_1 \cup C_2$ contains two distinct circuits, and (8.3) is not satisfied. Thus, the circuits as specified by the collection $\mathscr{I}$ satisfy the circuit-elimination property [axiom (c3) in Chapter 2, Proposition 2.2.4]. □

The key fact used in proving that M^+ is a matroid is the following lemma:

8.2.10. Lemma. *Let I be an independent set of M, and let its nullity relative to $M \to N$ be k. Then there exists a unique minimal subset $I' \subseteq I$ such that the nullity of I' relative to $M \to N$ equals k.*

Proof. We first prove that if a and b are elements in I such that $n(I - a) = k$ and $n(I - b) = k$, then $n(I - \{a, b\}) = k$. By the submodular inequality in the matroid N,

$$-r_N(I - a) - r_N(I - b) \le -r_N(I - \{a, b\}) - r_N(I).$$

Because I is independent in $M, n(J) = |J| - r_N(J)$ for any subset J of I. Adding $|I| - 2$ to both sides of the inequality, we obtain

$$n(I - a) + n(I - b) \le n(I - \{a, b\}) + n(I).$$

Because n is an increasing set function,

$$k = n(I - a) + n(I - b) - n(I) \le n(I - \{a, b\}) \le n(I) = k.$$

This proves our assertion. Using it repeatedly, we conclude that the intersection I' of all the subsets of I having nullity k also has nullity k. This subset I' is the unique minimal subset of I having nullity k. $\square$

To prove that M^+ as defined earlier is a matroid, we need to check through six cases. Let I be an independent set of M^+, and a an element of $I \cup E$ such that $I \cup a$ is dependent. The independent sets of M^+ are of the following form:

    (a) A: $A \subseteq S$, and A is independent in M.
    (b) J: J is any subset of E.
    (c) $A \cup J$: A is an independent set of M, and $|J| < n + 1 - n(A)$.

The element a can be from

    (1) S, or
    (2) E.

The six cases are checked as follows:

    (a) and (1): Obvious.
    (a) and (2): A is independent and has nullity n. Use Lemma 8.2.10.
    (b) and (1): $\{a\}$ has nullity 1, $J = E$, and $E \cup a$ is itself a minimal dependent set.
    (b) and (2): This is impossible.
    (c) and (1): $n(A \cup a) = n + 1 - |J|$, and $n(A) = n - |J|$. Use Lemma 8.2.10.
    (c) and (2): $|J| = n - n(A)$. Again, use Lemma 8.2.10.

This concludes the proof of Theorem 8.2.8.

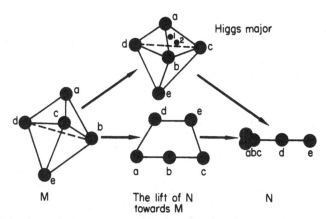

FIGURE 8.1. A quotient map, with its Higgs factorization and Higgs major. *Note:* The $(M \to N)$-closed sets are $\varnothing$, *abc*, *d*, *e*, and *abcde*; the $(M \to N)$-independent sets are $\varnothing$, *a*, *b*, *c*, *d*, *e*, *ad*, *ae*, *bd*, *be*, *cd*, *ce*, and *de*. The points 1 and 2 are added in general position on the plane *abc*.

The matroid M^+ constructed in the proof of Theorem 8.2.8 is called the *Higgs major* of the quotient map $M \to N$.

Consider now the sequence of matroids M_i, where

$$M_i = M^+ - \{i + 1, \ldots, n\}/\{1, \ldots, i\}.$$

Because contraction and deletion commute, we have

$$M_{i+1} = M_i \cup \{i + 1\}/(i + 1),$$

where $M_i \cup \{i + 1\}$ is the extension of M_i by the element $i + 1$ given by

$$M_{i+1} \cup \{i + 1\} = M^+ - \{i + 2, \ldots, n\}/\{1, \ldots, i\}.$$

In fact, M_{n-i} is the ith Higgs lift of N toward M, as discussed in Proposition 8.2.5, and we have recreated the factorization of $M \to N$ into elementary quotient maps

$$M = M_0 \to M_1 \to \cdots \to M_{n-1} \to M_n = N.$$

We end this section with the example in Figure 8.1.

8.3. ELEMENTARY QUOTIENT MAPS

An *elementary quotient map* $M \to N$ is a quotient map that decreases rank by at most 1; N is said to be an *elementary quotient* of M. Note that because the relative nullity is an increasing function, being elementary is equivalent to the weaker condition $r(M) \geq r(N) \geq r(M) - 1$. The importance of elementary quotient maps lies in the fact that they generate all quotient maps (Proposition 8.2.5). In addition, because they are relatively simple, their structure can be described explicitly.

As a special case of the factorization theorem, we have the following:

8.3.1. Proposition. *There is a one-to-one correspondence between elementary quotient maps with domain $M(S)$, or, equivalently, elementary quotients of M, and rank-preserving extensions of $M(S)$ by a single element e. This correspondence is given by the Higgs major M^+.*

Now, $M^+/e = M$ if and only if e is a loop or an isthmus in the extension M^+. Hence, an elementary quotient N of M is equal to M if and only if $r(M) = r(N)$; when this is the case, the map $M \to N$ is said to be *trivial*.

Let us now recall the definition of modular cut from Proposition 7.3.2. We prefer to include along with the closed sets of a modular cut all of their spanning subsets; this is clearly equivalent information. Thus, we define a *modular cut* $\mathcal{M}$ in $M(S)$ to be a collection of subsets of S satisfying the following:

MF1: If $A \subseteq B$ and A is in $\mathcal{M}$, then B is also in $\mathcal{M}$.

MF2: If the closure $\bar{A}$ is in $\mathcal{M}$, then A is also in $\mathcal{M}$.

MF3: If A and B are a modular pair of flats in $\mathcal{M}$, then their intersection $A \cap B$ is also in $\mathcal{M}$.

The *rim* of a modular cut is the collection of closed sets X not in $\mathcal{M}$ such that X is covered by a closed set Y in $\mathcal{M}$. (This closed set Y is unique, by MF3.)

An elementary quotient map can be described "internally" by a modular cut.

8.3.2. Theorem. *Let $M(S)$ be a matroid. There are one-to-one correspondences between the elementary quotients of M, rank-preserving extensions of M by a single element e, and nonempty modular cuts in M. These correspondences are as shown in Figure 8.2.*

Proof. The base of the triangle is Proposition 7.3.2. The left side is Proposition 8.3.1. Going down the right side is an easy exercise using the submodular inequality in N. To go up the right-hand side, let $M \cup e$ be the single-element extension associated with the modular cut $\mathcal{M}$. In the contraction $(M \cup e)/e$, a closed set X in the rim of $\mathcal{M}$ is no longer closed; indeed, its closure in the contraction is the unique closed set in $\mathcal{M}$ covering it. The proof can now be completed by checking (via the rank function) that the other closed sets in $\mathcal{M}$ not in the rim remain closed in the contraction. $\square$

The modular cut associated with $M \to N$ in the theorem is denoted by $\mathcal{M}(M \to N)$.

Once again, we end this section with an example, Figure 8.3.

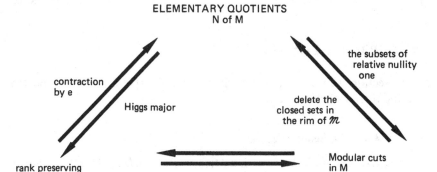

FIGURE 8.2. Correspondences between elementary quotients, rank-preserving extensions, and modular cuts.

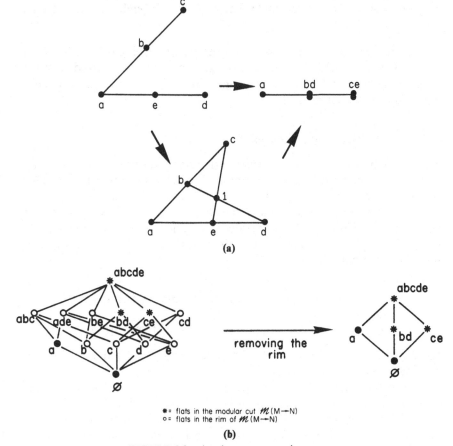

FIGURE 8.3. An elementary quotient map.

8.4. FURTHER TOPICS

In this section we present (mostly without proofs) several topics connected with the theory of strong maps.

8.4.A. The Scum Theorem

8.4.1. The scum theorem. *Let $M(S)$ be a matroid, and let M' be a minor of M. Then there exists an upper interval $[u, \hat{1}]$ in $L(M)$ such that the lattice of flats $L(M')$ can be embedded into $[u, \hat{1}]$.*

Proof. The lattice $L(M')$ can be embedded into some interval $[y, z]$ in $L(M)$. Let u be a minimal relative complement of z in $[y, \hat{1}]$. Then the map $\sigma^*; [y, z] \to [u, \hat{1}]$ given by $t \mapsto t \vee u$ is a strong map (check Definition 8.1.2). Moreover, the two intervals are of the same rank. Hence, by Exercise 8.10, σ^* is an embedding. We conclude that $L(M')$ can be embedded into $[u, \hat{1}]$.
□

The scum theorem is useful in questions involving excluded minors (see Proposition 7.4.13). Roughly speaking, it says that if a matroid contains a minor isomorphic to a given "configuration," then it contains that "configuration" in an upper minor.

8.4.B. Bimatroid Multiplication

A strong map shares many properties with a linear transformation and may fairly claim to be the combinatorial analogue of a linear transformation. This analogy can be made very precise using the notion of a bimatroid.

The theory of bimatroid abstracts the notion of a nonsingular minor in a matrix. A *bimatroid* B on the row set T and the column set S can be specified by a *birank function* $r(A, B)$ defined from pairs of subsets $A \subseteq S$, $B \subseteq T$ to the nonnegative integers satisfying

BR1: $r(\emptyset, \emptyset) = 0$.
BR2: $r(A, B) \le r(A \cup a, B) \le r(A, B) + 1$, and $r(A, B) \le r(A, B \cup b) \le r(A, B) + 1$.
BR3: $r(A \cup C, B \cap D) + r(A \cap C, B \cup D) \le r(A, B) + r(C, D)$.

These axioms are abstracted from the rank function (= maximum size of a nonsingular minor) of a matrix.

With a bimatroid is associated a collection of matroids $M(U)$, one for each subset $U \subseteq T$ on the column set S. The matroid $M(U)$ is called the *column matroid of B restricted to the rows U*, and its rank function is given by $r_{M(U)}(A) = r(A, U)$. The matroid $M(T)$ is called the column matroid of B. Similarly, we also have a collection of *row matroids* on T.

8.4.2. Lemma. *If $U \subseteq V$, then $M(U)$ is a quotient of $M(V)$.*

Now let B be a bimatroid between the row set E and column set S, and C a bimatroid between the row set T and column set E. By analogy with matrix multiplication, we define the *product bimatroid* $C \circ B$ of C and B to be the bimatroid with birank function

$$r(X, U) = \max_{A \subseteq E} \{|A|: r_B(X, A) = r_C(A, U) = |A|\}.$$

That is, it is the maximum size of a subset A of E that is independent in both the column matroid of C and the row matroid of B. By the matroid intersection theorem,

$$r(X, U) = \min_{A \cup B = E} \{r_B(X, A) + r_C(B, U)\}. \tag{8.4}$$

Using this, we can prove that the product $C \circ B$ is indeed a bimatroid.

8.4.3. Theorem. *The column matroid of $C \circ B$ is a quotient of the column matroid of B. Moreover, given a quotient map $M \to N$, there is a bimatroid B with column matroid M and another bimatroid C such that N is the column matroid of the product $C \circ B$.*

Proof. Use (8.4) and Proposition 8.1.6, part h, for the first part and the factorization theorem for the second. $\qquad\square$

8.4.C. Comaps and Retracts

There is a third notion of morphism for the theory of matroids, the notion of comaps. For the theory of comaps, it is technically easier to work with geometric lattices.

Let K and L be geometric lattices. A *comap* from K to L is a function $\gamma: K \to L$ satisfying the following:

CM1: If $x \downarrow y$, then $\gamma(x) \downarrow \gamma(y)$.
CM2: If x and y form a modular pair of flats in K, then

$$\gamma(x \wedge y) = \gamma(x) \wedge \gamma(y).$$

(Here, $\downarrow$ is the "covers or equals" relation.) If, in addition, γ satisfies

CM3: $\gamma(\hat{0}) = \hat{0}$,

γ is said to be a *normalized* comap. The composition of two comaps is a comap.

There are two basic kinds of comaps, embeddings and retractions. Rewording the definition given in Section 8.1, an *embedding* is an injective function $i: K \to L$ mapping $\hat{0}$ to $\hat{0}$, points to points, and preserving suprema.

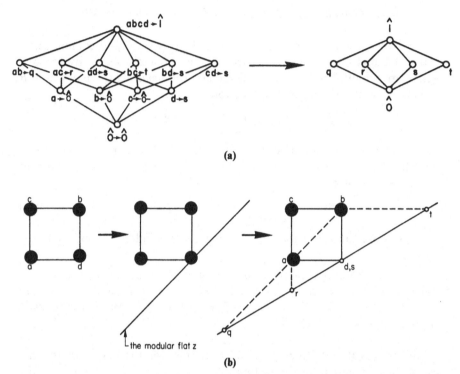

(a)

(b)

FIGURE 8.4. A comap, factored as an embedding followed by a retraction to a modular flat.

The *retraction of K to a modular flat z* is the function $\rho: K \rightarrow [\hat{0}, z], x \mapsto x \wedge z$. (Retractions to nonmodular flats are never comaps.) These two kinds of comaps generate all normalized comaps.

       8.4.4. Theorem. *Any normalized comap can be factored into an embedding followed by a retraction to a modular flat.*

      We omit the proof (Kung 1983). Instead, we present the example of Figure 8.4.

HISTORICAL NOTES

The foundations of the theory of strong maps were laid down by Henry Crapo (1965, 1967) and Denis Higgs (1968). In particular, the factorization theorem is due to Higgs (1968). Those were heady days, around 1965, as Crapo (1971b) testifies.

      Most of the results in Section 8.1 can be found in the works of Crapo and Higgs cited earlier. The content of Section 8.2.A is "lifted" from Crapo's

Bowdoin lecture notes (1971b); indeed, some readers (including the present author) may prefer the spontaneity of the *urtext*. The idea of Section 8.2.B, to construct the Higgs major directly, has occurred to Cheung and Crapo (1973), Kennedy (1975), Kung (1977), Las Vergnas (1980), and doubtless others. The proof given here has not appeared elsewhere. Section 8.2.c is due mainly to Crapo. In Section 8.3, the scum theorem is due to Higgs (1966a), bimatroids to Schrijver (1979) and, independently, but later, Kung (1978b), the connection of strong maps with bimatroid multiplication to Kung (1978b), and comaps to Crapo (1967/68, 1971b).

EXERCISES

8.1. Let $M(S)$ be a matroid, and $T \subseteq S$. Consider the embedding $i: M(T) \to M(S)$. Show that $i^*: L[M(T)] \to L[M(S)]$ is one-to-one.

8.2. Let $M(S)$ be a matroid. The *simplification* of M is the simple matroid M^s defined on the set of points (or rank-1 flats) of M with rank function given by the height function of the geometric lattice $L(M)$. (This concept is also considered in Section 7.4.)

   (a) Show that $L(M) \cong L(M^s)$ and that the function $a \mapsto \bar{a}$ is a strong map from M to M^s.

   (b) What happens if we apply the procedure described in the text to make M^s into a quotient of M?

   (c) Two elements a and b are *in series* in M if they are parallel in the orthogonal dual M^*. Show that the matroid N obtained by identifying any two elements in series equals $[(M^*)^s]^*$ and that $M \to N$ is a strong map. (*Hint:* N is obtainable from M by a sequence of contractions.)

8.3. Let S be a set of vectors in a vector space V, and M the vector matroid on S specified by linear dependence. Let $\lambda: V \to W$ be a linear transformation from V to a vector space W, and let N be the matroid on S specified as follows: A set $\{a_1, \ldots, a_m\}$ is independent if the vectors $\lambda(a_1), \ldots, \lambda(a_m)$ are distinct and linearly independent. Show that N is a quotient of M.

*8.4. (Kung 1977) Show that N is a quotient of M if and only if for any pair of independent sets I and J with the same M-closure, the fact that I is dependent in N implies that J is also dependent in N.

*8.5. (Crapo 1967) A quotient can be thought of as a matroid defined on a geometric lattice. Let N be a quotient of M. The function $x \mapsto x^{\#}$, where $x^{\#} = id^{\#}(x)$, is a closure operator on the geometric lattice $L(M)$ satisfying the following exchange condition.

   Let a and b be points and x any flat in $L(M)$ such that $a, b \nleq x^{\#}$. Then $a \leq (x \vee b)^{\#}$ if and only if $b \leq (x \vee a)^{\#}$.

[Here, the suprema are taken in $L(M)$.] Conversely, such a closure operator defines a quotient of M. Fill in the details.

*8.6. (Crapo and Rota 1970) Let $M(S)$ and $R(T)$ be matroids on disjoint sets. The *direct sum* $M \oplus R$ is the matroid on the union $S \cup T$ with the following rank function: For $A \subseteq S$ and $B \subseteq T$,

$$r_{M \oplus R}(A \cup B) = r_M(A) + r_R(B).$$

Show that the direct sum is the coproduct in the category of matroids and strong maps. That is, given strong maps $\sigma_1: M(S) \to Q(E)$ and $\sigma_2: R(T) \to Q(E)$, there exists a unique strong map $\sigma: (M \oplus R)(S \cup T) \to Q(E)$ such that

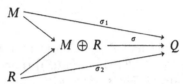

commutes. In particular, M and R are both quotients of $M \oplus R$.

*8.7. (Kung, private communication) The join of two matroids is another notion of summing. It originated in the theory of transversal matroids. Let M and N be matroids on the same set S. Consider the collection $\mathscr{I}$ of subsets of S defined as follows:

> A subset I is in $\mathscr{I}$ if and only if it can be partitioned into two blocks I_M and I_N such that I_M is independent in M and I_N is independent in N.

Prove that $\mathscr{I}$ is the collection of independent sets of a matroid. We call this matroid $M \vee N$ the *join* (other common names are *sum* and *union*) of the two matroids M and N. Prove that the join satisfies the following property:

> Let $M \to M'$ and $N \to N'$ be quotient maps. Then $M \vee N \to M' \vee N'$ is a quotient map.

Conclude that M and N are quotients of $M \vee N$. Finally, write out the orthogonally dual version of this exercise.

* 8.8. Suppose that $\sigma: M(S) \to N(T)$ is a surjective strong map, $L^i(N)$ is the ith Higgs lift, and τ is the quotient map from M to $L^i(N)$. Show that for $A \subseteq S$,

$$n_\tau(A) = \begin{cases} 0 & \text{if } n_\sigma(A) < i, \\ n_\sigma(A) - i + 1 & \text{Otherwise.} \end{cases}$$

*8.9. *An alternative approach to the factorization theorem* (Ingleton, in Welsh 1976, p. 315) First, a definition: Let r_1 and r_2 be two submodular functions. Then r_1 *grows faster than* r_2 if for all sets $A \subseteq B$,

$$r_1(B) - r_1(A) \ge r_2(B) - r_2(A).$$

For example, if $M \to N$ is a quotient map, r_M grows faster than r_N (Proposition 8.1.6, part h).

(a) Show that if r_1 and r_2 are two submodular functions such that r_1 grows faster than r_2, then $r = \min(r_1, r_2)$ is also submodular, and r grows faster than r_2.

Now prove the factorization theorem following these steps. Let $\sigma: M(S) \to N(T)$ be a surjective strong map of nullity n. Let $R(E)$ be any matroid of rank n. Let s_1 be the rank function in the direct sum $M \oplus R$, and let

$$s_2 = r_{N \oplus Z} + n,$$

where Z is the matroid of rank zero on E, and n is the constant function n.

(b) Show that s_1 grows faster than s_2 (remember: $M \oplus R \to N \oplus Z$ is strong), and conclude that $r = \min(s_1, s_2)$ is submodular and increasing.

(c) Show that r is the rank function of a matroid (the only condition missing is unit increase).

(d) Show that r coincides with r_M on S and that for $A \subseteq S$, $r_N(A) = r(A \cup E) - n$.

*8.10. (Higgs 1968) Let $\sigma: M(S) \to N(T)$ be a strong map such that $r(M) = r(N)$. Show that σ^* is one-to-one, and hence we can relabel the elements of S so that σ is an embedding. Conclude that if $M \to N$ is a quotient map and $r(M) = r(N)$, then $M = N$.

*8.11. (Higgs 1968) Let $\sigma: M(S) \to N(T)$ be a surjective strong map. Show that every flat in N has a preimage of equal rank and that this flat is σ-independent.

8.12. Go through the proof of Theorem 8.2.8 with an arbitrary matroid $R(E)$ of rank n instead of the free matroid on $\{1, 2, \ldots, n\}$.

8.13. Using the factorization theorem, prove Proposition 8.1.6, part f.

*8.14. *Elementary factorizations and majors* (Cheung and Crapo 1973; Kennedy 1975; Kung 1977; and others) Let $M \to N$ be a quotient map of nullity n. An *elementary factorization of $M \to N$ of length p* is a sequence (M_i) of elementary quotient maps

$$M = M_0 \to M_1 \to M_2 \to \cdots \to M_{p-1} \to M_p = N.$$

Note that $p \geq n$ and that exactly $p - n$ of these elementary quotient maps are trivial.

(a) Show that there is a matroid M^+ on the set $S \cup E$, where $E = \{1, 2, \ldots, p\}$, such that

$$M_i = M^+ - \{i + 1, \ldots, p\}/\{1, \ldots, i\}. \tag{8.5}$$

[*Hint:* Consider the matroid M^+ on $S \cup E$ whose dependent sets are all sets containing a subset of the following type: (a) A: A is dependent in M, and (b) $J \cup A$: A is independent in M, $j = \max\{i: i \in$

$J\}$, and $|J| = j + 1 - n_j(A)$. Here, $n_j(A)$ is the nullity of A relative to $M \to M_j$.]

The matroid M^+ is called the *major* of the elementary factorization (M_i). We write

$$M^+ = \text{Major}(M_i).$$

Conversely, given any matroid M^+ on the set $S \cup E$ extending M and satisfying $M^+/E = N$, we can define an elementary factorization of $M \to N$ by (8.5). We write

$$(M_i) = \text{Fact } M^+.$$

The set of all matroids M^+ on $S \cup E$ satisfying $M^+(S) = M$ and $M^+/E = N$ can be made into a partially ordered set $\mathbf{M}$ by imposing the weak order. In the *weak order*, $M_1 \leq M_2$ if and only if every independent set of M_1 is also independent in M_2. Similarly, the set of all elementary factorizations can be made into a partially ordered set $\mathbf{F}$ under the following order relation: $(M_i) \leq (M_i')$ if and only if for every i, $M_i \leq M_i'$ in the weak order.

(b) Show that the functions Fact and Major form a Galois coconnection between $\mathbf{M}$ and $\mathbf{F}$. More precisely, show the following:
   (1) Major and Fact are order-preserving maps.
   (2) Fact[Major(M_i)] = (M_i).
   (3) Major(Fact M^+) $\geq M^+$.
   Show also that part (3) is the best possible result.

(c) Show that $\mathbf{F}$ has a maximum, namely, the sequence

$$M \to M \to M \to \cdots \to M \to L^{n-1} \to L^{n-2} \to \cdots \to L^1 \to N$$
$$\leftarrow \quad\quad p - n \text{ times} \quad\quad \rightarrow$$

where L^i is the ith Higgs lift of N toward M. Show that the major of this elementary factorization is the direct sum of the Higgs major (with $E = \{p + 1, \ldots, p + n\}$) and $p - n$ isthmuses.

(d) Show that orthogonal duality acts contravariantly (i.e., reverses all the arrows) on the foregoing situation. [*Hint:* Use Proposition 8.1.6, part f, and the fact that $M \leq N$ if and only if $M^* \leq N^*$ (Lemma 9.3.2).]

An example illustrating these notions is given in Figure 8.5.

*8.15. *The drop construction* (Cheung and Crapo 1973)
   (a) Let $\sigma: M(S) \to N(T)$ be a surjective strong map of nullity n. Show that the subsets in S of σ-nullity n form a modular cut in $M(S)$. Note, however, that the subsets of nullity i, $i < n$, do not necessarily form a modular cut. Similarly, the subsets of strictly positive nullity do not necessarily form a modular cut.
   (b) Now let $M \to N$ be a quotient map of nullity n. Let $\mathcal{M}(M \to N)$ be the modular cut of all subsets of relative nullity n. (This notation

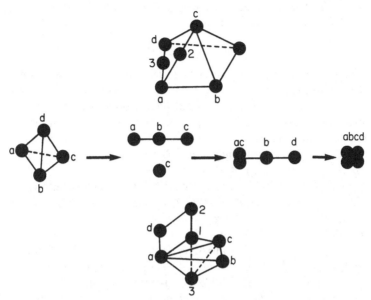

FIGURE 8.5. An extension of the free matroid on $\{a, b, c, d\}$, the elementary factorization, and the major.

is consistent with the notation in the text for elementary-quotient maps.) The (*first*) *drop* Drop(M) of M *toward* N is the elementary quotient defined by

$$\mathcal{M}[M \to \text{Drop}(M)] = \mathcal{M}(M \to N).$$

The *i*th *drop* is defined recursively by $\text{Drop}^0(M) = M$ and $\text{Drop}^{i+1}(M) = \text{Drop}[\text{Drop}^i(M)]$. Show that the *i*th drop of M toward N is the $(n - i)$th lift of N toward M; that is,

$$\text{Drop}^i(M) = \text{Lift}^{n-i}(N).$$

(c) Show that $\mathcal{M}(M \to N)$ is the largest closed cut $\mathcal{N}$ of subsets in S satisfying the following: If A is in $\mathcal{N}$ and is M-closed, then A is also N-closed. [*Note:* A *closed cut* $\mathcal{N}$ *of subsets* is a collection of subsets satisfying MF1 and MF2 (see Section 8.3).]

*8.16. *Modular cuts and elementary factorizations.* Let

$$M = M_0 \to M_1 \to \cdots \to M_{n-1} \to M_n = N$$

be a factorization of $M \to N$, and let $\mathcal{M}_i$ be the modular cut $\mathcal{M}(M_{i-1} \to M_i)$.

(a) (Cheung and Crapo 1973) $\mathcal{M}(M \to N) = \bigcap_{i=1}^n \mathcal{M}_i$.

(b) (Dowling and Kelly 1974) For a subset A, let

$$f(A) = |\{i: A \in \mathcal{M}_i\}|.$$

Show that A is a closed set in N if and only if, for every subset B that strictly contains A,

$$f(B) - f(A) \nleqq r_N(B) - r_N(A).$$

*8.17. *Characterizations of the Higgs factorization* (Cheung and Crapo 1973; Kelly and Kennedy 1978) Let $M \to N$ be a quotient map of nullity n. The elementary factorization $[\text{Lift}^{n-i}(N)]$ obtained from the lift construction is called the *Higgs factorization* of $M \to N$. Show that the following are equivalent to (M_i) being the Higgs factorization:
(a) M_i is the $(n-i)$th Higgs lift of N toward M.
(b) M_i is the ith drop of M toward N.
(c) (M_i) is the maximum in the weak order on factorizations.
(d) The modular cuts $\mathcal{M}_i$, where $\mathcal{M}_i = \mathcal{M}(M_{i-1} \to M_i)$, are nested; that is, $\mathcal{M}_1 \subseteq \mathcal{M}_2 \subseteq \cdots \subseteq \mathcal{M}_n$.
(e) $\mathcal{M}_i = \{A : n_{M \to N}(A) > n - i\}$.
    [*Hint:* Prove this auxiliary lemma: $L = \text{Lift}(N)$ if and only if $\mathcal{M}(L \to N)$ is the collection of $(M \to N)$-dependent sets if and only if $\mathcal{M}(L \to N)$ contains all the $(M \to L)$-dependent sets.]

*8.18. *Essential nullity relative to a quotient map* (Kelly and Kennedy 1978) The *essential nullity of a subset A relative to the quotient map $M \to N$* is defined by

$$n^e_{M \to N}(A) = n_{M \to N}(A) - \max\{n_{M \to N}(B) : B \text{ is a flat covered by } A\}.$$

Show that the essential nullity of A equals the number of modular cuts $\mathcal{M}_i$ in the Higgs factorization in which $\bar{A}^N$ occurs as a minimal closed set.

*8.19. *Factorizations of the canonical map*
    (a) *Higgs lift and truncations* (Dowling and Kelly 1976) Let $M(S)$ be a matroid of rank r. The *truncation* $T(M)$ of M is the matroid on S whose rank function is given by

$$r_{T(M)}(A) = \begin{cases} r_M(A) - 1 & \text{if } r_M(A) = r, \\ r_M(A) & \text{otherwise.} \end{cases}$$

    Equivalently, $T(M)$ is the drop of M toward the matroid Z of rank zero on S. The closed sets of $T(M)$ are all the closed sets of M except the copoints, and hence $T(M)$ is an elementary quotient of M. What is the modular cut $\mathcal{M}[M \to T(M)]$? Let $\text{Lift}_B(M)$ be the Higgs lift of M toward the free matroid B on S. Show that

$$\text{Lift}_B(M) = T(M^*)^*.$$

    Conclude that the Higgs factorization of the canonical map $B \to M$ is the orthogonal dual of

$$M^* \to T(M^*) \to T^2(M^*) \to \cdots \to T^{n-r}(M^*) = Z.$$

[*Hint:* Use the fact that $\text{Lift}_B^{n-i}(M)$ is the maximum in the partial order **F**.]

(b) *Essential nullity* (Kelly and Kennedy 1978) Let $n^e(A)$ be the essential nullity of A relative to the canonical map $B \to M$. Show that $n^e(A) > 0$ if and only if $\bar{A}$ is a *cyclic flat*; that is, $\bar{A}$ is a union of circuits, or, equivalently, the matroid restricted to $\bar{A}$ has no isthmuses.

(c) Let

$$B = M_0 \to M_1 \to \cdots \to M_{n-1} \to M_n = M$$

be an elementary factorization of the canonical map $B \to M$, where M is a matroid of nullity n. Let $\mathscr{E}_i = \{E_{i1}, \ldots, E_{ik_i}\}$ be an irredundant set of generators for the modular cut $\mathscr{M}(M_{i-1} \to M_i)$. [A modular cut $\mathscr{M}$ is *generated* in the matroid $M(S)$ by the sets $E_1, \ldots, E_k$ if it is the intersection of all the modular cuts in $M(S)$ containing $E_1, \ldots, E_k$. We write $\mathscr{M} = \langle E_1, \ldots, E_k \rangle$. The collection $E_1, \ldots, E_k$ is *irredundant* if for all i, $\mathscr{M} \neq \langle E_1, \ldots, E_{i-1}, E_{i+1}, \ldots, E_k \rangle$.] Show that a subset A is independent in M if and only if for every choice of subsets $\mathbf{E} = \{E_{1j_1}, E_{2j_2}, \ldots, E_{nj_n}\}$, one from each $\mathscr{E}_i$, the complement of A contains a system of distinct representatives (or transversal) of $\mathbf{E}$.

*8.20. *The antilift* (Cheung and Crapo 1973) Let $M \to N$ be a quotient map. The *antilift of N away from M* is the elementary quotient A of N defined as follows: $\mathscr{M}(N \to A)$ is the modular cut generated by the $(M \to N)$-dependent sets. Show that if $N \neq A$, then N is the lift of A toward M. Find an example for which $\text{Antilift}(N) = N$.

*8.21. *The degree of a modular cut* (Cheung and Crapo 1973) Let $\mathscr{M}$ be a modular cut in $M(S)$. The *degree* of $\mathscr{M}$ is defined to be the maximum relative nullity of a quotient N of M such that $\mathscr{M}(M \to N) = \mathscr{M}$. Find a construction that will yield a quotient with nullity the degree of $\mathscr{M}$. (*Remark:* Repeating the antilift construction as far as it will go does not work.)

*8.22. *Modular ideals* (Dowling and Kelly 1976) Describe as in Theorem 8.3.2 all the elementary quotient maps with codomain N. (*Hint:* $M \to N$ is elementary if and only if $N^* \to M^*$ is elementary. Use this to construct an orthogonal analogue of modular cut.)

*8.23. *Principal maps.* An elementary quotient $M \to N$ is a *principal map* if the modular cut $\mathscr{M}(M \to N)$ can be generated by a single set E.

(a) Show that any elementary quotient map with domain a modular matroid is principal. A quotient map $M \to N$ has a *principal factorization* if there exists a factorization (M_i) such that for every i, $M_{i-1} \to M_i$ is principal. A matroid is *principal* if the canonical map $B \to M$ has a principal factorization.

(b) (Brown 1974; Dowling and Kelly 1974; Kung 1978a) Show that a matroid is principal if and only if it is the orthogonal dual of a

transversal matroid. (For a description of transversal matroids, see
Chapter 1. One way to prove this is to use Exercise 8.19, part c;
another way is to consider the free matrix representation of a transversal matroid.)

(c) (Oxley, Prendergast, and Row 1982) Show that the Higgs factorization of the canonical map of a principal matroid is, in general,
not a principal factorization. Show that the collection of matroids
whose Higgs factorizations are principal is closed under minors
and orthogonal duality, but not direct sums. Characterize this collection by excluded minors.

*8.24. *The alpha function.* Let $M(S)$ be a matroid with nullity function n. The
alpha function is a function defined on the subsets of S recursively by

$$\alpha(A) = n(A) - \sum \{\alpha(F): F \text{ is a flat strictly contained in } A\}.$$

(a) (Mason 1972; Kung 1978a) Prove that a matroid is principal (i.e.,
the orthogonal dual of a transversal matroid) if and only if for all
$A \subseteq S$, $\alpha(A) \geq 0$. [*Hint:* To prove the implication, investigate how
α changes under an elementary quotient map by looking at the
associated modular cut. To prove the converse, observe that,
counting with multiplicity $\alpha(F)$, there are exactly n flats (where $n =$
nullity of M) with $\alpha(F) > 0$. List those flats $F_1, \ldots, F_n$. Show that
the factorization (M_i), with $\mathcal{M}(M_{i-1} \to M_i) = \langle F_i \rangle$, works. This
proof also yields the "maximal presentation" of the orthogonally
dual transversal matroid.]

(b) (Kung 1978a) By replacing nullity with nullity relative to a quotient
map, define a relative alpha function. Show that $M \to N$ has a
principal factorization if and only if the relative alpha function is
nonnegative.

*8.25. *Quotient bundles and extensions* (Cheung and Crapo 1973) Let $R(S \cup E)$
be a matroid on the disjoint union $S \cup E$, and let $M = R(S)$ and $N =
R(E)$. We say that R is an *extension of* $M(S)$ *by* $N(E)$. Show that if
$r(R) = r(M)$, the extension R determines an *M-quotient bundle* on the
lattice $L(N)$ of flats of N; that is, R determines a family $\{Q(x): x \in L(N)\}$
of quotients of M indexed by flats of $L(N)$ satisfying

QB1: $Q(\hat{0}) = M$.
QB2: If x covers y, then $Q(y) \to Q(x)$ is an elementary quotient map.
QB3: For x, y in $L(N)$,

$$\mathcal{M}[Q(x \wedge y) \to Q(x)] \cap \mathcal{M}[Q(x \wedge y) \to Q(y)] = \mathcal{M}[Q(x \wedge y) \to Q(x \vee y)].$$

Conversely, every M-quotient bundle over $L(N)$ determines an extension R of M by N. As an example, construct the M-quotient bundle
over the free matroid on E determined by the Higgs factorization.
Finally, do Exercise 8.14, part a, using quotient bundles.

8.26. *Coordinatization.* Consider the canonical map $B \to M$, where M is a
matroid coordinatizable over a field **k**. Show that the Higgs lift of M

toward B and the Higgs major of $B \to M$ are both coordinatizable over some extension field of **k**.

*8.27. (Crapo 1968; Las Vergnas 1977/78, 1980) Develop a Tutte-Grothendieck theory for strong maps.

*8.28. (Faigle 1980) Develop a theory of strong maps for geometries on partially ordered sets.

*8.29. (Sachs 1971, 1972) A *geometric map* is a strong map mapping closed sets to closed sets. Investigate geometric maps and their relationships with affine geometries.

*8.30. The embedding theorem of Kantor (1974) for locally projective geometric lattice says the following: Let L be a geometric lattice of rank n, where $n \geq 5$. Suppose that every upper interval $[x, \hat{1}]$, where x is a flat of corank 4, is isomorphic to the lattice of subspaces of a projective geometry of dimension 3 over a field **k**. Then L can be embedded in the lattice of subspaces of the projective geometry of dimension $n - 1$ over the field **k**. Prove Kantor's embedding theorem for the finite fields GF(2) and GF(3) using the scum theorem and the fact that coordinatizability over GF(2) and GF(3) can be characterized by a finite set of excluded minors.

8.31. Let $\mathscr{H}$ be a collection of geometric lattices closed under taking minors. Show that $\mathscr{H}$ can be characterized by a finite set of excluded minors (i.e., there exists a finite set $E_1, \ldots, E_n$ of geometric lattices such that $G \in \mathscr{H}$ if and only if for all i, E_i is not a minor of G) if and only if there exists an integer c such that if rank$(G) \geq c$, and for every point p in G, $G/p \in \mathscr{H}$, then $G \in \mathscr{H}$.

The last five exercises are from Crapo 1968, 1971a).

*8.32. Let K be a geometric lattice, $Q_m(K)$ the (nongeometric) lattice obtained from K by identifying all the flats of rank less than or equal to m, and $D_m(K)$ the Dilworth completion of $Q_m(K)$ (see Lemma 7.7.1). Show that the composition $K \to Q_m(K) \to D_m(K)$ is a comap.

*8.33. *The Crapo join.* For $j = 1$ and 2, let K_j be a geometric lattice with rank function r_j, and δ_j a normalized comap from K_j to a geometric lattice L. Impose the following rank function on the product lattice $K_1 \times K_2$:

$$r(x_1, x_2) = \left[\!\!\left[\sum_{i=1}^{2} \{r_i(x_i) - r_L(\delta_i(x_i))\} \right]\!\!\right] + r_L(\delta_1(x_1) \vee \delta_2(x_2)).$$

Show that the closed flats (i.e., the maximal flats with a given rank) form a geometric lattice J. The lattice J (or its associated geometry) is called the *Crapo join of K_1 and K_2 across the normalized comaps δ_1 and δ_2* and is denoted by $K_1 \square K_2$. Prove that $J \to L$, defined by $(x_1, x_2) \mapsto \delta_1(x_1) \vee \delta_2(x_2)$, is a comap.

*8.34. Let $\Delta(U)$ and $\Gamma(V)$ be graphs on disjoint vertex sets U and V, and $i: U \to \Omega$ and $j: V \to \Omega$ be injective maps into an index set Ω. On the disjoint union $U + V$, impose the equivalence relation $u \sim v$ if $i(u) = j(v)$;

call the set of equivalence classes $U \square V$, and denote an equivalence class by $\tilde{u}$ (or $\tilde{v}$). Let $\Delta \square \Gamma$ be the graph on the vertex set $U \square V$ obtained from Δ and Γ by identifying the vertices in each equivalence class; i.e., the edge-set of $\Delta \square \Gamma$ is the collection $(\tilde{w}_1, \tilde{w}_2)$, where (w_1, w_2) is an edge of Δ or Γ. Show that the lattice of closed subgraphs of $\Delta \square \Gamma$ is the Crapo join of the lattices $L(\Delta)$ and $L(\Gamma)$ of closed subgraphs of Δ and Γ across the embeddings $\delta \colon L(\Delta) \to L(K_U)$ and $\gamma \colon L(\Gamma) \to L(K_V)$. Here, K_U and K_V are the complete graphs on the vertex sets U and V.

*8.35. *Single-element extensions.* Let $\mathcal{K}$ be a modular cut of the geometric lattice L (i.e., $\mathcal{K}$ is a collection of flats satisfying MF1 and MF3 in Section 8.2.C), $\mathbf{2}$ the lattice $\{\hat{0}, \hat{1}\}$, $\gamma \colon L \to \mathbf{2}$ the comap $\gamma(x) = \hat{1}$ if $x \in \mathcal{K}$ and $\hat{0}$ otherwise, and $\delta \colon \mathbf{2} \to \mathbf{2}$ the identity comap. Show that the single-element extension of the geometric lattice L determined by $\mathcal{K}$ equals the Crapo join of L and $\mathbf{2}$ across γ and δ.

*8.36. *Geometric duality between comaps and strong maps.* Let $\gamma \colon K \to L$ be a comap.

(a) Prove that if $\gamma(\hat{0}) = \hat{0}$, then the rank function

$$r_P(x) = r_K(x) - r_L[\gamma(x)]$$

defines a matroid P on the points of K.

(b) Show that $id \colon M_K \to P$, where M_K is the matroid on the points of K determined by K, is a quotient map.

(c) Show that if x covers y in K, then $\gamma(x) = \gamma(y)$ if and only if $id^*(x) \neq id^*(y)$.

REFERENCES

Brown, T. J. (1974). Transversal theory and F-products. *J. Combin. Theory Ser. A* **17**: 290–8.

Cheung, A. L. C. (1974). Compatibility of extensions of a combinatorial geometry. Thesis, University of Waterloo, Ontario, Canada.

Cheung, A. L. C., and Crapo, H. H. (1973). On relative position in extensions of combinatorial geometries. Preprint, University of Waterloo.

Cordovil, R. (1983). Sur la compatibilité des extensions ponctuelles d'un matroïde. *J. Combin. Theory Ser. B* **34**: 209–23.

Crapo, H. H. (1965) Single-element extensions of matroids. *J. Res. Nat. Bur. Standards Sect. B* **69B**: 55–65.

(1967). Structure theory for geometric lattices. *Rend. Sem. Mat. Univ. Padova* **38**: 14–22.

(1967/68). The joining of exchange geometries. *J. Math. Mech.* 17: 837–52.

(1968). Möbius inversion in lattices. *Archiv der Math. (Basel)* **19**: 595–607.

(1971a). Orthogonal representations of combinatorial geometries, in *Atti del Convegno di Geometria Combinatoria e sue Applicazioni*, pp. 175–86. Instituto di Matematica, Università di Perugia.

(1971b). Constructions in combinatorial geometry. Notes for N.S.F. Advanced Science Seminar in Combinatorial Theory, Bowdoin College, Brunswick, Me.

Crapo, H. H., and Rota, G.-C. (1970). *Combinatorial Geometries*, preliminary edition. M.I.T. Press, Cambridge, Mass.

Dowling, T. A., and Kelly, D. G. (1974). Elementary strong maps and transversal geometries. *Discrete Math.* 7: 209–24.

(1976). Elementary strong maps between combinatorial geometries, in *Colloquia Internazionale sulle Teorie Combinatorie, Tomo II*, pp. 121–52. Accad. Naz. Lincei, Rome.

Faigle, U. (1980). Ueber Morphismen halbmodularer Verbände. *Aequationes Math.* 21: 53–67.

Higgs, D. A. (1966). Maps of geometries. *J. London Math. Soc.* 41: 612–18.

(1966a). Geometry. Lecture notes, University of Waterloo.

(1968). Strong maps of geometries. *J. Combin. Theory* 5: 185–91.

Kantor, W. M. (1974). Dimension and embedding theorems for geometric lattices. *J. Combin. Theory Ser. A* 17: 173–95.

Kelly, D. G., and Kennedy, D. (1978). The Higgs factorization of a geometric strong map. *Discrete Math.* 22: 139–46.

Kennedy, D. (1975). Majors of geometric strong maps. *Discrete Math.* 12: 309–40.

Kung, J. P. S. (1977). The core extraction axiom for combinatorial geometries. *Discrete Math.* 19: 167–75.

(1978a). The alpha function of a matroid. I. Transversal matroids. *Stud. Appl. Math.* 58: 263–75.

(1978b). Bimatroids and invariants. *Adv. Math.* 30: 238–49.

(1983). A factorization theorem for comaps of geometric lattices. *J. Combin. Theory Ser. B* 34: 40–7.

Las Vergnas, M. (1977/78). Acyclic and totally cyclic orientations of combinatorial geometries. *Discrete Math.* 20: 51–61.

(1980). On the Tutte polynomial of a morphism of matroids. *Combinatorics 79, Part I, Ann. Discrete Math.* 8: 7–20.

Mason, J. H. (1972). On a class of matroids arising from paths in graphs. *Proc. London Math. Soc. (3)* 25: 55–74.

(1977). Matroids as the study of geometrical configurations, in *Higher Combinatorics*, edited by M. Aigner, pp. 133–76. Reidel, Dordrecht.

Ore, O. (1962). *Theory of Graphs*. Amer. Math. Soc. Colloq. Publ., Vol. 38, Amer. Math. Soc. Providence, R.I.

Oxley, J. G., Prendergast, K., and Row, D. (1982). Matroids whose ground sets are the domains of functions. *J. Australian Math. Soc. Ser. A* 32: 380–7.

Sachs, D. (1971). Geometric mappings on geometric lattices. *Canad. J. Math.* 23: 22–35.

(1972). A note on geometric mappings. *Rend. Sem. Mat. Padova* 47: 23–8.

Schrijver, A. (1979). Matroids and linking systems. *J. Combin. Theory Ser. B* 26: 349–69.

Welsh, D. J. A. (1976). *Matroid Theory*. Academic Press, New York.

Wilde, P. J. (1977). Matroids with given restrictions and contractions. *J. Combin. Theory Ser. B* 22: 122–30.

CHAPTER 9

Weak Maps

Joseph P. S. Kung and Hien Q. Nguyen

9.1. THE WEAK ORDER

The geometric intuition underlying the notion of weak maps is the idea of *special position*. Roughly speaking, if A and B are two sets of points (labeled by the same index set) in a vector space, then B is said to be *more specially positioned* than A if the points in B are "more constrained" than the points in A. For example, if we admit constraints that are algebraic equations, the sets in Figure 9.1b–c are more specially positioned than the set in Figure 9.1a.

In matroid theory, we are concerned only with constraints that are *linear* equations. Thus, the notion of special position can be rendered combinatorially as follows: B is *more specially positioned* than A if the family of linearly dependent sets in B contains the family of linearly dependent sets in A.

When the two sets A and B are not necessarily labeled in the same way, we have the notion of a weak map. To take into account possible linear equations of the form $\mathbf{x} = 0$, we need to add a zero vector. We adjoin a zero element to any matroid M to obtain a matroid M_o, as in Chapter 8.

9.1.1. Definition. *A weak map τ from the matroid $M(S)$ to the matroid $N(T)$ is a function $\tau: S \cup o \to T \cup o$ mapping o to o and satisfying the following*

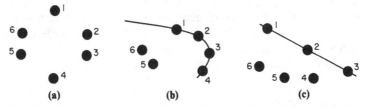

FIGURE 9.1. Special position (points 1–4 in part b lie on an algebraic curve).

condition:

> *If I is a subset of S such that τ restricted to I is one-to-one and $\tau(I)$ is independent in N_o, then I is independent in M.*

Because the rank of a subset A is the maximum size of an independent set contained in A, it is immediate from the definition that a function τ is a weak map if and only if for all subsets A contained in S,

$$r_M(A) \geq r_{N_o}[\tau(A)].$$

Thus, by Lemma 8.1.7, every strong map is also a weak map. Other examples are given in the exercises.

As with strong maps, we can, by relabeling, obtain from any weak map an equivalent weak map between matroids on the same set. Let τ be a weak map from $M(S)$ to $N(T)$. Let $\tilde{N}$ be the matroid defined on S as follows: For $A \subseteq S$,

$$r_{\tilde{N}}(A) = r_{N_o}[\tau(A)].$$

The identity map *id* on $S \cup o$ is a weak map from $M(S)$ to $\tilde{N}(S)$, and the lattice of flats $L(\tilde{N})$ is isomorphic to the lower interval $[\hat{0}, \overline{\tau(S)}]$ in $L(N)$. Thus, except for what happens outside the image of τ, the matroid $\tilde{N}(S)$ contains all the information in the weak map τ. Henceforth we shall restrict our attention to weak maps whose underlying function is the identity.

9.1.2. Proposition. *Let M and N be matroids on the same set S. The following are equivalent:*

(a) *The identity function id: $S \to S$ is a weak map from M to N.*
(b) *Every independent set in N is also independent in M.*
(c) *Every dependent set in M is also dependent in N.*
(d) *Every circuit of M contains a circuit of N.*
(e) *For every subset A in S, $r_M(A) \geq r_N(A)$.*

The proof is easy and is left as an exercise.

From the equivalence of (a) and (e), say, we see that a partial order is defined on the collection of all matroids on the set S as follows:

$$M \geq N \text{ if the identity function is a weak map from } M \text{ to } N.$$

This partial order is called the *weak order*. The maximum of the weak order is the free matroid (or Boolean algebra) on S, and the minimum is the zero matroid, the matroid on S of rank zero. As a partially ordered set, the weak order is rather unstructured: It is not a lattice when $|S| \geq 5$ and does not satisfy the Jordan-Dedekind chain condition when $|S| \geq 4$.

9.2. WEAK CUTS

Unlike the situation for strong maps or comaps, there are no factorization theorems for weak maps. Consequently, the structure of weak maps cannot be described as neatly as the structure of strong maps or comaps. This is not surprising, for the model for a weak map is a specialization of coordinates (see Exercise 9.2), and this is an algebraic notion rather than a geometric notion. Nevertheless, we can obtain an "internal" description of a weak map in terms of weak cuts that is analogous to the description of an elementary-quotient map in terms of a modular cut (Theorem 8.3.2). This description yields an algorithm for constructing all the weak-map images of a given matroid.

To describe a weak cut, it is best to begin with the following crypto-morphism.

9.2.1. Cryptomorphism. *A family $\mathcal{D}$ of subsets of a set S is the family of dependent sets of a matroid M on S if and only if $\mathcal{D}$ satisfies the following conditions:*

(dwl) $\varnothing \notin \mathcal{D}$.

(dw2) *$\mathcal{D}$ is an ascending family; that is, if $A \in \mathcal{D}$, $A \subseteq B$, then $B \in \mathcal{D}$.*

(dw3) *Let I be a subset of S, and let a and b be two distinct elements of S not in I. Suppose that $I \notin \mathcal{D}$, $I \cup a \in \mathcal{D}$, and $I \cup b \in \mathcal{D}$. Then, for any element c in I, $(I - c) \cup \{a, b\}$ is in $\mathcal{D}$.*

Proof. We shall first prove that Cryptomorphism 9.2.1 is implied by the circuit-elimination axioms [axioms (c1)–(c3), Proposition 2.2.4]. We only need to show (dw3). Because I is independent, the dependent set $I \cup a$ contains a circuit C_a containing a. Similarly, $I \cup b$ contains a circuit C_b containing b. Thus, if $c \notin C_a \cap C_b$, $(I - c) \cup \{a, b\}$ contains either C_a or C_b or both and is dependent. On the other hand, if $c \in C_a \cap C_b$, the circuit-elimination axiom (c3) implies that $C_a \cup C_b - c$, and hence $(I - c) \cup \{a, b\}$ contains a circuit.

In order to prove the converse, we first prove the following lemma.

9.2.2. Lemma. *Let $\mathcal{D}$ be a family of subsets of S satisfying (dw2) and (dw3). Then $\mathcal{D}$ also satisfies the following conditions for $n \geq 2$:*

(dw3)$_n$ *Let I be a subset of S, and let $a_1, \ldots, a_n$ be n distinct elements in S not in I. Suppose that $I \notin \mathcal{D}$, and for every i, $1 \leq i \leq n$, $I \cup \{a_1, \ldots, a_{i-1}, a_{i+1}, \ldots, a_n\} \in \mathcal{D}$. Then, for every $c \in I$, $(I - c) \cup \{a_1, \ldots, a_n\} \in \mathcal{D}$.*

Proof. We proceed by induction on n. If $n = 2$, (dw3)$_n$ is just (dw3). Now assume that (dw3)$_{n-1}$ holds. Let $I^+ = I \cup \{a_1, \ldots, a_n\}$, and consider the subsets $I^+ - \{a_i, a_j\}$. If there exists some $\{a_i, a_j\}$ such that $I^+ - \{a_i, a_j\} \notin \mathcal{D}$, then we can apply (dw3)$_2$ to the set $I^+ - \{a_i, a_j\}$ and the elements a_i and a_j to conclude that $(I - c) \cup \{a_1, \ldots, a_n\} \in \mathcal{D}$ for every $c \in I$. Thus, we can suppose that $I^+ - \{a_i, a_j\} \in \mathcal{D}$. In particular, we can assume that $I \cup \{a_1, \ldots, a_{i-1}, a_{i+1}, \ldots, a_{n-1}\} \in \mathcal{D}$ for every i, $1 \le i \le n - 1$. Applying (dw3)$_{n-1}$ to I and the $n - 1$ elements $a_1, \ldots, a_{n-1}$, we conclude that $(I - c) \cup \{a_1, \ldots, a_{n-1}\} \in \mathcal{D}$. By (dw2), $(I - c) \cup \{a_1, \ldots, a_n\} \in \mathcal{D}$. $\qquad \square$

Now consider the minimal subsets in $\mathcal{D}$. These minimal subsets satisfy axiom (c1) by (dw1) and axiom (c2) by minimality. Let C_1 and C_2 be distinct minimal subsets. Then $C_1 \cap C_2 \notin \mathcal{D}$. Applying (dw3)$_n$ for a suitable n to the set $C_1 \cap C_2$ and the elements in $(C_1 \cup C_2) - (C_1 \cap C_2)$, we conclude that $C_1 \cup C_2 - c$ is in $\mathcal{D}$ for every $c \in C_1 \cap C_2$. This proves axiom (c3). $\qquad \square$

Now suppose that M and N are matroids on S and $M \ge N$ in the weak order. Then the family $\mathcal{D}$ of dependent sets of M is contained in the family $\mathcal{D}'$ of dependent sets of N. The *weak cut* $\mathcal{W}$ of the weak map $id: M(S) \to N(S)$ is the family of subsets of S independent in M but dependent in N; that is, $\mathcal{W} = \mathcal{D}' - \mathcal{D}$.

9.2.3. Theorem. *Let M be a matroid on S, and let $\mathcal{W}$ be a subfamily of the family $\mathcal{I}$ of independent sets of M. Then $\mathcal{W}$ is the weak cut of a weak map $id: M(S) \to N(S)$ for some matroid $N(S)$ if and only if the following conditions hold:*

(wc1) *$\varnothing \notin \mathcal{W}$.*

(wc2) *$\mathcal{W}$ is an ascending family within $\mathcal{I}$; that is, if $A \in \mathcal{W}, A \subseteq B$, $B \in \mathcal{I}$, then $B \in \mathcal{W}$.*

(wc3) *Let I be an independent subset in M not in $\mathcal{W}$, and let a and b be two distinct elements of S not in I. Suppose that $I \cup a \notin \mathcal{I} - \mathcal{W}$ and $I \cup b \notin \mathcal{I} - \mathcal{W}$. Then, for any element c in I, $(I - c) \cup \{a, b\} \notin \mathcal{I} - \mathcal{W}$.*

Proof. The only condition for $\mathcal{W}$ to be a weak cut is that $\mathcal{D} \cup \mathcal{W}$ (where $\mathcal{D}$ is the family of dependent sets in M) is the family of dependent sets of a matroid. Thus, the theorem follows immediately from Cryptomorphism 9.2.1. $\qquad \square$

Let $\mathcal{A}$ be a nonempty subfamily of independent sets of a matroid $M(S)$. A weak cut $\mathcal{W}$ containing $\mathcal{A}$ is said to be *minimal (relative to containing $\mathcal{A}$)* if there is no weak cut $\mathcal{V}$ such that $\mathcal{A} \subseteq \mathcal{V} \subsetneq \mathcal{W}$. Because the intersection of two weak cuts is, in general, not a weak cut (see Example 9.2.8), there may be more than one minimal weak cut containing a given subfamily. We now give an algorithm for constructing a minimal weak cut containing a given subfamily. To do so, we need to rephrase Theorem 9.2.3 in a more useful form. We use the following notation: If $\mathscr{A}$ is a family of subsets, then $\mathscr{A}_k$ is the subfamily of subsets in $\mathscr{A}$ of size k.

9.2.4. Proposition. *A subfamily $\mathscr{W}$ of the family $\mathscr{I}$ of independent sets of a matroid M on the set S is a weak cut of M if and only if for any integer k, $0 \leq k \leq r_M(S)$, the following conditions hold:*

(wc$_k$1) $\varnothing \notin \mathscr{W}_k$.

(wc$_k$2) *If $I \in \mathscr{I}_k$ and there exists $a \in T$ such that $I - a \in \mathscr{W}_{k-1}$, then $I \in \mathscr{W}_k$.*

(wc$_k$3) *Let $I \in \mathscr{I}_{k-1} - \mathscr{W}_{k-1}$, and let a and b be distinct elements not in I. Suppose that both $I \cup a$ and $I \cup b$ are not in $\mathscr{I}_k - \mathscr{W}_k$. Then any subset in $\mathscr{I}_k$ contained in $I \cup \{a, b\}$ is in $\mathscr{W}_k$.*

Using this result we obtain the following algorithm:

9.2.5. Algorithm. *Input: A nonempty subfamily $\mathscr{A}$ of independent sets of a matroid $M(S)$. Procedure: Set $\mathscr{W}_0 = \varnothing$. For $0 < k \leq r_M(S)$, construct $\mathscr{W}_k$ inductively as follows:*

$\mathscr{W}_k$ = *intersection of all the subfamilies of $\mathscr{I}_k$ containing $\mathscr{A}_k$ and satisfying* (wc$_k$1), (wc$_k$2), *and* (wc$_k$3).

Output: $\mathscr{W} = \bigcup_k \mathscr{W}_k$, *a minimal weak cut in $M(S)$ containing $\mathscr{A}$.*

The weak cut $\mathscr{W}$ is said to be *generated* by the subfamily $\mathscr{A}$.

To check that the algorithm is correct, we need the following result:

9.2.6. Lemma. *Let the subfamilies $\mathscr{A}$ and $\mathscr{W}_{k-1}$ be given. If $\mathscr{U}_k$ and $\mathscr{V}_k$ are two subfamilies of $\mathscr{I}_k$ containing $\mathscr{A}_k$ and satisfying* (wc$_k$1)–(wc$_k$3), *then their intersection $\mathscr{U}_k \cap \mathscr{V}_k$ also contains $\mathscr{A}_k$ and satisfies* (wc$_k$1)–(wc$_k$3).

The proof is routine and is omitted.

Using Proposition 9.2.4 and Lemma 9.2.6, the family constructed by the algorithm is a weak cut of $M(S)$. In addition, because the subfamilies $\mathscr{W}_k$ are minimal (note that they are intersections), the family $\mathscr{W}$ is also minimal.

In practice, the construction of $\mathscr{W}_k$ will proceed by starting off with the subsets in $\mathscr{A}_k$, adding those independent sets I such that for some $a \in I$, $I - a \in \mathscr{W}_{k-1}$, and finally adding the independent subsets needed in order for $\mathscr{W}_k$ to satisfy (wc$_k$3). To implement the final step, the following consequence of (wc$_k$3) is often useful:

9.2.7. Lemma. *Let $\mathscr{W}$ be a weak cut of the matroid $M(S)$. If I and J are independent sets with the same closure in M and $I \notin \mathscr{W}$, then for all $a \in S$, $I \cup a \in \mathscr{W}$ implies $J \cup a \in \mathscr{W}$.*

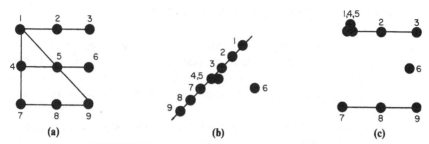

FIGURE 9.2. A matroid and two weak-map images.

The proof is left as an exercise.

We end this section with an example illustrating Algorithm 9.2.5.

9.2.8. Example. Let $M(S)$ be the matroid given by Figure 9.2a, and let $\mathscr{A} = \{45, 235\}$. (In this example, subsets of S are written as sequences; thus, $\{2, 3, 5\}$ is written as 235.) Applying Algorithm 9.2.5, we obtain

$$\mathscr{W}_0 = \varnothing,$$

$$\mathscr{W}_1 = \varnothing,$$

$$\mathscr{W}_2 = \{45\}.$$

To construct $\mathscr{W}_3$, we first add 235. Then, to satisfy (wc$_3$2), we add 145, 245, 345, 457, 458, and 459.

To indicate briefly applications of (wc$_3$3) and Lemma 9.2.7, we shall use the notation

$$I \cup a, I \cup b \,|\, I \Rightarrow (I - c) \cup \{a, b\},$$

$$I \cup a \Rightarrow J \cup a.$$

The highlights of the remaining portion of the algorithm are

(a) $345, 235 \,|\, 35 \Rightarrow 234;\quad 234 \Rightarrow 134;\quad 124, 147 \,|\, 14 \Rightarrow 127.$
(b) $235 \Rightarrow 125;\quad 125, 159 \,|\, 15 \Rightarrow 129;\quad 127, 129 \,|\, 12 \Rightarrow 179.$
(c) For $i, j \neq 4, 5, 6,\; 4ij,\quad 45i \,|\, 4i \Rightarrow 5ij.$

From these deductions and Lemma 9.2.7, we conclude that any independent subset of size 3 in 12345789 is in $\mathscr{W}_3$. Hence, because the subfamily

$$\{I : I \in \mathscr{I}_3, I \subset 12345789\}$$

satisfies (wc$_3$3), this subfamily is in fact $\mathscr{W}_3$. Because $r_M(S) = 3$, we are done. The weak-map image corresponding to $\mathscr{W}$ is shown in Figure 9.2b.

This weak cut is not the only minimal weak cut of $M(S)$ containing $\{45, 235\}$. Another weak cut containing $\{45, 235\}$ is the one corresponding to the weak-map image shown in Figure 9.2c. This weak cut is generated by $\{14, 15\}$.

9.3. RANK-PRESERVING WEAK MAPS

An approach to describing weak maps, suggested by the existence of elementary factorizations for strong maps (see Theorem 8.2.7) is to study how weak maps can be constructed by composing together more basic weak maps. As a first step toward this, we consider rank-preserving weak maps and truncations.

A weak map id: $M(S) \to N(S)$ is said to be *rank-preserving* if rank$(M) =$ rank(N). We abbreviate this situation by

$$M \overset{\text{rp}}{\to} N.$$

If $M(S)$ is a matroid of rank m, the kth *truncation* $T^k(M)$ of M is the matroid of rank $m - k$ whose rank function r' is given as follows: For $A \subseteq S$,

$$r'(A) = \begin{cases} m - k & \text{if } r_M(A) \geq m - k, \\ r_M(A) & \text{otherwise.} \end{cases}$$

If id: $M(S) \to N(S)$ is a weak map with rank$(M) = m$ and rank$(N) = n$, then it is evident that id: $T^{m-n}[M(S)] \to N(S)$ is a rank-preserving weak map. Thus, we have the following:

9.3.1. Lemma. *Every weak map* id: $M(S) \to N(S)$ *can be factored uniquely into a truncation followed by a rank-preserving weak map.*

Unlike strong maps, rank-preserving weak maps are *covariant* under orthogonal duality.

9.3.2. Lemma. $M \overset{\text{rp}}{\to} N$ *if and only if* $M^* \overset{\text{rp}}{\to} N^*$.

Proof. $M \overset{\text{rp}}{\to} N$ if and only if every basis of N is also a basis of M. Now observe that the bases of M^* are exactly the complements of bases of M. $\qquad\square$

More intriguing are the connections of rank-preserving weak maps with Whitney numbers. This will give us a brief preview of several topics from the next volume.

Let $M(S)$ be a finite matroid. The *Tutte polynomial* (Tutte 1947; Brylawski 1972) $t(M; z, x)$ is the polynomial defined inductively as follows:

tp1. If a is neither a loop nor an isthmus, then

$$t(M; z, x) = t(M - a; z, x) + t(M/a; z, x).$$

tp2. $t(M \oplus N; z, x) = t(M; z, x)t(N; z, x)$.

tp3. If M is the matroid of rank 1 on a one-element set, $t(M; z, x) = z$.

tp4. If M is the matroid of rank zero on a one-element set, $t(M; z, x) = x$.

The *Whitney numbers $w_k(M)$ of the first kind* are defined by

$$w_k(M) = (-1)^k \sum_{x:\, r(x)=k} \mu(\hat{0}, x),$$

where the sum is over all flats x in $L(M)$ of rank k, and μ is the Möbius function of $L(M)$. It can be proved that

$$w_k(M) = \frac{1}{k!} \sum_{i=k}^{\text{rank}(M)} \binom{i}{k} \left[\frac{\partial^i}{\partial z^i} t(M; z, x) \right]_{x=z=0}$$

The *Whitney numbers $W_k(M)$ of the second kind* are defined by

$$W_k(M) = \textit{number of flats in } L(M) \textit{ of rank } k.$$

9.3.3. Proposition. *Let $M \overset{rp}{\to} N$. Then $W_k(M) \geq W_k(N)$.*

Sketch of proof. We begin with a lemma, for which the proof is omitted.

9.3.4. Lemma. *There exists a one-to-one function i from the family $\mathscr{C}_N$ of circuits of N into the family $\mathscr{C}_M$ of circuits of M such that for any $C \in \mathscr{C}_N$, $C \subseteq i(C)$ and $r_N[i(C)] = |i(C)| - 1$.*

Applying this lemma to $M^* \overset{rp}{\to} N^*$ and using the fact that the bonds (or complements of copoints) of M^* are exactly the circuits of M, we obtain a one-to-one function of the copoints of N into the copoints of M. This proves the inequality when $k = \text{rank}(M) - 1$. Now repeat this argument for the rank-preserving weak maps $T^r(M) \overset{rp}{\to} T^r(N)$ obtained by truncating M and N. $\qquad \square$

9.3.5. Proposition. *Let $M \overset{rp}{\to} N$, and let z be a positive real number Then $t(M; z, 0) \geq t(N; z, 0)$, the inequality being strict if M has no loops.*

Sketch of proof. We first observe that if $M \overset{rp}{\to} N$ and a is not an isthmus of N, then a can be deleted from M and N to obtain $M - a \overset{rp}{\to} N - a$; similarly, if a is not a loop of N, we can contract a to obtain $M/a \overset{rp}{\to} N/a$. The proposition now follows by induction, using the defining properties of the Tutte polynomial and the fact that for M a matroid of rank 1,

$$t(M; z, 0) = z > 0. \qquad \square$$

A similar argument shows the following:

9.3.6. Proposition. *Let $M \overset{rp}{\to} N$, and let z be a nonnegative real number.* Then

$$\left[\frac{\partial^k t(M; \zeta, x)}{\partial^k \zeta} \right]_{\zeta=z,\, x=0} \geq \left[\frac{\partial^k t(N; \zeta, x)}{\partial^k \zeta} \right]_{\zeta=z,\, x=0}$$

9.3.7. Corollary. *Let* $M \xrightarrow{\text{rp}} N$. *Then* $w_k(M) \geq w_k(N)$.

By taking orthogonal duals and using the fact that

$$t(M^*; z, x) = t(M; x, z),$$

we obtain the following variants of Propositions 9.3.5 and 9.3.6:

9.3.8. Corollary. *Let* $M \xrightarrow{\text{rp}} N$, *and let* x *be a positive real number. Then* $t(M; 0, x) \geq t(N; 0, x)$, *the inequality being strict if* M *has no isthmuses.*

9.3.9. Corollary. *Let* $M \xrightarrow{\text{rp}} N$, *and let* x *be a nonnegative real number. Then*

$$\left[\frac{\partial^k t(M; z, \xi)}{\partial \xi^k}\right]_{z=0,\ \xi=x} \geq \left[\frac{\partial^k t(N; z, \xi)}{\partial \xi^k}\right]_{z=0,\ \xi=x}$$

9.4. SIMPLE WEAK MAPS OF BINARY MATROIDS

A weak map *id*: $M(S) \rightarrow N(S)$ is said to be *simple* if M covers or equals N in the weak order. Because every weak map whose underlying function is the identity is trivially a composition of simple weak maps, knowing the structure of simple weak maps would be a key step toward understanding weak maps. Apart from the easy result that a simple weak map that strictly decreases rank is a first truncation, not much is known about simple weak maps in general.

When the matroids involved are assumed to be binary, a complete description of simple weak maps is known (Lucas 1975).

9.4.1. Theorem. *Let* M *be a binary matroid on the set* S, *and let* $M \xrightarrow{\text{rp}} N$ *be simple. Then there exists a subset* $F \subseteq S$ *such that* $N = M/F \oplus M(F)$.

Proof. The proof requires several lemmas:

9.4.2. Lemma. *If* $M \xrightarrow{\text{rp}} N$ *and* M *is binary, then* N *is also binary.*

Proof. Use Exercise 9.13 and the fact that a matroid is binary if and only if it does not contain the four-point line as a minor.   □

Now let $M \xrightarrow{\text{rp}} N$. Because rank$(M) = $ rank$(N) = r$, say, there exists a basis $B = \{b_1, \ldots, b_r\}$ of M that remains a basis of N. Coordinatize M and N as column vectors over GF(2) so that the element b_i is coordinatized by the standard column vector all of whose coordinates are zero, except for the ith coordinate, which equals 1. Let $\hat{M}$ be a matrix with rows indexed by $\{1, \ldots, r\}$ and columns indexed by S whose $i a$th entry m_{ia} equals the ith coordinate of the column vector coordinatizing a as an element of M. The matrix $\hat{N} = (n_{ia})$ is defined similarly.

9.4.3. Lemma. *Let $A \subseteq \{1, \ldots, r\}$ and $B \subseteq S$ be two subsets of row and column indices such that $|A| = |B|$. If the submatrix obtained by restricting $\hat{N}$ to A and B is nonsingular, then the submatrix obtained by restricting $\hat{M}$ to A and B is also nonsingular.*

The proof of Lemma 9.4.3 is easy and is omitted.

The main step in the proof of Theorem 9.4.1 is to show the following:

9.4.4. Lemma. *Let M be a binary matroid on S, $M \xrightarrow{\text{rp}} N$, and $M \neq N$. Then N is separable.*

Proof. If M is separable, then by Exercise 9.12, N is also separable. We can thus assume that M is connected. When $|S| = 2$, the only rank-preserving weak map $M \xrightarrow{\text{rp}} N$ with M connected is $\circ\circ \xrightarrow{\text{rp}} \circ (\circ)$, where the domain is the matroid of rank 1 with two parallel elements and the codomain is the direct sum of an isthmus and a loop. The lemma holds in this case.

We now proceed by induction. If there exists a loop or isthmus in N, then we are done. Thus, we can assume that there exists an element a in S that is neither a loop nor an isthmus of N. By deleting and contracting a, we obtain two rank-preserving weak maps $M - a \xrightarrow{\text{rp}} N - a$ and $M/a \xrightarrow{\text{rp}} N/a$. There are four cases to consider:

Case I: $M - a = N - a$ and $M/a = N/a$. This implies $M = N$ and therefore cannot happen.

Case II: $M - a \neq N - a$ and $M/a \neq N/a$. By induction, both $N - a$ and N/a are separable. Hence, by a theorem of Tutte (1966) to the effect that if a matroid N is connected, then at least one of $N - a$ or N/a is connected, we conclude that N is separable.

Case III: $M - a = N - a$ and $M/a \neq N/a$. Let $\hat{M}$ and $\hat{N}$ be the matrices with entries in GF(2) coordinatizing M and N. Because $M - a = N - a$, the matrices $\hat{M}$ and $\hat{N}$ are identical when restricted to the columns indexed by $S - a$. However, in the column indexed by a, some of the nonzero coordinates in $\hat{M}$ become zero in $\hat{N}$. Consider the subsets of row and column indices defined inductively by

$$R_0 = \{i: n_{ia} = m_{ia} = 1\},$$
$$R_1 = \{i: n_{ia} = 0, \text{ and } m_{ia} = 1\},$$
$$L_1 = \{b \in S - a: n_{ib} = 1 \text{ for some } i \in R_1\},$$

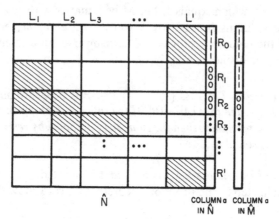

FIGURE 9.3. A matrix coordinating M and N over GF(2).

and, for $k \geq 2$,

$$R_k = \left\{ i: i \notin \bigcup_{j=0}^{k-1} R_j, \text{ and } n_{ib} = 1 \text{ for some } b \in L_{k-1} \right\},$$

$$L_k = \left\{ b \in S - a: b \notin \bigcup_{j=1}^{k-1} L_j, \text{ and } n_{ib} = 1 \text{ for some } i \in R_k \right\}.$$

In addition, define R' to be the set of row indices not in any of the sets R_i, and define L' to be the set of column indices not in any of the sets L_i. The idea behind these definitions is illustrated in Figure 9.3, where the rows and columns are permuted to make the pattern clearer.

In Figure 9.3 there are exactly two shaded rectangles in the submatrix indexed by $\{1, \ldots, r\} \times L_k$. In the upper shaded rectangle there is at least one 1 in each column. In the lower shaded rectangle there is at least one 1 in each row. There are no conditions on the entries in the shaded rectangles $R_0 \times L'$ and $R' \times L'$. Finally, all the entries in the unshaded rectangles are zero. Except for the rectangles $R_0 \times L_k$, this follows immediately from the definition. To see why the entries in $R_0 \times L_1$ are zero, let $(i, b) \in R_0 \times L_1$. By definition of L_1, there exists $j \in R_1$ such that $n_{jb} = 1$. Consider the submatrix indexed by $\{i, j\} \times \{b, a\}$. It equals

$$\begin{bmatrix} n_{ib} & 1 \\ 1 & 0 \end{bmatrix}$$

in $\hat{N}$ and hence is nonsingular. In $\hat{M}$, however, it equals

$$\begin{bmatrix} n_{ib} & 1 \\ 1 & 1 \end{bmatrix}$$

and, by Lemma 9.4.3, is nonsingular. Hence, $n_{ib} = 0$. The argument that the entries in $R_0 \times L_k$ are zero proceeds by induction on k. Let $(i, b) \in R_0 \times L_k$.

By definition of L_k, there exists j_k in R_k with $n_{j_k b} = 1$. Similarly, by definition of R_k, there exists b_{k-1} in L_{k-1} with $n_{j_k b_{k-1}} = 1$. Repeating this process, we obtain the row indices $j_k, j_{k-1}, \ldots, j_1$ and the column indices $b_{k-1}, \ldots, b_1$.

Consider the submatrix indexed by $\{j_1, \ldots, j_k, i\} \times \{b_1, \ldots, b_{k-1}, b, a\}$. Using the induction hypothesis, it equals

$$
\begin{bmatrix}
0 & 0 & 0 & \cdots & & n_{ib} & 1 \\
1 & 0 & 0 & & & & * \\
1 & 1 & 0 & & & & 0 \\
0 & 1 & 1 & & & & 0 \\
& & & \ddots & & & \\
0 & & & & 1 & 1 & 0
\end{bmatrix},
$$

where the entry * equals 0 in $\hat{N}$ and 1 in $\hat{M}$. Computing determinants, we see that this submatrix is nonsingular in $\hat{N}$. Because it must also be nonsingular in $\hat{M}$, by Lemma 9.4.3, we conclude that $n_{ib} = 0$.

To finish this case, observe that on permuting rows and columns, the matrix $\hat{N}$ can be put into the form

$$
\begin{bmatrix}
\hat{N}_1 & 0 \\
0 & \hat{N}_2
\end{bmatrix},
$$

where $\hat{N}_1$ is the submatrix indexed by $(R_0 \cup R') \times L'$, and $\hat{N}_2$ is the submatrix indexed by $(R_1 \cup R_2 \cup \ldots) \times (L_1 \cup L_2 \cup \ldots)$. We conclude that the matroid N is the direct sum $N - (L_1 \cup L_2 \cup \ldots) \oplus N - L'$.

Case IV: $M - a \neq N - a$ and $M/a = N/a$. This case is orthogonally dual to Case III and can be reduced to it by using Lemma 9.3.2. $\square$

We also need the following "interpolation" lemma:

9.4.5. Lemma. *Let M be a binary matroid on S, and $M \overset{rp}{\to} N$. Suppose that $F \subseteq S$ is a separator of N, such that $r_N(F) = r_M(F)$. Then*

$$
M \overset{rp}{\to} M/F \oplus M(F) \overset{rp}{\to} N.
$$

The proof consists of checking that the independent sets of N are independent in $M/F \oplus M(F)$ and that the independent sets of $M/F \oplus M(F)$ are independent in M.

The final lemma shows that separators of the kind needed to apply Lemma 9.4.5 exist.

9.4.6. Lemma. *Let M be a binary matroid on S, and $M \overset{rp}{\to} N$. Then there exists a connected component $F \subseteq S$ of N such that the restriction $N(F)$ equals the restriction $M(F)$.*

Proof. Once again, the proof is by induction. As in the proof of Lemma 9.4.4, the lemma is true when $|S| = 2$. If N is connected, then by Lemma 9.4.4, $M = N$, and the lemma holds trivially. Let $N = N(F_1) \oplus N(F_2) \oplus \cdots \oplus N(F_s)$ be a decomposition of N into its connected components. We distinguish three cases:

Case I: $F_i = \{a\}$ for some i. If a is an isthmus of N, then $N(\{a\}) = M(\{a\})$. If a is a loop, then $M - a \overset{\text{rg}}{\to} N - a$, and by induction, $N - a$ has a connected component $F_j, j \neq i$, such that $(N - a)(F_j) = (M - a)(F_j)$. But F_j is also a connected component of N.

Case II: Every connected component F_i of N consists of two parallel elements. Let $B = \{b_1, \ldots, b_s\}$ be a basis of M that remains a basis in N, let b_i' be the element parallel to b_i, and let $\hat{M}$ and $\hat{N}$ be the matrices defined in the proof of Lemma 9.4.4 coordinatizing M and N. Because b_i' is parallel to b_i, the submatrix of $\hat{N}$ indexed by $\{1, \ldots, s\} \times \{b_1', \ldots, b_s'\}$ is the identity matrix. In $\hat{M}$, some of the zero entries in this submatrix are changed to 1's. Now, if there exists a column b_i' in $\hat{M}$ that still contains exactly one nonzero entry, then $N(\{b_i, b_i'\}) = M(\{b_i, b_i'\})$. Thus, we are left with the case in which every column b_i' in $\hat{M}$ has at least two 1's. But, by Exercise 9.14, this implies that a subset of $\{b_1', \ldots, b_s'\}$ is dependent in M, contradicting $M \overset{\text{rg}}{\to} N$.

Case III: Some connected component, say F_1, contains at least three elements. We use the following lemma:

9.4.7. Lemma. *Let M be a connected matroid with at least three elements. Then there exist two elements a and b such that either $M - a$ and $M - b$ are both connected or $M^* - a$ and $M^* - b$ are both connected.*

Proof. Let c_1, c_2, and c_3 be three distinct elements in M. Because M is connected, by a theorem of Tutte (1966), at least one of $M - c_i$ or M/c_i is connected. Among the possibilities, either two of the $M - c_i$'s are connected or to of the M/c_i's (and hence $M^* - c_i$'s) are connected.   □

By taking the orthogonal dual and using the fact that $M^* \overset{\text{rg}}{\to} N^*$ if necessary, we can assume that there exist two elements a and b in F_1 such that $N(F_1 - a)$ and $N(F_1 - b)$ are both connected. Now the lemma holds for $M - a \overset{\text{rg}}{\to} N - a$ by induction, and hence, one of the components $F_1 - a, \ldots,$ F_s of $N - a$ equals its preimage in M. If this is the case for a component F_i, $i \geq 2$, then we are done. Thus, we can assume that $M(F_1 - a) = N(F_1 - a)$. Similarly, we can assume that $M(F_1 - b) = N(F_1 - b)$. Because $N(F_1)$, $N(F_1 - a)$, and $N(F_1 - b)$ are connected,

$$r_N(F_1) = r_N(F_1 - a) = r_N(F_1 - b) = r_N(F - \{a, b\}).$$

Because the corresponding restrictions for M are also connected (use Exercise 9.11), these rank equalities also hold in M. But $M(F_1 - \{a, b\}) =$

$N(F_1 - \{a, b\})$, and so F_1 and $F_1 - \{a, b\}$ have the same rank in M and N; thus, $M(F_1) \overset{\text{rp}}{\to} N(F_1)$. Because $N(F_1)$ is connected, we conclude from Lemma 9.4.4 that $M(F_1) = N(F_1)$. $\qquad\square$

We are now ready to prove Theorem 9.4.1. First suppose that M is connected. By Lemma 9.4.6 there exists a connected component F of N such that $M(F) = N(F)$. By Lemma 9.4.5, the simple weak map $M \overset{\text{rp}}{\to} N$ factors into

$$M \overset{\text{rp}}{\to} M/F \oplus M(F) \overset{\text{rp}}{\to} N.$$

Because M is connected, $M \neq M/F \oplus M(F)$. Hence, $N = M/F \oplus M(F)$.

If M is not connected, let $F_1, \dots, F_s$ be the connected components of M. Because $M \overset{\text{rp}}{\to} N$ is simple, $M(F_i) = N(F_i)$ for all but one, say F_s, of the components. Applying the first part to the connected matroid $M(F_s)$, we obtain a connected component $F \subseteq F_s$ of $N(F_s)$ such that $N(F_s) = [M(F_s)]/F \oplus M(F)$. Hence,

$$N = N(F_1) \oplus \cdots \oplus N(F_{s-1}) \oplus [M(F_s)]/F \oplus M(F) = M/F \oplus M(F). \quad\square$$

We end this section with several consequences of Theorem 9.4.1.

9.4.8. Corollary. *Let M be a connected binary matroid on S. There exists a one-to-one correspondence between simple rank-preserving weak maps $M \overset{\text{rp}}{\to} N$ and subsets $F \subseteq S$ such that M/F and $M(F)$ are both connected.*

9.4.9. Corollary. *Let $\# M$ be the number of connected components of the matroid M. If M is a binary matroid, then $M \overset{\text{rp}}{\to} N$ is simple if and only if $\# N = \# M + 1$.*

9.4.10. Corollary. *Let M be a binary matroid. Then $M \overset{\text{rp}}{\to} N$ can be factored into $\# N - \# M$ nontrivial simple weak maps, and every such factorization involves exactly $\# N - \# M$ simple weak maps.*

The proofs of these corollaries are left as exercises.

HISTORICAL NOTES

Weak maps (or maps, as they were first called) were discovered by Denis Higgs (1966b). Dean Lucas (1975) developed the theory of rank-preserving weak maps and simple weak maps of binary matroids presented in Sections 9.3 and 9.4. Weak cuts were discovered by Nguyen (1979a) and, in a slightly different version, by Kung (1980). The proof of Theorem 9.2.3 presented here is new and is based on a suggestion by J. Mason.

EXERCISES

9.1. A weak map $\tau: M(S) \to N(T)$ induces a function $\tau^*: L(M) \to L(N)$ as follows: If x is a flat in $L(M)$ with underlying set X, then $\tau^*(x)$ is the flat in $L(N)$ with underlying set $\overline{\tau(X)}$. Show by an example that in contrast to the situation with strong maps, * does not preserve composition; that is, $(\theta \circ \tau)^* \neq \theta^* \circ \tau^*$ in general.

9.2. *Specialization of coordinates.* Let $M(S, R)$ be a matroid on S coordinatized over an integral domain R (i.e., the independent sets are exactly the linearly independent sets in the field of fractions of R), and let P be a prime ideal in R. Show that the matroid $M(S, R/P)$ obtained from $M(S, R)$ by regarding the coordinates as elements in the quotient ring R/P is a weak-map image of $M(S, R)$.

9.3. *Retracts.* Let $M(S)$ be a matroid, and $X \subseteq S$. Show that the function $\tau: S \cup o \to X \cup o$, defined by $\tau(o) = o$, $\tau|_X = id$, and $\tau(S - X) = \{o\}$ is a weak map. Such weak maps are called *retracts.* If X is a closed set and x is the flat in $L(M)$ with underlying set X, show that for any flat $u \in L(M)$, $\tau^*(u) = u \wedge x$.

9.4. Let $M(S)$ be a matroid, and $F \subseteq S$. Show that the direct sum $M/F \oplus F$ is a weak-map image of M.

9.5. The weak order can be modified to take into account isomorphisms of matroids. Consider the relation defined on the collection of all matroids on a set S as follows: $M \geq N$ if there exists a permutation α of S that is a weak map from M to N. This relation is a quasi order (antisymmetry may not hold), and the equivalence classes under the equivalence relation $M \simeq N$ if $M \geq N$ and $N \geq M$ are exactly the isomorphism classes of matroids defined on S. Thus, the relation $\geq$ defines a partial order, also called the *weak order* on the collection of isomorphism classes of matroids on S. Draw the Hasse diagram of this partial order when $|S| \leq 5$.

*9.6. *The lexicographic order* (Higgs 1966a) Let M be a matroid on S, and let

$$R_i(M) = \{A \subseteq S: r_M(A) = i\}.$$

The *lexicographic order* $\geq_l$ on the collection of all matroids on S is defined as follows:

$M >_l N$ if there exists an integer k such that $R_i(M) = R_i(N)$ for $i = 0, 1, \ldots, k - 1$, and $R_k(M) \subsetneq R_k(N)$.

(a) Show that $\geq_l$ defines a lattice on the collection of all matroids on S. [*Hint:* The infimum of the matroids M and N is the matroid P given by

$$R_i(P) = R_i(M) \cap R_i(N), \qquad i = 0, 1, \ldots, k,$$

$$R_{k+1}(P) = \text{all the other subsets of } S,$$

where k is the least positive integer such that $R_k(M) \neq R_k(N)$.]

(b) (Lucas 1975) Show that if $M \geq N$ in the weak order, then $M \geq_l N$ in the lexicographic order.

9.7. Let $\mathcal{M}$ be a modular cut in the matroid $M(S)$. Show that $\mathcal{M} \cap \mathcal{I}$, the family of independent sets contained in $\mathcal{M}$, is a weak cut of $M(S)$ and that the weak-map image of $M(S)$ corresponding to the weak cut is the elementary quotient of $M(S)$ corresponding to the modular cut $\mathcal{M}$. As a special case, deduce that the family $\mathcal{B}$ of all bases of $M(S)$ is a weak cut whose corresponding weak-map image is the truncation $T(M)$ of M.

*9.8. *Projections onto a closed set* (Nguyen 1978) Let E be a subset of S, X a closed set of the matroid $M(S)$, and I an independent set in M spanning the closed set X. Let $\mathcal{W}$ be the weak cut generated by the family.

$$\{I \cup a: a \in E, \text{ but } a \notin X\}$$

of independent sets, and let P be the weak-map image corresponding to $\mathcal{W}$. The rank-preserving weak map $M \to P$ is called the *projection of E onto X.*

(a) Show that $r_p(X \cup E) = r_M(X)$.

(b) Show that if N is a matroid on S such that $r_N(X \cup E) = r_M(X)$ and $M \geq N \geq P$, then $N = P$.

(c) Characterize the simple projections $M \to P$. (This problem has been solved only when X is a copoint of M.)

*9.9. (Nguyen 1979a) Show that a weak cut is generated by the subfamily of minimal independent sets contained in it. (However, it can be generated by other subfamilies as well.)

*9.10. (Knuth 1975; Bixby 1978; Nguyen 1979a) Algorithm 9.2.5 constructs the family of dependent sets of the weak-map image N of M corresponding to the weak cut $\mathcal{W}$. Develop an algorithm for constructing the family of closed sets of M. Compare this algorithm with the "random matroid" algorithm of Knuth (1975).

*9.11. *Erections* (Crapo 1970; Roberts 1974, 1975; Las Vergnas 1976; Nguyen 1978) Let $M \geq N$ in the weak order. A matroid E is said to be a *nontrivial erection of N toward M* if $M \geq E \geq N$ and $T(E)$, the first truncation of E, equals N.

(a) We use the same notation as in Proposition 9.2.4. Let $\text{rank}(N) = n$, let $\mathcal{W}$ be the weak cut of M corresponding to N, $\mathcal{I}$ the family of independent sets of M, and $\mathcal{E}_{n+1}$ a subfamily of $\mathcal{I}_{n+1}$. Show that $\mathcal{E} = (\mathcal{W} - \mathcal{W}_{n+1}) \cup \mathcal{E}_{n+1}$ is a weak cut of M if and only if the following hold:

(ec1) If $I \in \mathcal{I}_{n+1}$ and for some $a \in I$, $I - a \in \mathcal{W}_n$, then $I \in \mathcal{E}_{n+1}$.

(ec2) Let $I \in \mathcal{I}_n - \mathcal{W}_n$, and let a and b be distinct elements not in I. Suppose that both $I \cup a$ and $I \cup b$ are not in $\mathcal{I}_{n+1} - \mathcal{E}_{n+1}$. Then any subset in $\mathcal{I}_{n+1}$ contained in $I \cup \{a,b\}$ is in $\mathcal{E}_{n+1}$.

Such a family $\mathcal{E}$ is called an *erection cut.*

(b) Show that $\mathcal{W}$ is an erection cut; it is called the *trivial* erection cut.

Evidently, there exists a nontrivial erection of N toward M if and only if there exists an erection cut that is not trivial.

(c) Show that there exists no nontrivial erection of the Fano plane F_7 toward the free matroid B on 7 elements.

(d) Show that if $\mathscr{E}_{n+1}$ and $\mathscr{E}'_{n+1}$ define erection cuts, then so does their intersection $\mathscr{E}_{n+1} \cap \mathscr{E}'_{n+1}$. Deduce that the collection of all non-trivial erections of N toward M is either empty or has a unique maximum in the weak order.

(e) Interpret Higgs's lexicographic order (Exercise 9.6) in terms of truncations and erections.

9.12. (Lucas 1975) Show that a rank-preserving weak map preserves separators.

*9.13. (Lucas 1975) Let $M(S) \overset{\text{rp}}{\to} N(S)$, and let K be a minor of M defined on a subset $E \subseteq S$. Show that there exists a minor L defined on E such that $K \overset{\text{rp}}{\to} L$.

9.14. Let $R = (r_{ij})$ be a 0-1 matrix with rows and columns indexed by $\{1, \ldots, k\}$. Suppose that $r_{ii} = 1$ for all i, and every column of R contains at least two 1's. Show that there exists a subset $A \subseteq \{1, \ldots, k\}$ such that the principal submatrix of R with rows and columns indexed by A has exactly two 1's in each column.

*9.15. (Lucas 1975) Let M be a binary matroid, and $M \overset{\text{rp}}{\to} N$. Show that if K is a connected minor of N, then K is also a connected minor of M. Hence, show that if $\mathscr{H}$ is a collection of binary matroids closed under minors and direct sums, then $\mathscr{H}$ is also closed under rank-preserving weak-map images.

REFERENCES

Bixby, R. (1978). The solution to a matroid problem of Knuth. *Discrete Math.* **21**: 87–8.

Brylawski, T. H. (1972) A decomposition for combinatorial geometries. *Trans. Amer. Math. Soc.* **171**: 235–82.

Crapo, H. H. (1970). Erecting geometries, in *Proceedings of the Second Chapel Hill Conference on Combinatorial Mathematics and Its Applications*, pp. 74–99. University of North Carolina Press, Chapel Hill, N.C.

Higgs, D. A. (1966a). A lattice order on the set of all matroids on a set. *Canad. Math. Bull.* **9**: 684–5.

(1966b). Maps of geometries. *J. London Math. Soc.* **41**: 612–18.

Knuth, D. E. (1975). Random matroids. *Discrete Math.* **12**: 341–58.

Kung, J. P. S. (1980). On specializations of matroids. *Stud. Appl. Math.* **62**: 183–7.

Las Vergnas, M. (1976). On certain constructions for matroids, in *Proceedings of the Fifth British Combinatorics Conference*, pp. 395–404. Utilitas Math., Winnipeg, Manitoba, Canada.

Lucas, D. (1974). Properties of rank-preserving weak maps. *Bull. Amer. Math. Soc.* **80**: 127–31.

(1975). Weak maps of combinatorial geometries. *Trans. Amer. Math. Soc.* **206**: 247–79.

Nguyen, H. Q. (1978). Projections and weak maps in combinatorial geometries. *Discrete Math.* **24**: 281–9.

(1979a). Weak cuts of combinatorial geometries. *Trans. Amer. Math. Soc.* **250**: 247–62.

(1979b). Constructing the free erection of a geometry. *J. Combin. Theory Ser. B* **27**: 216–24.

Roberts, L. (1974). Characterization of a pregeometry by its flats, in *Combinatorial Mathematics, Lecture Notes in Mathematics, Vol. 403*, edited by D. A. Holton, pp. 101–4. Springer-Verlag, Berlin.

(1975). All erections of a combinatorial geometry and their automorphism groups, in *Combinatorial Mathematics III, Lecture Notes in Mathematics, Vol. 452*, edited by A. P. Street and W. D. Wallis, pp. 210–13. Springer-Verlag, Berlin.

Tutte, W. T. (1947). A ring in graph theory. *Proc. Cambridge Philosophical Society* **43**: 26–40.

(1966). Connectivity in matroids. *Canad. J. Math.* **18**: 1301–24.

CHAPTER 10

Semimodular Functions

Hien Q. Nguyen

It was shown in Chapter 2 that the notion of rank function provides a crypto-morphic theory of matroids. The semimodularity property of the rank is essentially equivalent to the basis-exchange or circuit-elimination axioms and thus is central to matroid theory.

In this chapter we shall consider a more general class of functions on subsets of a finite set E that are semimodular and nondecreasing, but not necessarily nonnegative, normalized, or unit-increased.

The relationships between semimodular functions and matroids have been known and studied since the very beginning of matroid theory. Many results have been found and sometimes independently rediscovered. This chapter presents a unifying theory that attempts to explain most of these known results as derived essentially from Dilworth's fundamental theorem about the embedding of a point lattice into a geometric lattice. Dilworth's original proof was based on a construction of a rank function from a semimodular function, which is a special case of what will be more generally defined in this chapter as expansions. The main thrust of the exposition is to study the properties and applications of expansions.

10.1. GENERAL PROPERTIES OF SEMIMODULAR FUNCTIONS

In the most general setting that will be of interest we shall consider a point lattice L and classes of integer-valued functions defined on L. Two notions of semimodularity can be defined for such functions:

(1) (Global semimodularity) for any elements x, y of L, $f(x) + f(y) \geq f(x \wedge y) + f(x \vee y)$.
(2) (Local semimodularity) for each element x and atoms p, q of L, $p \not< x$, $q \not< x$, $f(x \vee p) + f(x \vee q) \geq f(x) + f(x \vee p \vee q)$.

The proof that the two definitions are equivalent is routine and is left as an exercise.

We define $\mathscr{C}(L)$ to be the set of integer-valued functions defined on L that are nondecreasing [if $x < y$, then $f(x) \leq f(y)$] and semimodular.

When the lattice L is the Boolean algebra $B(E)$ of some finite set E, for simplicity we shall write $\mathscr{C}(E)$ instead of $\mathscr{C}[B(E)]$.

Common examples of elements of $\mathscr{C}(E)$ are constant functions, cardinality, and rank functions.

10.1.1 Proposition. *A function $f \in \mathscr{C}(E)$ is a rank function if and only if f is normalized, and for any $a \in E$, $f(a) \leq 1$.*

10.1.2. Proposition. *If f and g are elements of $\mathscr{C}(E)$, then so is $b_0 + b_1 f + b_2 g$, with b_0, b_1, b_2 integers and b_1, b_2 nonnegative.*

The most notable elements of $\mathscr{C}(E)$ are certainly rank functions, because they define matroid structures on E. More precisely, if r is a rank function, the geometric-closure operator associated to r is

$$A \subset E \to \bar{A} = \{a \,|\, a \in E, r(A \vee a) = r(A)\}.$$

In the case of a more general element of $\mathscr{C}(E)$, it is interesting to see that there is a weaker structure induced on E that bears some analogies to a matroid. For any $f \in \mathscr{C}(E)$, the operator defined by

$$A \subset E \to \bar{A} = \{a \,|\, a \in E, f(A \vee a) = f(A)\}$$

can be shown to be a closure operator, called the f-closure.

10.1.3. Proposition. *Any closure operator defined on E is the f-closure for some $f \in \mathscr{C}(E)$.*

Proof. Suppose $A \to \bar{A}$ is a closure operator on E. The set of closed sets is a subsemilattice L of $B(E)$ (i.e., L is closed under set intersection, and if A, $B \in L$, $A \wedge B = \overline{A \cap B} = A \cap B$).

It is then possible to construct an element $f \varepsilon \mathscr{C}(L)$ that is strictly increasing on L: For example, set $f(E) = 0$, and for any element A of L covered by $E, f(A) = -1$. Then f is defined inductively as follows:

For all $X \subseteq E, f(X) = \inf\{f(A) + f(B) - f(A \vee B) | A > X, B > X\}$.

Thus, f is semimodular by construction. We have to show that f is strictly increasing. With f being defined by induction, and being strictly increasing at the outset, suppose that, until a certain step in the construction, f has been strictly increasing. Let $f(A)$ be the next value to be defined. If f is not strictly increasing, for some $B \in L, B > A, f(B) \leq f(A)$. Let C be any other element of L covering A. By construction,

$$f(A) \leq f(B) + f(C) - f(B \vee C).$$

Thus, $f(C) \geq f(A) - f(B) + f(B \vee C) \geq f(B \vee C)$, which is a contradiction. So f remains strictly increasing throughout the construction.

Consider the extension of f to all subsets of E defined as follows:

For all $A \subseteq E, f(A) = \inf\{f(U) | U \in L, A \subseteq U\} = f(\bar{A})$.

To prove that $f \in \mathscr{C}(E)$, consider $A \subseteq E$ and $a, b \in E$ and the expression

$$f(A \cup a) + f(A \cup b) - f(A) - f(A \cup a \cup b) = f(\overline{A \cup a}) + f(\overline{A \cup b})$$
$$- f(\bar{A}) - f(\overline{A \cup a \cup b}).$$

By semimodularity of f on L,

$$f(\overline{A \cup a}) + f(\overline{A \cup b}) \geq f(\overline{A \cup a \cup b}) + f[(\overline{A \cup a}) \cap (\overline{A \cup b})]$$
$$\geq f(\overline{A \cup a \cup b}) + f(\bar{A}).$$

We can then check that the f-closure is identical with the given closure: If A is closed (i.e., $A = \bar{A}$), then the following holds: For any $a \notin A, f(A \vee a) = f(\overline{A \cup a}) > f(\bar{A}) = f(A)$; so A is f-closed. Conversely, if A is f-closed, the following holds: For any $a \notin A, f(A \cup a) > f(A)$; so $\bar{A} \neq \overline{A \cup a}$, and A is closed. Proposition 10.1.3 is thus proved. $\square$

Given $f \in \mathscr{C}(E)$, a subset $I \subseteq E$ is said to be f-independent if, for any element $a \in I, f(I - a) < f(I)$. (Note that in the case of f being a rank function, we have the usual notion of independent sets.) A subset that is not f-independent is f-dependent; $\varnothing$ is by convention f-independent.

10.1.4. Proposition. *For any $f \in \mathscr{C}(E)$ we have the following:*

(a) *The family of f-independent sets is an order ideal of $B(E)$.*

(b) *The function f is completely determined by its values on f-independent sets:*

For every $A \subseteq E, f(A) = max\{f(I) | I$ f-independent$, I \subseteq A\}$.

Proof. (a) Let I be f-independent and $J \subseteq I$. If J is f-dependent, then for some $a \in J$, we have $f(J - a) = f(J)$, and $f(I - a) + f(J) \geq f(J - a) + f(I)$.

This implies $f(I - a) \geq f(I)$, and then $f(I - a) = f(I)$, which is a contradiction. Thus, J is f-independent.

(b) If A is f-independent, the formula is obviously true. If A is f-dependent, we have

$$f(A) \geq \max\{f(I)|I f\text{-independent}, I \subseteq A\}.$$

On the other hand, for some $a \in A$, $f(A - a) = f(A)$, and we can construct a descending chain

$$A = A_0 \supset A_1 = A - a \supset A_2 \supset \cdots \supset A_k,$$

where A_i, $1 \leq i \leq k$, is f-dependent and $f(A_i) = f(A_{i-1})$. Let A_k be the minimal subset with such a property. Clearly A_k is f-independent and $f(A) = f(A_k)$, and so

$$f(A) \leq \max\{f(I)|I f\text{-independent}, I \subseteq A\},$$

and the formula is proved.   □

10.1.5. *A set I is f-independent if and only if for any element $a \in I$, the f-closure of $I - a$ is strictly contained in the f-closure of I.*

The proof is left as an exercise.

10.2. EXPANSIONS AND DILWORTH'S EMBEDDING

One simple way of constructing semimodular functions that are not rank functions is to consider a matroid $M(E)$ and a partition $\mathscr{P} = (P_1, P_2, \ldots, P_m)$ of E. The restriction of the rank function of $M(E)$ to the sublattice of $B(E)$ generated by $P_1, P_2, \ldots, P_m$ is then a semimodular function on the set $\{P_1, P_2, \ldots, P_m\}$ that is normalized and nondecreasing, but usually not unit-increasing.

Example. Let us consider the matroid $M(E)$ represented in Figure 10.1, where $E = \{1, 2, 3, 4, 5, 6, 7, 8\}$. Let $\mathscr{P}$ be the partition $(\{1, 2, 3\}, \{4, 5, 6\}, \{7, 8\})$ of E. The sublattice of $B(E)$ generated by $\mathscr{P}$ consists of the following subsets:

    rank 0: $\varnothing$,
    rank 1: $\{1, 2, 3\}$, $\{4, 5, 6\}$, $\{7, 8\}$,
    rank 2: $\{1, 2, 3, 4, 5, 6\}$, $\{1, 2, 3, 7, 8\}$, $\{4, 5, 6, 7, 8\}$,
    rank 3: $\{1, 2, 3, 4, 5, 6, 7, 8\}$.

The restriction of the rank function of $M(E)$ to this sublattice gives the following function:

    $f(\varnothing) = 0$,
    $f(\{1, 2, 3\}) = 2, f(\{4, 5, 6\}) = 3, f(\{7, 8\}) = 2,$

FIGURE 10.1. A matroid $M(E)$.

$$f(\{1,2,3,5,6\}) = 3, f(\{1,2,3,7,8\}) = 3, f(\{4,5,6,7,8\}) = 4,$$
$$f(\{1,2,3,4,5,6,7,8\} = 4.$$

Conversely, an interesting question to investigate is the following: Given a normalized function $f \in \mathscr{C}(T)$ for some set T, is it possible to find a matroid $M(E)$ whose rank function coincides with f when restricted to a sublattice of $B(E)$ isomorphic to $B(T)$?

For most of our applications we shall need a more general setting, and instead of considering f given in $\mathscr{C}(T)$ for some set T, we shall assume that $f \in \mathscr{C}(L)$ for some general point lattice L. Also, unless otherwise stated, we shall consider only normalized functions in this section.

10.2.1. Definition. *Given a point lattice L, and $f \in \mathscr{C}(L)$, a rank function r defined on a set X [or the corresponding matroid $M(X)$] is an expansion of f if and only if the following hold:*

(a) *There is a sublattice of $B(X)$ isomorphic to L [the isomorphism being denoted $\phi: A \in L \leftrightarrow \phi(A) \in B(X)$].*
(b) *For every $A \in L$, $r[\phi(A)] = f(A)$.*

Constructing an expansion of f necessitates the construction of a set X and a matroid on X. The isomorphism ϕ between L and $B(X)$ will also be denoted as $A \in L \leftrightarrow X_A \in B(X)$ for simplicity of notation.

Another way of interpreting expansions is to consider the mapping ϕ as an embedding of the lattice L into the geometric lattice corresponding to $M(X)$. This was the original approach taken by Dilworth (Crawley and Dilworth 1973), where he proved that any finite point lattice can be embedded into a geometric lattice. Dilworth's construction in this proof gives an example of a special expansion that will be called the free expansion.

If $f(a) = 0$ for some atom a of L, the set X_a is composed exclusively of loops in any expansion of f. If we consider the lattice L' obtained from L by deleting a, and the function f' induced on it by f, any expansion of f is obtained by adding some loops to an expansion of f'. Without loss of generality we shall thus assume that for any atom a of L, $f(a) > 0$, and we shall consider expansions that have no loops.

Let $\mathscr{A}$ be the set of atoms of L, and r the rank function of $M(X)$. For any $a, b \in \mathscr{A}$, we have $X_a \cap X_b = X_{a \wedge b} = X_0 = \varnothing$, and for any $a \in \mathscr{A}$,

$r(X_a) = f(a)$. Thus,

$$|X| \geq \sum_{a \in \mathscr{A}} |X_a| \geq \sum_{a \in \mathscr{A}} r(X_a) = \sum_{a \in \mathscr{A}} f(a).$$

We shall show that, in fact, we can restrict our attention to expansions defined on a set X such that $|X| = \sum_{a \in \mathscr{A}} f(a)$.

10.2.2. Proposition. *Given $f \in \mathscr{C}(L)$, if $M(X)$ is an expansion of f, then there is an expansion $M'(X')$ of f such that the following hold:*

(a) $X' \subset X$ and $|X'| = \sum_{a \in \mathscr{A}} f(a)$.
(b) M' is a submatroid of M.

Proof. Because r is the rank function of $M(X)$, $|X_a| \geq r(X_a) = f(a)$ for all $a \in \mathscr{A}$.

For any $a \in \mathscr{A}$, take a maximal independent set X'_a of X_a and consider the set $X' = \bigcup_{a \in \mathscr{A}} X'_a$: We claim that the submatroid of $M(X)$ defined on X' is also an expansion of f.

For any $A \in L$, to the subset X_A we associate the subset $X'_A = \bigcup_{a \leq A} X'_a$. Clearly the lattice $\{X_A | A \in L\}$ is isomorphic to the lattice $\{X'_A | A \in L\}$.

Furthermore, for any $a \in \mathscr{A}$, $r(X'_a) = r(X_a)$, so that for any $A \in L$, $r(X'_A) = r(\bigcup_{a \leq A} X'_a) = r(\bigcup_{a \leq A} X_a) = f(A)$. $\square$

In the following, for any $f \in \mathscr{C}(L)$, we shall consider only expansions of f defined on a minimal set X, $|X| = \sum_{a \in \mathscr{A}} f(a)$. X is the disjoint union $\bigcup_{a \in \mathscr{A}} X_a$, where $|X_a| = f(a)$. The isomorphism ϕ is then given by $A \in L \to \mathscr{C}(A) = X_A = \bigcup_{a \in \mathscr{A}} X_a$. One of the main results of this section is that for any normalized function f of $\mathscr{C}(L)$, an expansion exists.

In any expansion of f, with rank function r, if $I \subseteq X$ is independent, then for any element $A \in L$, we must have $|I \cap X_A| \leq r(X_A) = f(A)$. Let us suppose the converse, that a set $D \subseteq X$ is dependent if and only if there exists an $A \in L$ such that $|D \cap X_A| > f(A)$. It turns out that this simple supposition leads to a particular expansion of f, called the *free expansion*. More precisely, let us consider the family $\{T | T \subseteq X, |T \cap X_A| > f(A)$ for some $A \in L\}$, and let H be the set of its minimal elements (for set inclusion).

10.2.3. Proposition. *H is the set of circuits of an expansion of f.*

Proof. We need only show that H has the circuit-elimination axiom. First, it is easy to see that if T is an element of H, then $T \cap X_A = f(A) + 1$ for some $A \in L$.

Consider $T, V \in H$ [we know then that there exist $A, B \in L$ such that $|T \cap X_A| = f(A) + 1$, $|V \cap X_B| = f(B) + 1$], and $x \in T \cap V$. We want to show that $T \cup V - x = W$ contains an element of H such that $|W \cap X_c| > f(C)$ for some element $C \in L$.

If $|W \cap X_{A \wedge B}| > f(A \wedge B)$ there is nothing to prove; so suppose that $|W \cap X_{A \wedge B}| \leq f(A \wedge B)$. We have

$$|W \cap X_{A \vee B}| \geq |W \cap (X_A \cup X_B)| = |(W \cap X_A) \cup (W \cap X_B)|$$
$$= |W \cap X_A| + |W \cap X_B| - |W \cap (X_A \cap X_B)|.$$

Now

$$|W \cap X_A| = |(T \cup V - v) \cap X_A| \geq |T \cap X_A| - 1 = f(A).$$

Similarly, $|W \cap X_B| \geq f(B)$. On the other hand, by hypothesis,

$$|W \cap (X_A \cap X_B)| = |W \cap X_{A \wedge B}| \leq f(A \wedge B).$$

Finally,

$$|W \cap X_{A \vee B}| \geq f(A) + f(B) - f(A \wedge B) \geq f(A \vee B).$$

If $|W \cap X_{A \vee B}| = f(A \vee B)$, *the following equalities hold:*

(a) $|W \cap X_A| = |(T \cup V - x) \cap X_A| = |T \cap X_A| - 1.$
(b) $|W \cap X_B| = |(T \cup V - v) \cap X_B| = |T \cap X_B| - 1.$
(c) $|W \cap X_{A \wedge B}| = f(A \wedge B).$

Equality (a) implies $x \in X_A$ and $V \cap X_A = T \cap V \cap X_A$, and, similarly, from (b), $x \in X_B$ and $T \cap X_B = T \cap V \cap X_B$. We have

$$|(T \cup V) \cap X_{A \wedge B}| = |(T \cup V - x) \cap X_{A \wedge B}| + 1 = |W \cap X_{A \wedge B}| + 1$$
$$= f(A \wedge B) + 1.$$

On the other hand,

$$(T \cup V) \cap X_{A \wedge B} = (T \cap X_A \cap X_B) \cup (V \cap X_A \cap X_B)$$
$$= T \cap V \cap X_A \cap X_B,$$

and

$$|(T \cup V) \cap X_{A \wedge B}| = |(T \cap V) \cap X_{A \wedge B}| \leq f(A \wedge B),$$

because $T \cap V \notin H$.

The hypothesis $|W \cap X_{A \vee B}| = f(A \vee B)$ being rejected, we have found an element of L, namely $A \vee B$, such that $|W \cap X_{A \vee B}| > f(A \vee B)$. H is thus the set of circuits of a matroid $M(X)$. The family of dependent sets is $\{D | D \subseteq X$, for all $A \in L, |D \cap X_A| > f(A)\}$; the family of independent sets is $\{I | I \subseteq X$, for all $A \in L, |I \cap X_A| \leq f(A)\}$. It remains to show that for all $A \in L$, if r is the rank function of $M(X)$, $r(X_A) = f(A)$ [clearly we have $r(X_A) \leq f(A)$].

10.2.4. For all subset T of X, $r(T) \geq \inf_{A \in L} \{f(A) + |T - X_A|\}$.

Proof. Let I be a maximal independent set contained in T: $T = I \cup x_1 \cup x_2 \cup \cdots \cup x_k$, where $x_i \notin I$. For all i, $1 \leq i \leq k$, $I \cup x_i$ is dependent: There is an element A_i of L such that $|(I \cup x_i) \cap X_{A_i}| > f(A_i)$. I being independent, we also have $|I \cap X_{A_i}| \leq f(A_i)$; thus, $|I \cap X_{A_i}| = f(A_i)$, and $x_i \in X_{A_i}$.

Using the fact that for any $A, B \in L$, such that $|I \cap X_A| = f(A)$ and $|I \cap X_B| = f(B)$, we have $|I \cap (X_A \cup X_B)| = |I \cap X_{A \vee B}| = f(A \vee B)$. [Proof:

$$f(A \vee B) \geq |I \cap X_{A \vee B}| \geq |I \cap (X_A \cup X_B)|$$
$$= |I \cap X_A| + |I \cap X_B| - |I \cap X_A \cap X_B|$$
$$\geq f(A) + f(B) - f(A \wedge B) \geq f(A \vee B).]$$

We can infer that

$$\left| I \cap \left(\bigcup_{i=1}^{k} X_{A_i} \right) \right| = \left| I \cap X_{\bigvee_{i=1}^{k} A_i} \right| = f\left(\bigvee_{i=1}^{k} A_i \right).$$

Then

$$f\left(\bigvee_{i=1}^{k} A_i \right) + \left| T - \bigcup_{i=1}^{k} X_{A_i} \right|$$
$$= f\left(\bigvee_{i=1}^{k} A_i \right) + \left| (I \cup x_1 \cup x_2 \cup \cdots \cup x_k) - \bigcup_{i=1}^{k} X_{A_i} \right|$$
$$= f\left(\bigvee_{i=1}^{k} A_i \right) + \left| I - \bigcup_{i=1}^{k} X_{A_i} \right| \quad \text{(because } x_i \in X_{A_i})$$
$$= \left| I \cap \bigcup_{i=1}^{k} X_{A_i} \right| + \left| I - \bigcup_{i=1}^{k} X_{A_i} \right| = |I| = r(T).$$

So we have $r(T) \geq \inf_{A \in L} \{f(A) + |T - X_A|\}$. Lemma 10.2.4 is thus proved.
□

It follows from Lemma 10.2.4 that for all $A \in L$,

$$r(X_A) \geq \inf_{B \in L} \{f(B) + |X_A - X_B|\} \geq f(A),$$

and thus $r(X_A) = f(A)$. The matroid $M(X)$ is an expansion of f.
□

10.2.5. If $f(1) = N$, where 1 is the top element of L, then the family of bases of $M(X)$ is

$$\mathscr{B} = \{B | B \subseteq X, |B| = N, \text{ and for every } A \in L, |B \cap X_A| \leq f(A)\}.$$

Given an element $f \in \mathscr{C}(L)$ that is normalized, the set of expansions of f is thus never empty. If we restrict our attention to expansions defined on the minimal cardinality set X, $|X| = \sum_{a \in \mathscr{A}} f(a)$, we shall denote the set of all those expansions by $\mathscr{E}(f, X)$. Thus, $\mathscr{E}(f, X)$ is a subset of the set of all pregeometries defined on X and is endowed with the induced weak order.

10.2.6. Proposition. $\mathscr{E}(f, X)$, *ordered by the weak order, has a largest element.*

Proof. Let $\mathscr{B}$ be defined as in Corollary 10.2.5, and let the matroid on X defined by $\mathscr{B}$ be denoted by $E(f, X)$.

If M is an element of $\mathscr{E}(f, X)$, any basis B of M must verify the conditions $|B| = N$ and for all $A \in L$, $|B \cap \phi(A)| \le f(A)$, so that $B \in \mathscr{B}$; M is thus a weak-map image of $E(f, X)$, and $E(f, X)$ is the largest element of $\mathscr{E}(f, X)$. $\square$

$E(f, X)$ will be called the free expansion of f. We shall also write $E(f)$ when there is no ambiguity, and r_E will be its rank function. As before, we shall write $\phi(A) = X_A$.

It will be left as an exercise for the reader to prove that the free expansion is the same construct as defined in Dilworth's proof of the lattice embedding theorem.

10.2.7. Proposition. *A set* X_A *is independent in* $E(f)$ *if and only if* $f(A) = \sum_{a \in \mathscr{A}, a \le A} f(a)$.

Proof. Suppose X_A is independent. Then, for all $B \in L$, $|X_A \cap X_B| \le f(B)$; in particular, for $B = A$,

$$\sum_{\substack{a \in \mathscr{A} \\ a \le A}} f(A) = \sum_{\substack{a \in \mathscr{A} \\ a \le A}} |X_a| = |X_A| \le f(A).$$

By semimodularity we must have $f(A) = \sum_{a \in \mathscr{A}, a \le A} f(a)$.

Conversely, if $f(A) = \sum_{a \in \mathscr{A}, a \le A} f(a)$, then for any subset $A' \subseteq A$, we also have

$$f(A') = \sum_{\substack{a \in \mathscr{A} \\ a \le A'}} f(a);$$

for all $B \in L$,

$$|X_A \cap X_B| = |X_{A \wedge B}| = \sum_{\substack{a \in \mathscr{A} \\ a \le A \wedge B}} f(a) = f(A \wedge B) \le f(B).$$

So X_A is independent in $E(f)$. $\square$

10.2.8. Proposition. *A set X_A is closed in $E(f)$ if and only if A is f-closed in L.*

Proof. If A is not f-closed in L, then for some $B \in L$, $A \leq B$ and $A \neq B$, we have $f(A) = f(B)$, $X_A \subset X_B$, and $r_E(X_A) = r_E(X_B)$, with $X_A \neq X_B$; X_A is not closed in $E(f)$.

Suppose now that A is f-closed in L: For all $a \in \mathscr{A}$, $a \leq A$, $f(A \vee a) > f(A)$. [This also implies that for all $B \leq A$, $f(B \vee a) > f(B)$.] To show that X_A is closed, we shall show that given any maximal independent set $I \subseteq X_A$, for any $x \in X$, $x \notin X_A$, $I \cup x$ is an independent set.

Because $x \notin X_A$, there is an atom $a \in \mathscr{A}$, $a \not\leq A$, such that $x \in X_a$. For any $B \in L$, we have the following: If $a \not\leq B$,

$$|(I \cup x) \cap X_B| = |I \cap X_B| \leq f(B).$$

If $a \leq B$,

$$|(I \cup x) \cap X_B| = |I \cap X_B| + 1 \leq r_E(X_A \cap X_B) + 1$$
$$= f(A \wedge B) + 1 \leq f[(A \wedge B) \vee a] \leq f(B). \qquad \square$$

10.2.9. Proposition. *The rank function of $E(f)$, r_E, is as follows:*

$$\text{For every } T \subseteq X, r_E(T) = \inf_{A \in L} \{f(A) + |T - X_A|\}.$$

Proof. Given $T \subseteq X$, let I be an independent set contained in T. For any $A \in L$,

$$|I| = |I \cap X_A| + |I - X_A| \leq f(A) + |I - X_A| \leq f(A) + |T - X_A|.$$

In particular, for a maximal independent set of T, we have the following:

$$\text{For all } A \in L, r_E(T) \leq f(A) + |T - X_A|.$$

On the other hand, by Lemma 10.2.4,

$$r_E(T) \geq \inf_{A \in L} \{f(A) + |T - X_A|\}. \qquad \square$$

Example. Let $E = \{a, b, c\}$, and let f be defined as follows:

$$f(\varnothing) = 0,$$
$$f(\{a\}) = 2, \ f(\{b\}) = 3, \ f(\{c\}) = 2,$$
$$f(\{a, b\}) = 3, \ f(\{a, c\}) = 3, \ f(\{b, c\}) = 4,$$
$$f(\{a, b, c\}) = 4.$$

$E(f)$ is the matroid depicted in Figure 10.2. It is easy to see that $\mathscr{E}(f, X)$ contains many expansions of f other than $E(f)$. A few examples are shown in Figure 10.3.

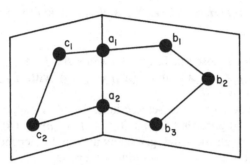

FIGURE 10.2. Example of a free expansion.

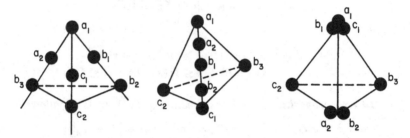

FIGURE 10.3. Examples of nonfree expansions.

10.3. REDUCTIONS

Whereas expansions of a normalized function $f \in \mathscr{C}(E)$ are usually matroids on a larger set than E, we shall now consider a class of constructions that associate to f a matroid on E itself.

    10.3.1. Definition. *Given a function $f \in \mathscr{C}(E)$, a reduction of f is a matroid on E whose rank function r satisfies the following: for any subset A of S, $r(A) \leq \max\{0, f(A)\}$.*

    The following example shows that usually there are many distinct reductions of a given element f of $\mathscr{C}(E)$.

Example. Let $E = \{a, b, c\}$, and let f be defined as

$$f(\varnothing) = 0,$$
$$f(\{a\}) = 2, \; f(\{b\}) = 3, \; f(\{c\}) = 2,$$
$$f(\{a, b\}) = 3, \; f(\{a, c\}) = 3, \; f(\{b, c\}) = 4,$$
$$f(\{a, b, c\}) = 4.$$

All the loopless reductions, up to permutations of the labels a, b, c, are

> the matroid of rank 1 on $\{a,b,c\}$,
> the matroid of rank 2 on $\{a,b,c\}$ with $\{a,b\}$ of rank 1,
> the three-point line on $\{a,b,c\}$,
> the free matroid on $\{a,b,c\}$.

Thus, in this particular example, all matroids on E with no loops are reductions of f.

If f has nonpositive values, we can define the set $E^1 = \{x \mid x \in E, f(x) > 0\}$ and consider the function g on E^1: For all $x \in E^1$, $g(x) = f(x)$. If $M(E^1)$ is a reduction of g, the matroid obtained by adding a set $E - E^1$ of loops to $M(E^1)$ is a reduction of f. Clearly, all reductions of f can be obtained this way; so without loss of generality, in the remainder of this section we shall consider only elements of $\mathscr{C}(E)$ that take positive values on points.

One way of representing reductions is to consider them as special submatroids of expansions of f. More precisely, let $D(f, X)$ be any expansion of f with rank function r_D on the set $X = \bigcup_{a \in E} X_a$, $|X_a| = f(a)$. In each subset X_a, we select arbitrarily a point p_a. For each $A \subseteq S$, we let $p_A = \{p_a \mid a \in A\}$. Let $M(p_E)$ be the restriction of $D(f)$ to p_E, and r_M its rank function. Because of the bijection $p_a \leftrightarrow a \in E$, $M(p_E)$ can be considered as a matroid $M(E)$ on E. It is easy to see that $M(E)$ is a reduction of f: For any subset A of E,

$$r_M(A) = r_M(p_A) = r_E(p_A)$$

$$= \inf_{T \subseteq E} \{f(T) + |p_A - X_T|\}$$

$$\leq f(A) + |p_A - X_A| = f(A).$$

In this fashion, any expansion of f can define several nonisomorphic reductions of f. However, not all reductions of f can be obtained that way, as can be seen from the following:

Example. Let $E = \{a,b,c\}$, and let f be defined as

$$f(\varnothing) = 0,$$
$$f(\{a\}) = 2, f(\{b\}) = 1, f(\{c\}) = 2,$$
$$f(\{a,b\}) = 2, f(\{a,c\}) = 3, f(\{b,c\}) = 2,$$
$$f(\{a,b,c\}) = 3.$$

The three-point-line matroid on $\{a,b,c\}$ is a reduction of f. On the other hand, in any expansion of f, $f(\{a,b\}) = 2$ and $f(\{b,c\}) = 2$ imply that the points a_1, a_2, b are on a line, as well as b, c_1, c_2. If the three-point line on a_1, b, c_1 is a submatroid, then necessarily all points a_1, a_2, b, c_1, c_2 must be on a line, which contradicts $f(\{a,b,c\}) = 3$.

If we apply the construction described earlier to the free expansion $E(f)$, we obtain a reduction whose independent sets can be described explicitly.

A set $p_A \subseteq p_E$ is independent in $M(p_E)$ if and only if it is independent in $E(f)$; that is, for any subset T of E, $|p_A \cap X_T| \le f(T)$. But

$$|p_A \cap X_T| = \left|p_A \cap \left(\bigcup_{a \subset T} \{p_a\}\right)\right| = |A \cap T|,$$

so that the condition is equivalent to the following:

p_A is independent in $M(p_E)$ if and only if for any subset A' of A, $|A'| \le f(A')$.

This condition is independent of the choices of the particular $p_a \in X_a$, and so there is one unique reduction of f that can be associated to the free expansion. This matroid will be denoted by $R(f)$.

10.3.2. Proposition. *Given an element* $f \in \mathscr{C}(E)$, *we have the following:*

(a) *The family* $\mathscr{I}$ *of subsets of* E,

$$\mathscr{I} = \{I \mid I \subseteq E, \text{ for every } I' \subseteq I, |I'| \le f(I')\},$$

is the family of independent sets of a matroid $R(f)$ *on* E.

(b) $R(f)$ *is the submatroid of the free expansion* $E(f, X)$ *defined on the set* $X' \subseteq X$ *such that for each element* a *of* E, $|X' \cap X_a| = 1$.

The rank function r of $R(f)$ is given by Proposition 10.2.9: For any subset T of E,

$$
\begin{aligned}
r(T) = r(X_T) &= \inf_{A \subseteq E} \{f(A) + |X_T - X_A|\} \\
&= \inf_{A \subseteq E} \{f(A) + |T - A|\} \\
&= \inf_{A \subseteq T} \{f(A) + |T - A|\}.
\end{aligned}
$$

10.3.3. Proposition. $R(f)$ *is the largest element in the set of reductions of* f *ordered by the weak-map order.*

Proof. Let $M(f)$ be any reduction of f with rank function r_M. If I is independent in $M(f)$, then for any subset I' of I, $|I'| = r_M(I') \le f(I')$ [because $M(f)$ is a reduction of f], which is the condition for I to be independent in $R(f)$.   □

$R(f)$ will be called the free reduction of f. In the rest of this section we shall study further properties of $R(f)$.

10.3.4. Proposition. *Given* $f \in \mathscr{C}(E)$, *a subset* $K \subseteq E$, $|K| \ge 2$, *is a circuit of* $R(f)$ *if and only if* $|K| = f(K) + 1$, *and for every* $K' \subseteq K, f(K') \ge |K'|$.

Proof. The condition means that K is dependent, and any subset K' of K is independent. Thus, K is a circuit of $R(f)$.

Conversely, let K be a circuit of $R(f)$. Because K is dependent, K has a subset K_0 such that $f(K_0) < |K_0|$ $(K_0 \neq \varnothing)$. However, for any subset $K' \subseteq K$, $K' \neq K$, K' is independent, and thus $|K'| \leq f(K')$ $(K' \neq \varnothing)$. Thus, $K_0 = K$, and $f(K) < |K|$.

On the other hand, for any element x of K, $f(K) \geq f(K - x) \geq |K - x| = |K| - 1 \geq f(K)$. We have equality all along, and thus $f(K) = |K| - 1$.

$\square$

We have to exclude the case $|K| = 1$; that is, K is a loop because the condition of Proposition 10.3.4 is not necessary then.

Example: Let $E = \{a, b\}$, and let f be defined as follows:

$f(\varnothing) = -4$,
$f(\{a\}) = 1, f(\{b\}) = -3$,
$f(\{a, b\}) = 1$;

b is a loop, and no relation of the foregoing type exists.

A set $A \subseteq E$ is said to be *normal* in $R(f)$ if and only if A is independent in $R(f)$, and $f(A) = |A|$. From the proof of Proposition 10.3.4 we see that if K is a circuit of $R(f)$ and $|K| \geq 2$, then for any $x \in K$, $K - x$ is a normal set in $R(f)$.

10.3.5. Proposition. *If a point $a \in E$ is contained in no normal subset of $R(f)$, then a is either a loop or an isthmus of $R(f)$.* $\square$

The proof is left as an exercise.

We have seen that any element $f \in \mathscr{C}(E)$ defines a unique matroid $R(f)$. It is, however, true that to a given matroid $M(E)$, there may be many elements of $\mathscr{C}(E)$ whose free reduction is $M(E)$.

Given a matroid $M(E)$, let $\mathscr{R}(M)$ be the subset of $\mathscr{C}(E)$ defined by

$$\mathscr{R}(M) = \{f \mid f \in \mathscr{C}(E), R(f) \equiv M(E)\}.$$

The set $\mathscr{R}(M)$ is characterized in the following way:

10.3.6. Proposition. *Given a matroid $M(E)$ with rank function r, an element $f \in \mathscr{C}(E)$ belongs to $\mathscr{R}(M)$ if and only if*

(a) $f(A) \geq r(A)$ *for any subset A of S, A not consisting exclusively of loops of M,*
(b) $f(K) = r(K)$ *for any circuit K of M, $|K| \geq 2$.*

Proof. Let $f \in \mathcal{R}(M)$. Let A be a nonempty subset of E. If A is independent, then for any nonempty subset A' of A, $f(A') \geq |A'|$; in particular, for $A' = A$, we have $f(A) \geq |A| = r(A)$.

If A is dependent, and if A is not a set of loops of M, then A contains a maximal independent set $I_A, I_A \neq \emptyset$:

$$r(A) = r(I_A) \leq f(I_A) \leq f(A).$$

Thus, (a) is proved to be necessary.

By Proposition 10.3.4, if K is a circuit of M, $|K| \geq 2$, $f(K) = |K| - 1 = r(K)$. The necessity of (a) and (b) is proved.

Suppose now that an element $f \in \mathcal{C}(E)$ satisfies (a) and (b). Let A be an independent set of M. For any subset A' of A, A' being independent in M, we have, by (a), $f(A') \geq r(A') = |A'|$. A is thus independent in $R(f)$.

Let A be dependent in M. A contains a circuit K of M. If $|K| = 1$, K is a loop of M and a loop of $R(f)$, and so A is dependent in $R(f)$. If $|K| \geq 2$, then $f(K) = r(K) = |K| - 1$: K is dependent in $R(f)$, and so is A.

Thus, M and $R(f)$ have the same independent sets. They are therefore identical. $\qquad\qquad\qquad\qquad\qquad\qquad\qquad\qquad\qquad\qquad\qquad\qquad\quad \square$

When we are interested in only nonnegative elements of $\mathcal{C}(E)$, we define $\mathcal{R}^+(M)$ as

$$\mathcal{R}^+(M) = \{f \mid f \in \mathcal{R}(M), f \geq 0\}.$$

10.3.7. Corollary. *Given a matroid $M(E)$, $f \in \mathcal{C}(E)$ belongs to $\mathcal{R}^+(M)$ if and only if*

(a) $f(A) \geq r(A)$ *for any subset A of S,*
(b) $f(K) = r(K)$ *for any circuit K of M.*

We say that $\mathcal{R}^+(M)$ is bounded if there is an integer $N \geq 0$ such that for any $f \in \mathcal{R}^+(M)$, $f(E) \leq N$. Clearly, $\mathcal{R}^+(M)$ has a finite number of elements if and only if $\mathcal{R}^+(M)$ is bounded.

10.3.8. Proposition. *Given a matroid $M(E)$ with rank function r, $\mathcal{R}^+(M)$ has a finite number of elements if and only if M has no isthmus.*

Proof. If M has no isthmus,

$$\bigcup_{\substack{K \text{ circuit} \\ \text{of } M}} K = E.$$

For any subset A of S and $f \in \mathcal{R}^+(M)$,

$$f(A) \leq \sum_{\substack{K \text{ circuit} \\ \text{of } M}} f(K) = \sum_{\substack{K \text{ circuit} \\ \text{of } M}} (|K| - 1).$$

$\mathcal{R}^+(M)$ is thus bounded and has a finite number of elements.

Conversely, suppose M has an isthmus a. Consider the function δ_a defined by:

$$\text{for } A \subseteq E, \delta_a(A) = 1 \quad \text{if } a \in A,$$
$$= 0 \quad \text{if } a \notin A.$$

Clearly, $\delta_a \in \mathscr{C}(E)$, and for any positive integer k, we want to show that $r + k\delta_a \in \mathscr{R}^+(M)$.

For any subset A of E, we have $(r + k\delta_a)(A) \geq r(A)$, and if K is a circuit of M, necessarily $a \notin K$, and $(r + k\delta_a)(K) = r(K)$. By Corollary 10.3.7, $r + k\delta_a \in \mathscr{R}^+(M)$, and $\mathscr{R}^+(G)$ has an infinite number of elements. $\quad\square$

Among the cases in which $\mathscr{R}^+(M)$ has a finite number of elements we can investigate the cases in which $|\mathscr{R}^+(M)|$ is minimal: If we suppose that $M(E)$ is loopless, we always have $|\mathscr{R}^+(M)| \geq 1$. Take the function f defined by $f(A) = r(A)$ for any nonempty subset A of E, and $f(\varnothing) = 1$.

10.3.9. Lemma. *If $M(E)$ is a loopless matroid, and if $|\mathscr{R}^+(M)| > 2$, there is an element $f \in \mathscr{R}^+(M)$ such that for some element q of $E, f(q) = 2$, and for any other element a of $E, f(a) = 1$.*

Proof. Suppose $|\mathscr{R}^+(M)| > 2$, and let f be an element of $\mathscr{R}^+(M)$ such that f is normalized and $f \neq r$, where r is the rank function of $M(S)$.

Let $E(f)$ be the free expansion of f, and r_E its rank function. We know that by taking arbitrarily a point p_a in each X_a for $a \in E$, the submatroid of $E(f)$ defined on $p_E = \{p_a | a \in E\}$ is isomorphic to the free reduction of f, in this case to M itself, as $f \in \mathscr{R}^+(M)$: For any subset A of $E, r_E(p_A) = r(A)$.

Let $q \in E$ be a point such that $f(q) > 1$ (if such a point does not exist, f will be identical with r), and $|X_q| > 1$. Let y be a point of X_q such that $y \neq p_q$. Let B be the Boolean algebra that is the sublattice of $B(X)$ generated by the subsets $\{p_a\}, a \in E, a \neq q$, and $\{p_q, y\}$. B is isomorphic to $B(E)$. The restriction of r_E to B is a function g that we can consider as being defined on $B(E)$. We have

$$g(q) = g(\{p_q, y\}) = r_E(\{p_q, y\}) = 2,$$

and for any element a of $E, a \neq q$,

$$g(a) = g(\{p_a\}) = r_E(\{p_a\}) = 1.$$

We claim that $g \in \mathscr{R}^+(M)$. For any subset A of E, we have the following:

If $q \notin A, g(A) = g(p_A) = r_E(p_A) = r(A)$.

If $q \in A, g(A) = g(\{p_q, y\} \cup p_{A-q}) = r_E(p_{A-q} \cup p_q \cup y)$
$$\geq r_E(p_{A-q} \cup p_q) = r_E(p_A) = r(A).$$

So $g(A) \geq r(A)$ for all subsets of A.

Let K be a circuit of M. If $q \notin K$, then as before, $g(K) = r(K)$. Suppose $q \in K$: $g(K) = r_E(p_{K-q} \cup p_q \cup y)$. Using the properties of the free expansion $E(f)$, we have

$$f(K) = r_E(X_K) \geq r_E(p_{K-q} \cup p_q \cup y) \geq r_E(x_K) = r(K).$$

However, $f \in \mathcal{R}^+(G)$ implies that $f(K) = r(K)$, and so we have equality all along:

$$g(K) = r_E(p_{K-q} \cup p_q \cup y) = r(K); \text{ so } g \in \mathcal{R}^+(M). \qquad \square$$

10.3.10. Proposition. *If $M(E)$ is a loopless matroid, $|\mathcal{R}^+(M)| = 2$ if and only if, for any element a of E, $\bar{a}$ (closure of a) belongs to the modular cut generated by the closures of all the circuits of M containing a.*

Proof. Suppose $|\mathcal{R}^+(M)| > 2$. By Lemma 10.3.9 there is an $f \in \mathcal{R}^+(M)$ such that for some element p of $E, f(p) = 2$, and for any other element a of E, $f(a) = 1$. Let $E(f)$ be the free expansion of f. $|X| = |E| + 1$. Let $X_p = \{p, q\}$. We know that the submatroid of $E(f)$ defined on $X - \{q\}$ is isomorphic to M: $E(f)$ can be considered as a single-point extension of M. Let K be any circuit of M not containing p,

$$R_E(K) = r(K) = f(K) = r_E(X_K) = r_E(K \cup q),$$

and so $q \in \bar{K}^E$

Thus, $\bar{K}^M$ is in the modular cut $\mathcal{M}$ defining the extension by q. Furthermore, if K is a circuit of G containing p, then $\{p, q\} \subseteq \bar{K}^E$ and $r_E(\{p, q\}) > 1$; $\bar{p}^M$ does not belong to $\mathcal{M}$.

Conversely, suppose that there is a point $p \in E$ such that $\bar{p}$ does not belong to the modular cut $\mathcal{M}$ generated by all the circuits of M containing p. Consider the single-element extension defined by $\mathcal{M}$: $H(X)$, where $X = E \cup q$. Define a function f on $B(E)$ as follows: For any subset A of E,

$$\text{if } p \notin A, f(A) = r_H(A);$$

$$\text{if } p \in A, f(A) = r_H(A \cup q).$$

We have $f(p) = 2$ and any other element a of E, $f(a) = 1$. We claim that $f \in \mathcal{R}^+(M)$. Let A be any subset of E;

$$\text{if } p \notin A, f(A) = r_H(A) = r(A);$$

$$\text{if } p \in A, f(A) = r_H(A \cup q) \geq r_H(A) = r(A).$$

Let K be a circuit of M;

$$\text{if } p \notin K, f(K) = r_H(K) = r(K);$$

$$\text{if } p \in K, f(K) = r_H(K \cup q) = r_H(K) = r(K)$$

(because $\bar{K}^M$ is in $\mathcal{M}$, and $\overline{K - q}^H = \bar{K}^H$).

Finally, $f \in \mathscr{R}^+(M)$, with $f > r$: $|\mathscr{R}^+(M)| > 2$. Proposition 10.3.10 is thus proved.   □

10.3.11. Corollary. *A necessary condition for having* $|\mathscr{R}^+(M)| = 2$ *is that for each element a of E, r* $[\bigcap \{\bar{K}^M | K$ *circuit of M, a* $\in K\} = 1$.

As illustrations for Proposition 10.3.10, we can say that $|\mathscr{R}^+(M)| = 2$ for the Fano plane and the cube in 3-space.

10.4. APPLICATIONS OF EXPANSIONS AND REDUCTIONS

10.4.A. Matroid Constructions

Most of the well-known matroid constructions, such as deletion, contraction, and truncation, are extensions to matroids of traditional constructions in graphs, projective geometries, or lattices. Expansions and reductions seem to be more original to matroids.

Two classic constructions can be defined in terms of reductions: Dilworth truncation and matroid join.

10.4.1 Definition. *Given a matroid M(E) with rank function r and an integer* $k \leq r(E)$, *let L be the lattice formed by identifying into a single lattice element all flats of M(E) of rank* $\leq k$ *and by keeping all other flats of M(E). The kth Dilworth truncation is the free reduction of the following function* $f \in \mathscr{C}(E)$ *defined for all subsets A of the set of flats of rank* $k + 1$:

$$f(A) = r(\sup_L A) - k.$$

10.4.2. Definition. *Given n matroids* $M_1(E)$, $M_2(E), \ldots, M_n(E)$, *with rank functions* $r_1, r_2, \ldots, r_n$, *the join of* $M_1(E)$, $M_2(E), \ldots, M_n(E)$ *is the free reduction of the function* $r = r_1 + r_2 + \cdots + r_n$.

Properties of both of these constructions have been studied in Chapter 7. In this section we show that expansions and reductions define a correspondence between semimodular, nondecreasing functions and matroids that gives rise to whole new classes of constructions.

More precisely, let $\mathscr{C}$ be the set of all integer-valued, normalized, nondecreasing, and semimodular functions defined on finite sets, and let $\mathscr{M}$ be the set of all finite matroids. Using free expansion and reduction, it is possible to associate to each map of $\mathscr{M}$ into itself a map of $\mathscr{C}$ into itself, and vice versa.

Whereas we are more interested in forming new constructions of matroids, let us first illustrate the idea by extending the usual matroid truncation to $\mathscr{C}$. The objective is to define the truncation of a function $f \in \mathscr{C}$.

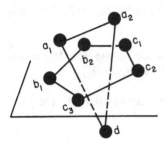

FIGURE 10.4. Free expansion of f.

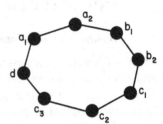

FIGURE 10.5. Truncation of $E(f)$.

Let $f \in \mathscr{C}$ be a function defined on a set $E = \{a, b, c, d\}$ in the following way:

$f(\varnothing) = 0$,
$f(\{a\}) = 2, f(\{b\}) = 2, f(\{c\}) = 3, f(\{d\}) = 1$,
f takes the value 4 on $\{a, b\}$, $\{a, c\}$, and $\{c, d\}$, and 3 on $\{a, d\}$, $\{b, c\}$, and $\{b, d\}$,
f takes the value 4 on all three-element subsets and on E.

The free expansion of f is $E(f)$, depicted in Figure 10.4. The truncation of $E(f)$ is shown in Figure 10.5. The restriction of its rank function to $B(\{a_1 a_2, b_1 b_2, c_1 c_2 c_3, d\})$ is the function f' defined the following way:

$f'(\varnothing) = 0$,
f' takes the same values as f on all points of E,
f' takes the value 3 on all other subsets of E.

By definition, f' is the truncation of f.

It is easy to show, in the general case, that given $f \in \mathscr{C}(E)$, the truncation of f is the following function g on E defined for each subset A of E:

$$g(A) = f(A) \quad \text{if } f(A) < f(E),$$

$$g(A) = f(A) - 1 \quad \text{if } f(A) = f(E).$$

In a similar way we can define, for any element $f \in \mathscr{C}$, its contractions, deletions, and so forth.

Conversely, if we start with a map of $\mathscr{C}$ into itself, we can expect to be able to derive some matroid constructions. A natural map to consider is $f \in \mathscr{C} \to k_0 + k_1 f$, with k_0, k_1 integers, $k_1 > 0$. We shall work out the case $k_0 = 0$ in some detail. Two constructions can be defined, E_k and R_k.

(1) Given a matroid $M(E)$ with rank function r, multiply r by a positive integer k, and form the free expansion of kr: E_k: $M(E) \to E_k(M) = E(kr)$.

(2) Similarly, R_k: $M(E) \to R_k(M) = R(kr)$.

We shall only mention, without proofs, some of the most interesting properties of E_k and R_k. For more details, the reader is referred to Nguyen (1978).

Given a matroid $M(E)$ with rank function r, $E_k(M)$ is defined on a set X such that $X = \bigcup_{a \in E} X_a$, and for each $a \in E$, $|X_a| = k$. $E_k(M)$ has rank $kr(E)$, so that $E_k(M)$ is a free matroid (i.e., a matroid of isthmuses) if and only if M is a free matroid. The independent sets of $E_k(M)$ are

$$\{I | I \subseteq X, \text{ for every } A \subseteq E, |I \cap X_A| \le kr(A)\}.$$

Its rank function is the following function r_k defined for each subset T of X:

$$r_k(T) = \inf_{A \subseteq E} \{kr(A) + |T - X_A|\}.$$

We shall call the canonical surjection of X onto E the point map s defined for each element x of X: $x \to s(x)$ such that $x \in X_{s(x)}$. Then, for any $T \subseteq X$, $s(T) = \{s(x) | x \in T\}$. Equivalently, $s(T)$ is the smallest subset A of E such that $T \subseteq X_A$.

In general, circuits and flats of free expansions are not easy to determine. In the case of $E_k(M)$, we have the following:

10.4.3. Proposition. *If C is a circuit of M, any subset $K \subseteq X_C$ such that $|K| = kr(C) + 1$ is a circuit of $E_k(M)$.*

10.4.4. Proposition. *If K is a circuit of $E_k(M)$, then $\bar{K}^{E_k} = X_A$, where $A = \overline{s(k)}^M$.*

10.4.5. Proposition. *A flat F of $E_k(M)$ is a cyclic flat if and only if $s(F)$ is a cyclic flat of M and $F = X_{s(F)}$.*

The most interesting properties of E_k correspond to the fact that E_k is a functor in the category of matroids and strong maps that preserves many classes of morphisms: epimorphisms, monomorphisms, contractions, truncations, duality, and direct sums.

In the case of R_k, we note the following properties:

10.4.6. Proposition. *$M(S)$ is a quotient of $R_k(M)$.*

10.4.7. Proposition. *For any matroid* $M(E)$ *and two integers* $k_1, k_2 \geq 1$, $R_{k_2}(M)$ *is a quotient of* $R_{k_1}(M)$.

Just like E_k, R_k is a functor of the category of matroids and strong maps into itself, with interesting properties of preserving special classes of morphisms.

10.4.B. Decomposition and Extremal Problems

This is another example of the use of the parallelism between matroid and semimodular functions, brought about by the ideas of expansion and reduction, in solving problems arising in contexts of matroids or of functions.

Join-irreducible matroids. A problem of interest is to characterize the matroids that cannot be obtained as the join of some nontrivial matroids. Such matroids will be called join-irreducible.

10.4.8. Proposition. *A matroid* $M(E)$ *is join-irreducible if and only if any element of* $\mathcal{R}(M)$ *has a connected free expansion.*

The proof is left as an exercise.

In general, this proposition may not be very useful if $\mathcal{R}^+(M)$ contains many elements, but when $M(E)$ satisfies the conditions of Proposition 10.3.10, $\mathcal{R}^+(M)$ contains essentially one element [i.e., the rank function of $M(E)$ itself], so that $M(E)$ is join-irreducible if and only if $M(E)$ is connected. For example, the Fano plane and the cube in 3-space are join-irreducible.

Extremal semimodular, nondecreasing functions. In this section, $\mathscr{C}(E)$ is the set of real-valued nondecreasing and semimodular functions defined on the finite set E.

Given a finite set E, any real-valued function f on the Boolean algebra $B(E)$ is determined by an element of $\mathbb{R}^{2^E}$ [some arbitrary lexicographic ordering of the elements of $B(E)$ being fixed], consisting of the values of f on subsets of E.

An element of $\mathbb{R}^{2^E}$ represents a function f on $B(E)$ that is semimodular and nondecreasing if and only if the following condition is satisfied for any subset X of E and elements x and y of E:

$$\Delta f(X, x, y) = f(X \cup x) + f(X \cup y) - f(X) - f(X \cup x \cup y) \geq 0.$$

Thus, $\mathscr{C}(E)$ is the subset of $\mathbb{R}^{2^E}$ defined by those inequalities for all possible choices of X, x, and y and is thus a convex polyhedron in $\mathbb{R}^{2^E}$.

It is clear that for any function $f \in \mathscr{C}(E)$ and real number $k \geq 0$, kf is also an element of $\mathscr{C}(E)$. $\mathscr{C}(E)$ is a convex cone. One problem is to characterize the extremal rays of $\mathscr{C}(E)$. More precisely, we have the following definition: An element x of $\mathscr{C}(E)$ is extremal if and only if the only way to write $x = x_1 + x_2$, with $x_1 \in \mathscr{C}(E)$, $x_2 \in \mathscr{C}(E)$, is when, for some real number t, $x_1 = tx$ and $x_2 = (1 - t)x$.

The following remarks will allow us to restrict our attention to a smaller subset of $\mathscr{C}(E)$ for searching for all extremal elements of $\mathscr{C}(E)$.

1. For any $f \in \mathscr{C}(E)$ and real number y, the function $f + y$ is also an element of $\mathscr{C}(E)$: f is extremal if and only if the function $f - f(\varnothing)$ is extremal. Without loss of generality, we shall consider normalized extremal elements of $\mathscr{C}(E)$ [i.e., $f(\varnothing) = 0$].

2. $\mathscr{C}(E)$ is defined by a finite number of inequalities: An extremal ray of $\mathscr{C}(E)$ is the solution of some system of equations consisting of a subset of those inequalities transformed into equalities. The values of an extremal function are the solution of a system of linear equations with integer coefficients. These values are thus rational. Thus, a function $f \in \mathscr{C}(E)$ is extremal if and only if there is an integer k such that the function kf is integer-valued and extremal; without loss of generality we can restrict our attention to integer-valued functions.

3. Given $f \in \mathscr{C}(E)$, let E' be the support of f,

$$E' = \{x \,|\, x \in E, f(x) \neq 0\},$$

and let f' be the restriction of f to $B(E')$.

We first need to show that f is completely determined by f':

For every $A \subseteq E$, if $A \subseteq E'$, then $f(A) = f'(A)$,

and if $A \nsubseteq E'$, then $f(A) \leq f(A \cap E') + f(A - E')$.

But for any element a of $A - E'$,

$$f(a) = 0; \text{ so } f(A - E') \leq \sum_{a \in A - E'} f(a) = 0,$$

and $f(A - E') = 0$. Thus, $f(A) \leq f(A \cap E') = f'(A \cap E')$. So $f(A) = f'(A \cap E')$, by the nondecreasing property. Finally, f is determined for any subset A of E:

$$f(A) = f'(A \cap E').$$

Furthermore, f is extremal if and only if f' is extremal.

As a result of these remarks, without loss of generality we shall consider only extremal elements of $\mathscr{C}(E)$ that are integer-valued, normalized, and positive on points. For simplicity, we shall call this subset $\mathscr{C}(E)$ as well.

The following proposition gives a characterization of extremality for an element of a convex cone that will often be useful.

10.4.9. Proposition. *C being a convex cone in $\mathbb{R}^n$, an element $x \in C$ is not extremal if and only if for some $\varepsilon \in \mathbb{R}^n$, $\varepsilon \neq 0$ and ε not proportional to x, both elements $x + \varepsilon$ and $x - \varepsilon$ are elements of C.*

The proof is left as an exercise.

We shall often determine the extremality of a given element $f \in \mathscr{C}(E)$

by trying to find a real-valued function ε such that $f + \varepsilon \in \mathscr{C}(E)$ and $f - \varepsilon \in \mathscr{C}(E)$. In these cases we have to solve the following system of inequations:

$$\text{For every } X \subseteq E, x, y \in E, \Delta(f + \varepsilon)(X, x, y) \geq 0,$$

$$\Delta(f - \varepsilon)(X, x, y) \geq 0.$$

The following simple lemma will prove useful.

10.4.10. Lemma. *Because f and ε are two functions of $C(E)$ such that $\Delta(f + \varepsilon) \geq 0$ and $\Delta(f - \varepsilon) \geq 0$, if for some $X \subseteq E$ and $x, y \in E$ we have $\Delta f(X, x, y) = 0$, then necessarily $\Delta\varepsilon(X, x, y) = 0$.*

Proof. We want to have $\Delta(f + \varepsilon)(X, x, y) \geq 0$;

$$\Delta(f + \varepsilon)(X, x, y) = \Delta f(X, x, y) + \Delta\varepsilon(X, x, y) = \Delta\varepsilon(X, x, y) \geq 0,$$

and also $\Delta(f - \varepsilon)(X, x, y) \geq 0$,

$$\Delta(f - \varepsilon)(X, x, y) = -\Delta\varepsilon(X, x\ y) \geq 0. \qquad \square$$

As a first step we shall concern ourselves with the smaller class of rank functions [i.e. elements f of $\mathscr{C}(E)$ such that $f(a) \leq 1$ for any element a of E]. Let r be a rank function on E and $M(E)$ be the corresponding matroid. We shall show that extremality of r is equivalent to connectedness of $M(E)$.

The following result is proved in Proposition 3.4.6. Given a matroid $M(E)$, we have the following:

(a) The relation R defined among elements of E by "$a, b \in E$, aRb if and only if there is a circuit of $M(E)$ containing a and b" is an equivalence relation.

(b) The equivalence classes are the separators of $M(E)$.

We shall use an equivalent formulation for defining R:

10.4.11. Lemma. *Two elements a and b of E are related via R, namely aRb, if and only if there is a basis B of $M(E)$ such that $a \in B$ and $(B - a) \cup b$ is a basis of $M(E)$.*

Proof. If there is a circuit K containing a and b, we can complete $K - b$ into a basis B of $M(E)$: $B = (K - b) \cup C$, with $C \cap (K - b) = \varnothing$. Then $r[(B - a) \cup b] = r[(K - a) \cup C] = r(K \cup C) = r(B)$, and $(B - a) \cup b$ is a basis.

Conversely, let B be a basis containing a such that $(B - a) \cup b$ is also a basis. Because the set $B \cup b$ is dependent, b belongs to a circuit K contained in $B \cap b$. If the circuit K does not contain a, then K is a subset of $(B - a) \cup b$, which is impossible because of the independence of $(B - a) \cup b$. Thus, K contains a, and aRb. $\qquad \square$

We shall use the following notation: To each point a of E we associate a real variable x_a, and for any family of subsets of E, $(A_i)_{i \in 1}$, $(\sum_{a \in A_i} x_a =)_{i \in 1}$ will stand for the system of equations

$$\sum_{a \in A_1} x_a = \sum_{a \in A_2} x_a = \cdots = \sum_{a \in A_i} x_a = \cdots.$$

The main result of this section is the following:

10.4.12. Proposition. *Given a matroid $M(E)$ whose family of bases is $\mathscr{B} = (B_i)_{i \in I}$, the following properties are equivalent:*
(a) *$M(E)$ is connected.*
(b) *The system $(\sum_{a \in B_i} x_a =)_{i \in I}$ is of rank $|E| - 1$.*
(c) *The rank function r of M is an extremal element of $\mathscr{C}(E)$.*

Proof. (a) $\Rightarrow$ (b) M being connected, the equivalence relation R defines only one equivalence class; hence, by Lemma 10.4.11, for all elements p and q of E, there is a basis B such that $p \in B$ and $(B - p) \cup q \in \mathscr{B}$. Thus, in the system $\sum \equiv (\sum_{a \in B_i} x_a =)_{i \in I}$ we have the equation

$$\sum_{a \in B} x_a = \sum_{a \in (B - p) \cup q} x_a \qquad \text{(i.e., } x_p = x_q\text{)}.$$

Thus, $\sum$ has the solution $x_a = x_b$, for all elements a and b of E. This solution is unique up to a multiplicative constant; $\sum$ being a system of homogeneous equations, this means that $\sum$ is of rank $|E| - 1$.

(b) $\Rightarrow$ (c) M being given with its rank function r, we want to look for a function $\varepsilon: B(E) \to \mathbb{R}$ such that $r \pm \varepsilon \in \mathscr{C}(E)$.

By Lemma 10.4.10 we know that for any $X \subseteq E$, $x, y, \in E$ such that $\Delta r(X, x, y) = 0$, we must also have $\Delta \varepsilon(X, x, y) = 0$. In particular, let $B = \{b_1, b_2, \ldots, b_k\}$ be a basis of M:

$$r(b_1 b_2) = r(b_1) + r(b_2) \text{ implies } \varepsilon(b_1 b_2) = \varepsilon(b_1) + \varepsilon(b_2),$$

$$\vdots$$

$$r(b_1 b_2 \ldots b_k) = r(b_k) + r(b_1 b_2 \ldots b_{k-1}) \text{ implies}$$

$$\varepsilon(b_1 b_2 \ldots b_k) = \varepsilon(b_k) + \varepsilon(b_1 \ldots b_{k-1}),$$

and thus for any basis $B = \{b_1, b_2, \ldots, b_k\}$ of M,

$$\varepsilon(B) = \sum_{b_i \in B} \varepsilon(b_i).$$

On the other hand, let B_1 and B_2 be two bases of M. There is a maximal chain $B_1 = E_0 \subset E_1 \subset E_2 \subset \cdots \subset E$ such that $r(B_1) = r(E_1) = r(E_2) = \cdots = r(E)$ implying $\varepsilon(B_1) = \varepsilon(E_1) = \varepsilon(E)$. For the same reason, $\varepsilon(B_2) = \varepsilon(E)$, and finally, for any two bases of M, B_1 and B_2, we must have

$$\varepsilon(B_1) = \varepsilon(B_2).$$

Thus, the function ε must satisfy the system $\left[\sum_{a \in B_i} \varepsilon(a) = \right]_{i \in I}$, which is $\sum$; $\sum$ being of rank $|E| - 1$, the system has a solution unique up to a multiplicative constant. The values $r(a)$, $a \in E$, being a solution, there is a real number λ such that $\varepsilon(a) = \lambda r(a)$ for all $a \in E$. We want to show that this condition implies that

$$\varepsilon(A) = r(A) \quad \text{for all subsets } A \text{ of } S.$$

For any independent set A of M, clearly,

$$r(A) = \sum_{a \in A} r(a), \text{ implying } \varepsilon(A) = \sum_{a \in A} \varepsilon(a) = \sum_{a \in A} \lambda r(a) = \lambda r(A).$$

For any dependent set A, let I_A be a maximal independent set of A: $r(A) = r(I_A)$, and hence $\varepsilon(A) = \varepsilon(I_A) = \lambda r(I_A) = \lambda r(A)$.

Finally, the only possible functions ε such that $r \pm \varepsilon \in \mathscr{C}(E)$ are given by λr, for any real number λ. Thus, r is an extremal element of $\mathscr{C}(E)$.

(c) $\Rightarrow$ (a) If r is an extremal element of $\mathscr{C}(E)$, clearly $M(E)$ is connected. If M is disconnected, say $M(E) = M_1(E_1) + M_2(E_2)$, we have $r = r_1 + r_2$, where r_1 and r_2 are not proportional to r, which is a contradiction.

The proof of Proposition 10.4.12 is thus complete. $\square$

The characterization of extremal rank functions as rank functions of connected matroids is a stepping-stone to the general solution.

We say that an expansion of f is properly disconnected if it is the direct sum of matroids that are not all themselves expansions of the same function.

10.4.13. *An element f of $\mathscr{C}(E)$ is not extremal if and only if some integer multiple of f has a properly disconnected expansion.*

The proof is left as an exercise.

HISTORICAL NOTES

The main concept in this chapter is the expansion of a semimodular, non-decreasing function, which was introduced by Nguyen (1978a). The main theorem concerning the existence of the free expansion is a version of Dilworth's embedding of a point lattice into a geometric lattice. It was found independently by Nguyen (1978a) and given a proof involving matroid arguments that is somewhat simpler than Dilworth's original proof. Proposition 10.3.2 was found by Edmonds and Rota (1966).

One motivation for considering the constructions E_k and G_k was Rota's conjecture that G_k is a functor of the category of matroids and strong maps. It is interesting to notice that no direct proof to settle the conjecture was available before the introduction of the notion of expansion.

Once E_k is considered, most properties of G_k can be derived easily from the fact that G_k is a submatroid of E_k.

REFERENCES

Choquet, G. (1953/54). Theory of capacities. *Ann. Inst. Fourier* **5**: 131–295.

Crapo, H. H., and Rota, G.-C. (1970). *Combinatorial Geometries*, preliminary edition. M.I.T. Press, Cambridge, Mass.

Crawley, P., and Dilworth, R. P. (1973). *Algebraic Theory of Lattices*. Prentice-Hall, Englewood Cliffs, N.J.

Edmonds, J. (1968). Matroid partition, in *Mathematics of the Decision Sciences (I)*, *Lectures in Applied Mathematics, Vol. II*, edited by G. Dantzig and A. Veinott, pp. 335–45. American Mathematical Society, Providence, R.I.

    (1969). Submodular functions, matroids and certain polyhedra, in *Combinatorial Structures and Their Applications*, edited by R. Guy et al., pp. 69–87. Gordon & Breach, New York.

Edmonds, J., and Rota, G.-C. (1966). Submodular set functions, in Abstracts of the Waterloo Combinatorics Conference, University of Waterloo, Ontario, Canada.

Higgs, D. A. (1966a). A lattice order on the set of all matroids on a set. *Canad. Math. Bull.* **9**: 684–5.

    (1966b). Maps of geometries. *J. London Math. Soc.* **41**: 612–18.

    (1968). Strong maps of geometries. *J. Combin, Theory* **5**: 185–91.

Ingleton, A. W. (1959). A note on independence functions and rank. *J. London Math. Soc.* **34**: 49–56.

    (1971). Conditions for representability and transversality of matroids. *Springer Lecture Notes* **211**: 62–7.

Nguyen, H. Q. (1978a). Semi-modular functions and combinatorial geometries. *Trans. Amer. Math. Soc.* **238**: 355–83.

    (1978b). n-Alternation deficiency of geometric rank functions. *Discrete Math.* **21**: 273–83.

Pym, J. S., and Perfect, H. (1970). Submodular functions and independence structures. *J. Math. Anal. Appl.* **30**: 1–31.

Rosenmuller, J., and Weidner, H. G. (1972). A class of extreme convex set functions with finite carrier. *Advances in Mathematics* **9**: 1–38.

Tutte, W. T. (1965). Lectures on matroids. *J. Res. Nat. Bur. Standards* **69B**: 1–47.

Welsh, D. J. A. (1970). On matroid theorems of Edmonds and Rado. *J. London Math. Soc.* **2**: 251–6.

    (1971). Related classes of set functions, in *Studies in Pure Mathematics*, edited by L. Mirsky, pp. 261–74. Academic Press, London.

Whitney, H. (1935). On the abstract properties of linear dependence. *Amer. J. Math.* **57**: 509–33.

Appendix of Matroid Cryptomorphisms

Thomas Brylawski

The following is a survey of 13 of the more common ways to describe matroids. In the first section, an axiomatization is given for each description. The axiom A0 is usually a nontriviality or normalization condition (such as, for bases, that the family is nonempty) to rule out degeneracy, and A1 describes a general mathematical structure (e.g., that bases form a clutter in that no two are comparable). Finally, A2 is the characteristic axiom (basis exchange in our example) that distinguishes the family of matroidal bases from other clutters (such as k-edge paths in a graph, maximal antichains in a poset, etc.).

As a comparison, consider point-set topology. There, for example, the characteristic axiom for the (topological) closure operator, cl, would be

CL2$_T$. $\mathrm{cl}(A \cup B) = \mathrm{cl}(A) \cup \mathrm{cl}(B)$

(whereas the normalization axiom would be that $\bar{\varnothing} = \varnothing$).

Similarly, for the family $\mathscr{F}$ of closed sets, the topological axioms would be

F0$_T$. X and the empty set are closed.
F2$_T$. Finite unions of closed sets are closed.

Primed axioms, in general, can replace their unprimed counterparts to give an equivalent axiomatization. Thus, for example, bases can be axiomatized by B0, B1, B2, as well as by B0, B1, B2$^{(5)}$, or by B0, B1', B2.

Note that, for example, B1 and B1′ are·not equivalent per se, but among families of subsets of X that obey B0 and B2, they describe the same subclass of families. (In other words, B0, B1, B2 together prove B1′, and B0, B1′, B2 prove B1.) Note that bases give the largest number of equivalent axiomatizations from the apparently modest B2 [basis exchange] to the apparently much stronger $B2^{(2)}$ [symmetric subset exchange] to the (new) axiom $B2^{(9)}$ [algorithmic duality]. We remark that B2 and $B2^{(2)}$ were proved equivalent for nonlinear matroids only in the 1970s. Note that the first three axiomatizations involve special structures (lattice, subset operator, and integer-valued function on subsets), and the final 10 all axiomatize special families of subsets of the groundset.

In the second section we give the matroid cryptomorphisms that relate one matroid structure in an invertible way with another. Some of these cryptomorphisms are quite general (such as those that relate a simplicial complex with its clutter of maximal elements, or a closure system in which each point is closed with a point lattice). These are the cryptomorphisms $f_{\mathcal{M} \to \mathcal{M}'}$ that can be generalized to those that pair structures axiomatized by A0 and A1 with structures axiomatized by A′0 and A′1. Axiom A2 is then interpreted in $\mathcal{M}'$ by axiom A′2. Other cryptomorphisms, however, are more subtle, so that, for example, general closure systems do not satisfy a chain condition, and thus a rank function such as that given by $f_{\mathrm{CL} \to r}$ would not, in general, be well defined if the closure system obeyed only axioms CL0 and CL1.

The reader is encouraged to show that primed axioms are equivalent to their unprimed counterparts or to prove that any of the cryptomorphism pairs $f_{\mathcal{M} \to \mathcal{M}'}$ and $f_{\mathcal{M}' \to \mathcal{M}}$ are inverses and do, indeed, prove one axiom system from the other. One can also develop one's own cryptomorphic theory of matroids [two recent examples have been by the family of nonspanning sets and by the boundary operator: $A \mapsto \mathrm{cl}(A) \cap \mathrm{cl}(X - A)$].

In conclusion, we interpret our cryptomorphic descriptions for four important classes of matroids (vector, affine, transversal, and graphic) and then give a sampling of when an axiom or cryptomorphism has a special version that characterizes binary matroids. Research problems here could include giving cryptomorphic descriptions for other classes (orientable matroids, gammoids, etc.) or describing a class of structures that obeys a particular weaker or stronger set of axioms. [A recent example of the former is the class of "greedoids" that satisfy a generalization to strings of $B2^{(7)}$. An example of the latter is the class of matroids that satisfy bijective subset exchange (see footnote 11).]

In the following, we remind the reader that $B \supsetneq A$ means that the subset B properly contains A. Further, B covers A in the family $\mathcal{M}$, denoted $B \succ A$ if $B \supsetneq A$, but for no member C of $\mathcal{M}$ do we have that $B \supsetneq C \supsetneq A$. A similar definition of cover is used for lattices (posets) L, where atoms are lattice elements that cover the minimum element $\hat{0} \in L$.

AXIOMATIZATIONS FOR THE MATROID $M(E)$

(1)   *Geometric lattice L:* $f: E' \to A(L)$

L0. f maps a subset E' (the *nonloops*) of E onto the atoms A of a lattice L.

L1.   [Point lattice]   All elements of L are suprema of atoms.

   L1'.   [Relative complementation]   For all lattice elements c, d, and e, with $c < d < e$, there is $d' \in L$ such that $d' \vee d = e$ and $d' \wedge d = c$.

   L1''.   No two join-irreducible elements of L are comparable ($c \in L$ is join-irreducible if $c = d \vee e$ implies $c = d$ or $c = e$).

L2.   [Semimodularity]   There is a rank function ρ on L [$\rho(\hat{0}) = 0$, and $\rho(d) = \rho(c) + 1$ whenever $d \succ c$] such that for all d, $e \in L$,

$$\rho(d) + \rho(e) \geq \rho(d \vee e) + \rho(d \wedge e).$$

   L2'.   [Birkhoff covering property]   If c and d both cover $c \vee d$, then $c \vee d$ covers both c and d.

   L2''.   For all b, c, and d, if c covers or equals b, then $c \vee d$ covers or equals $b \vee d$.

   L2'''.   [Atom modularity]   If $a \succ \hat{0}$, then

$$\rho(b) + \rho(a) = \rho(b \vee a) + \rho(b \wedge a).[1]$$

(2)   *Closure operator:* cl: $2^E \to 2^E$

CL1.   [Closure axioms]
   (a)   [Increasing] $A \subseteq \text{cl}(A)$.
   (b)   [Monotone]   If $A \subseteq B$, then $\text{cl}(A) \subseteq \text{cl}(B)$.
   (c)   [Idempotent] $\text{cl}[\text{cl}(A)] = \text{cl}(A)$.
      CL1'.   (a) $A \subseteq \text{cl}(A)$.
               (b)   If $A \subseteq \text{cl}(B)$, then $\text{cl}(A) \subseteq \text{cl}(B)$.
      CL1''. $A \cup \text{cl}[\text{cl}(A)] \subseteq \text{cl}(A \cup B)$.

CL2.   [MacLane-Steinitz exchange]   If $x \in \text{cl}(A \cup \{y\}) - \text{cl}(A)$, then $y \in \text{cl}(A \cup \{x\})$.

   CL2'.   If $\text{cl}(A) \subsetneq B \subseteq \text{cl}(A \cup \{x\})$, then $\text{cl}(B) = \text{cl}(A \cup \{x\})$.

(3)   *Rank function:* $r: 2^E \to \mathbb{Z}$

R0.   [Normalization] $r(\varnothing) = 0$.

R1.   [Unit rank increase] $r(A \cup \{x\}) = r(A)$ or $r(A) + 1$.

   R1'. $r(\{x\}) = 0$ or 1 for all $x \in E.[2]$

---

[1] We can add a converse to L2''' and obtain one axiom equivalent to both L1 and L2:
$d \succ c$ if and only if $d = c \vee a$ for any element in the nonempty set of atoms
$\{a: a \leq d, a \not\leq c\}$.

[2] R1 is implied by R0, R1', and R2 (or R2''), not by R0, R1', and R2'.

R2. [Semimodularity] $r(A) + r(B) \geq r(A \cup B) + r(A \cap B)$.
R2′. [Local semimodularity] If $r(A) = r(A \cup \{x\}) = r(A \cup \{y\})$, then $r(A \cup \{x, y\}) = r(A)$.
R2″. If $A \supseteq B$, then $r(A) - r(B) \geq r(A \cup C) - r(B \cup C)$.[3]

Families of Subsets of E

(4) *Closed sets* or *flats:* $\mathscr{F}$
F0. $E \in \mathscr{F}$.
F1. [Closed-set family] If $F_1, F_2 \in \mathscr{F}$, then $F_1 \cap F_2 \in \mathscr{F}$.
F2. If $F_1, \ldots, F_k$ is the family of closed sets that cover F (i.e., each F_i contains F properly with no closed set between), then $F_1 - F, \ldots, F_k - F$ partition $E - F$.
F2′. $\cup\{F' : F' \succ F\} = E$.

(5) *Hyperplanes:* $\mathscr{H}$
H0. $E \notin \mathscr{H}$.
H1. [Clutter; Incomparability] One hyperplane cannot properly contain another.
H2. [Weak inclusion] For all distinct $H_1, H_2 \in \mathscr{H}$ and $x \in E$ there is a hyperplane H such that $(H_1 \cap H_2) \cup \{x\} \subseteq H$.
H2′. [Strong inclusion] For all distinct hyperplanes H_1, H_2, $x \notin H_1 \cup H_2$, and $y \in H_1 - H_2$, there exists $H \in \mathscr{H}$ such that $(H_1 \cap H_2) \cup \{x\} \subseteq H$ and $y \notin H$.

(6) *Circuits:* $\mathscr{C}$
C0. $\varnothing \notin \mathscr{C}$.
C1. [Incomparability] One circuit cannot properly contain another.
C2. [Weak elimination] $(C_1 \cup C_2) - \{x\}$ contains a circuit for all distinct circuits C_1 and C_2, and $x \in E$.

---

[3] The rank function is easily shown to be cryptomorphic to an *edge-labeled Boolean algebra*, where the edge (of the Hasse diagram) $\begin{smallmatrix} A \cup x \\ \nwarrow \\ A \end{smallmatrix}$ is labeled by $r(A \cup x) - r(A)$. R1 is equivalent to having all squares labeled as follows:

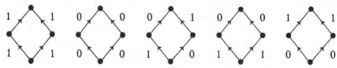

R2 is equivalent to further eliminating

C2'. If $I \subseteq E$ contains no circuit, then $I \cup \{x\}$ contains at most one circuit.

C2''. [Strong elimination] For all circuits C_1 and C_2 with $x \in C_1 \cap C_2$ and $y \in C_1 - C_2$, there is a circuit C containing y such that $C \subseteq (C_1 \cup C_2) - \{x\}$.[4]

C2'''. For all circuits $C_1, \ldots, C_k$ such that, for all i, $C_1 \not\subseteq \bigcup_{j=1}^{i-1} C_j$, and for all $E' \subseteq E$ with $|E'| < k$, there is a circuit C such that $C \subseteq (\bigcup_{j=1}^{k} C_j) - E'$.[5]

(7) *Bonds:* $\mathscr{C}^*$ are axiomatized the same as entry (6), Circuits.

(8) *Open sets:* $\mathcal{O}$

O0. $\varnothing \in \mathcal{O}$.

O1. Unions of open sets are open.

O2. For all open sets O_1 and O_2, and $x \in O_1 \cap O_2$, there is an open set O such that

$$(O_1 \cup O_2) - \{x\} \supseteq O \supseteq (O_1 \cup O_2) - (O_1 \cap O_2).$$

O2'. (a) Every open set is a union of minimal (nonempty) open sets, and (b) for all $O_1 \neq O_2$, $x \in O_1 \cap O_2$, there is a nonempty open set $O \subseteq (O_1 \cup O_2) - \{x\}$.[6]

O2''. If $O_1, \ldots, O_k$ is the family of open sets covered by O then $\bigcap_{j=1}^{k} O_j = \varnothing$.

(9) *Cycles:* $\mathcal{O}^*$ are axiomatized the same as entry (8), Open sets.

(10) *Spanning sets:* $\mathscr{S}$

S0. $E \in \mathscr{S}$ (or: S0'. $\mathscr{S} \neq \varnothing$).

S1. [Order filter] If $S \in \mathscr{S}$ and $S' \supseteq S$, then $S' \in \mathscr{S}$.

S2. If S and S' are spanning sets with $|S| = |S'| + 1$, then there is an element $x \in S - S'$ such that $S - \{x\}$ spans.[7]

S2'. Minimal spanning sets containing a fixed subset $E' \subseteq E$ are equicardinal.

(11) *Independent sets:* $\mathscr{I}$

I0. $\varnothing \in \mathscr{I}$ (or: I0'. $\mathscr{I} \neq \varnothing$).

---

[4] Weak elimination implies its strong counterpart directly only by a rather involved induction argument. However, the cryptomorphisms of the next section provide an alternate method of showing that C2 implies C2''. In particular, C0, C1, and C2 imply the independent-set axioms (using induction on $|I' - I|$ to prove I2). We then easily prove the rank-function axioms R0, R1, and R2. Finally, we use R0, R1, and R2'' (an easy consequence of R2) to prove strong circuit elimination.

[5] An important analogue of C2 that is a consequence of the circuit axioms (but does not imply C2) is *circuit transitivity*:
  If $x \in C_1 - C_2$, $y \in C_2 - C_1$, and $z \in C_1 \cap C_2$, then there is a circuit $C \subseteq C_1 \cup C_2$ that contains both x and y. (It may or may not contain z.)

[6] The family $\{\varnothing, x, y, xy, xyz\}$ shows that axiom O2'(b) alone does not imply O2.

[7] Complementing the family in footnote 8, we see that we cannot remove the restriction that x be not in S'.

I1. [Order ideal; Simplicial complex] If $I \in \mathscr{I}$ and $I' \subseteq I$, then $I' \in \mathscr{I}$.

I2. If I and I' are independent sets, with $|I| = |I'| + 1$, then there is an element $x \in I - I'$ such that $I' \cup \{x\}$ is independent.[8]

I2'. [Pure subcomplexes] For all $E' \subseteq E$, the maximal independent subsets of E' are equicardinal.

(12) *Dependent sets:* $\mathscr{D}$

D0. $\varnothing \notin \mathscr{D}$.

D1. [Order filter] If $D \in \mathscr{D}$ and $D' \supseteq D$, then $D' \in \mathscr{D}$.

D2. If D_1 and D_2 are dependent sets, then either $D_1 \cap D_2$ is dependent or $(D_1 \cup D_2) - \{x\}$ is dependent for all $x \in E$.

D2'. If $I \notin \mathscr{D}$ but $I \cup \{x\}$ and $I \cup \{y\}$ are both in $\mathscr{D}$, then all $(|I| + 1)$-element subsets of $D = I \cup \{x, y\}$ are dependent.[9]

(13) *Bases:* $\mathscr{B}$

B0. $\mathscr{B} \neq \varnothing$.

B1. [Incomparability] One basis cannot properly contain another.

B1'. [Equicardinality] All bases have the same size.

B2. [Weak exchange] For all bases B and B' and $x \in B$, there is a $y \in B'$ such that $(B - x) \cup \{y\}$ is a basis.

B2'. [Symmetric exchange] For all $B, B' \in \mathscr{B}$ and $x \in B$, there is a $y \in B'$ such that both $(B - x) \cup \{y\}$ and $(B' - y) \cup \{x\}$ are bases.[10]

B2$^{(2)}$. [Symmetric-subset exchange; Matroidal Laplace expansion] For all $B, B' \in \mathscr{B}$ and $A \subseteq B$, there is a subset $A' \subseteq B'$ such that $(B - A) \cup A'$ and $(B' - A') \cup A$ are both bases.

B2$^{(3)}$. [Bijective exchange] For all $B, B' \in \mathscr{B}$, there is a bijection $f: B \to B'$ such that $(B - x) \cup \{f(x)\}$ is a basis for all $x \in B$.[11]

B2$^{(4)}$. [Dual exchange] For all $B, B' \in \mathscr{B}$, $y \in B' - B$, there is an $x \in B - B'$ such that $(B - x) \cup \{y\}$ is a basis.[12]

---

[8] Weakening the stipulation $x \in I - I'$ in I2 by the statement $x \in I$ results in an axiom that does not imply I2, as is shown by the family $\{\varnothing, w, x, y, z, wx, wy, wz, xy, xz, yz, wxy\}$.

[9] D2' is D2 with the restriction on the hypothesis that $|D_1 - D_2| = |D_2 - D_1| = 1$.

[10] For both B2 and B2', we get an equivalent axiom by replacing "... $x \in B$ ... $y \in B'$ ..." with "... $x \in B - B'$ ... $y \in B' - B$ ...."

[11] Axiom B2$^{(3)}$ is, in fact, equivalent to both B1 and B2. In general, we cannot combine B2' and B2$^{(3)}$ to get bijective symmetric exchange {where both $(B - x) \cup \{f(x)\}$ and $[B' - f(x)] \cup \{x\}$ are bases}. The matroid of the complete graph K_4 does not satisfy bijective symmetric exchange. On the other hand, transversal matroids and their minors (*gammoids*) satisfy the stronger combination of B2$^{(2)}$ and B2$^{(3)}$: *bijective subset exchange*, where there exists a bijection $f: B \to B'$ such that $(B - A) \cup f(A)$ is a basis for all $A \subseteq B$.

[12] Axiom B2$^{(4)}$, with "... $y \in B' - B$ ... $x \in B - B'$ ..." replaced by $\overline{B2}^{(4)}$ ("... $y \in B'$ ... $x \in B$ ...") is too weak. For example, the equicardinal family $\{xyu, xyv, xyw, uvx, uvy, uvw\}$ satisfies B2$^{(4)}$ but not B2.

B2$^{(5)}$. [Basis interpolation or Middle basis axiom] For all
$I \subseteq B' \in \mathcal{B}$ and $S \supseteq B'' \in \mathcal{B}$, with $I \subseteq S$, there is a basis
B such that $I \subseteq B \subseteq S$.

B2$^{(6)}$. For any total order $<$ on E ($|E| = n$), associate the n-ary
relation $\mathcal{B}^<$ with the family $\mathcal{B}$, where $(x_1, x_2, \ldots, x_k,$
$x_k, \ldots, x_k) \in \mathcal{B}^<$ if $\{x_1, \ldots, x_k\} \in \mathcal{B}$ and $x_i > x_{i+1}$ for
all i. Then the lexicographically maximum member of
$\mathcal{B}^<$ is also the componentwise maximum.[13]

B2$^{(7)}$. The "greedy algorithm" gives the optimal member of $\mathcal{B}$:
If $w: E \to \mathbb{R}$ is any assignment of weights giving $w(x_1) \geq$
$w(x_2) \geq \cdots \geq w(x_n)$, and if we define $w(B) = \sum_{x \in B} w(x)$,
then $w(I_n) \geq w(B)$ for all $B \in \mathcal{B}$, where $I_0 = \varnothing$, and, for
all i,

$$I_i = \begin{cases} I_{i-1} \cup \{x_i\} & \text{if this subset is contained in a basis,} \\ I_{i-1} & \text{otherwise.}^{14} \end{cases}$$

B2$^{(8)}$. The "stingy algorithm" gives the optimal member of $\mathcal{B}$:
If $w: E \to \mathbb{R}$ is any assignment of weights giving $w(x_1) \geq$
$\cdots \geq w(x_n)$, then $w(S_0) \geq w(B)$ for all $B \in \mathcal{B}$, where $S_n = E$,
and, for $i = n - 1, n - 2, \ldots, 0$,

$$S_i = \begin{cases} S_{i+1} - \{x_{i+1}\} & \text{if this subset contains a basis,} \\ S_{i+1} & \text{otherwise.} \end{cases}$$

B2$^{(9)}$. [Algorithmic duality] For all one-to-one assignments
of weights, the greedy algorithm and the stingy algorithm
return the same subset.[15]

CRYPTOMORPHISMS

In this section, for any matroid M described by the axiom system $\mathcal{M}$, we
show how to get a structure that satisfies $\mathcal{M}'$. Using this *cryptomorphism* we
can prove the axioms for $\mathcal{M}'$ from the axioms for $\mathcal{M}$. Often, but not always,

[13] $(x_1, \ldots, x_n) < (x_1', \ldots, x_n')$ in the lexicographic order if for some k ($1 \leq k \leq n$), $x_k < x_k'$, while
$x_i = x_i'$ for all $i < k$. $(x_1, \ldots, x_n)$ is a componentwise maximum if $x_i \geq x_i'$ for all $(x_1', \ldots, x_n') \in$
$\mathcal{B}^<$ and all i. The least element of each basis is repeated to "fill out" the n-tuple. This
convention is not necessary under the equicardinality assumption B1'.

[14] If w is allowed to take on negative values, B2$^{(7)}$ will imply B1 as well as B2. An equivalent
axiom would result if we considered only one-to-one weight functions, in which case we
would have the conclusion that $w(I_n) > w(B)$ for all $B \in \mathcal{B}, B \neq I_n$. This axiomatization, at the
heart of algorithmic matroid theory, shows that if we have weighted subsets over a family
known to be the bases of a matroid, we can find the maximum in time $O\{n[\log n + I(n)]\}$,
where we sort in time $O(n \log n)$ and check independence in time $I(n)$.

[15] Axiom B2$^{(9)}$ is easily shown to imply the same axiom for the family $\mathcal{B}^*$ of complementary
subsets and is therefore a good way to prove matroid duality. Similar "self-dual" axioms
are B2' and B2$^{(5)}$.

the ith axiom for $\mathcal{M}$ implies the ith axiom for $\mathcal{M}'$ ($i = 0, 1, 2$). Sometimes, however, the axioms for $\mathcal{M}$ are used to make the cryptomorphism well-defined, and once it is, the axioms for $\mathcal{M}'$ follow easily.

Note that although it has little intuitive appeal, the rank function gives straightforward descriptions for all other axiomatizations.

We do not give cryptomorphic descriptions from open sets, spanning sets, or dependent sets, because matroids are seldom defined by these families. In addition, we do not present any cryptomorphisms to lattices, because most are quite awkward. [The only two natural ones are in terms of closed sets (L is obtained by ordering the members of $\mathcal{F}$ by inclusion) or hyperplanes (L is the infimum subsemilattice of the Boolean algebra 2^E generated by $\mathcal{H}$).] For cryptomorphisms in terms of the geometric lattice L, we assume that the matroid is a combinatorial geometry on the atoms of L.

The statement "(use $\bar{\mathcal{M}}$)" in the $(\mathcal{M}, \mathcal{M}')$ coordinate in Table A.1 means that there is no natural way to go directly from $\mathcal{M}$ to $\mathcal{M}'$ without using an intermediate cryptomorphism: $\mathcal{M} \to \bar{\mathcal{M}} \to \mathcal{M}'$. The most straightforward dual cryptomorphisms are mentioned in brackets: $[* \ldots]$. Hence, [*Dual set] in the $(7.\mathscr{C}^*, 6.\mathscr{C})$ coordinate means that the circuit family of M^* coincides with the bond family of M. Although these are not strictly cryptomorphisms, in that they do not relate one structural description of a matroid with another description of the same matroid, they do, for example, explain why circuits and bonds have the same axiom system.

PROTOTYPICAL EXAMPLES

We describe, where natural, each cryptomorphic description for the classes of vector matroids, affine matroids, transversal matroids, and graphical matroids. A check means that this example furnished the name of the cryptomorphism (such as circuits of graphs), and an arrow means that the corresponding axiom system is relatively easy to verify on the class. All matroids are on the set E.

Vector (or Linear)

E is the set of column vectors of an $r \times n$ matrix M' over the field F, where M' has linearly independent rows. $\hat{F}$ is a sufficiently large extension field of F.

1. L: $L(M)$ is the supremum semilattice of $PG(r-1, F)$ generated by the one-dimensional subspaces spanned by members of E.

$\to \checkmark$ 2. cl: Linear closure $[\mathbf{x} \in \text{cl}(A)$ if $\mathbf{x} = \sum f_i \mathbf{a}_i, f_i \in F, \mathbf{a}_i \in A]$.

$\checkmark$ 3. r: Linear rank $[r(A)$ is the size of a largest square nonsingular submatrix contained in the submatrix $A]$.

$\checkmark$ 4. $\mathcal{F}$: Linearly closed subsets; subspaces of F^r intersected with E.

TABLE A.1

In terms of:	2. Closure operator cl For all $A \subseteq E$, $cl(A) =$	3. Rank function $r(A) =$	4. Closed sets $F \in \mathscr{F}$ if	5. Hyperplanes $H \in \mathscr{H}$ if	6. Circuits $C \in \mathscr{C}$ if $C \neq \emptyset$ and
1. Lattice L (with atoms E)	$\{x \in E: x \leq \vee\{y: y \in A\}\}$	$\rho(\vee\{y: y \in A\})$, where ρ is the lattice rank (= length of a maximal chain from $\hat{0}$).	F is (the set of atoms below) some element in L	H is (the set of atoms below) some element covered by $\hat{1}$	The supremum semilattice generated by C is a (once) truncated Boolean algebra
2. Closure operator cl		(By recursion) if $A = B \cup \{x\}$, $r(B) + \begin{cases} 0 & x \in cl(B), \\ 1 & \text{otherwise.} \end{cases}$	$cl(F) = F$	$H \neq E$, and $x \notin H$ if and only if $cl(H \cup \{x\}) = E$	For $C' \subseteq C$, $cl(C') = C$ if and only if $|C - C'| = 1$
3. Rank function r	$\{x: r(A) = r(A \cup x)\}$		$r(F \cup \{x\}) = r(F) + 1$ for all $x \notin F$	$r(H) = r(E) - 1 = r(H \cup \{x\}) - 1$ for all $x \notin H$	$r(C) = |C| - 1 = r(C - \{x\})$ for all $x \in C$
4. Closed sets $\mathscr{F}$	$\cap\{F: F \supseteq A, F \in \mathscr{F}\}$	Maximum size of chain of closed sets, all properly contained in A		H is a maximal proper closed set	(Use cycles: add "C is minimal such that" in $4 \to 9$)
5. Hyperplanes $\mathscr{H}$	$\cap\{H: H \supseteq A, H \in \mathscr{H}\}$ (where the empty intersection equals E)	$c(\emptyset) - c(A)$, where for $A \subseteq E$, $c(A) = \max\{k: \text{there exist } H_1, \ldots, H_k \in \mathscr{H} \text{ where for all } j, H_j \supseteq A, \text{ and } H_j \supseteq H_1 \cap \ldots \cap H_{j-1}\}$	F is an intersection of hyperplanes		(Use cycles: add "C is minimal such that" in $5 \to 9$) [*Dual complement]
6. Circuits $\mathscr{C}$	$\{x: x \in A, \text{ or there is a } C \in \mathscr{C} \text{ with } x \in C \subseteq A \cup \{x\}\}$	$|A| - \max\{k: \text{there are } k \text{ circuits } C_1, \ldots, C_k, \text{ such that for all } j, C_j \subseteq A, \text{ and } C_j \not\subseteq C_1 \cup \ldots \cup C_{j-1}\}$	$|C - F| \neq 1$ for all $C \in \mathscr{C}$	(Use closed sets)	
7. Bonds $\mathscr{C}^*$	$E - \cup\{B: B \in \mathscr{C}^*, B \cap A = \emptyset\}$	(Use hyperplanes)		H is the complement of a bond	(Use cycles: add "C is minimal such that" in $7 \to 9$) [*Dual set] C is a minimal cycle
9. Cycles $\mathscr{O}^*$	$\{x: x \in A, \text{ or there is a } C \in \mathscr{O}^* \text{ with } x \in C \subseteq A \cup \{x\}\}$	$|A| - \max\{k: \emptyset \subsetneqq C_1 \subsetneqq C_2 \subsetneqq \ldots \subsetneqq C_k \subseteq A\}$	$|C - F| \neq 1$ for all $C \in \mathscr{O}^*$ [*Dual complement]	(Use closed sets)	
11. Independent sets $\mathscr{I}$	$E - \{x: x \notin A \text{ and } I \cup \{x\} \in \mathscr{I} \text{ for all } I \in \mathscr{I}, I \subseteq A\}$	$\max\{|I|: I \in \mathscr{I}, I \subseteq A\}$	$I \cup \{x\} \in \mathscr{I}$ whenever $I \subseteq F$, $I \in \mathscr{I}$, and $x \notin F$	(Use bases)	(Use dependent sets: add "C is minimal such that" in $11 \to 12$)
13. Bases $\mathscr{B}$	$\max\{|B \cap A|: B \in \mathscr{B}\}$	(Use independent sets)	(Use independent sets)	H is maximal with respect to containing no basis	(Use dependent sets: add "C is minimal such that" in $13 \to 12$)

	7. Bonds $B \in \mathscr{C}^*$ if $B \neq \emptyset$ and	8. Open sets $O \in \mathscr{O}$ if	9. Cycles $C \in \mathscr{C}$* if $C = \emptyset$ or	10. Spanning sets $S \in \mathscr{S}$ if	11. Independent sets $I \in \mathscr{I}$ if	12. Dependent sets $D \in \mathscr{D}$ if	13. Bases $B \in \mathscr{B}$ if
1.	(Use hyperplanes)	(Use closed sets)	$x \leq \vee \{y; y \in C - \{x\}\}$ for all $x \in C$	$\hat{1} = \vee \{x; x \in S\}$	Suprema of subsets of I form a Boolean algebra	There is a redundancy among the suprema of subsets of D	Suprema of subsets of B form a Boolean algebra that includes $\hat{1}$
2.	B is minimal such that $cl(E - B) \neq E$	$cl(E - O) = E - O$	$cl(C - x) = cl(C)$ for all $x \in C$	$cl(S) = E$	$x \notin cl(I - \{x\})$ for all $x \in I$	$x \in cl(D - \{x\})$ for some $x \in D$	$cl(B) = E \neq cl(B - x)$ for all $x \in B$
3.	B is minimal such that $r(E - B) = r(E) - 1$	$r(E - O) = r[(E - O) \cup x] - 1$ for all $x \in O$	$r(C - x) = r(C)$ for all $x \in C$	$r(S) = r(E)$	$r(I) = |I|$	$r(D) < |D|$	$r(B) = |B| = r(E)$
4.	B is a minimal complement of a closed set	O is a complement of a closed set	$|C - F| \neq 1$ for all $F \in \mathscr{F}$ [*Dual complement]	S is contained in no proper closed set	For every $x \in I$, $I - F = \{x\}$ for some $F \in \mathscr{F}$	There is an $x \in D$ such that $D - F \neq \{x\}$ for all $F \in \mathscr{F}$	(Use spanning sets: add "B minimal such that")
5.	B is a hyperplane complement	(Use bonds)	$|C - H| \neq 1$ for all $H \in \mathscr{H}$	S is contained in no hyperplane	For every $x \in I$, $I - H = \{x\}$ for some $H \in \mathscr{H}$	There is an $x \in D$ such that $D - H \neq \{x\}$ for all $H \in \mathscr{H}$	(Use spanning sets: add "B minimal such that")
6.	(Use open sets: add "B is minimal such that" in 6 → 8) [*Dual set]		C is a union of circuits	For all $x \notin S$, there is a circuit C such that $x \in C \subseteq S \cup \{x\}$	I contains no circuit	D contains some circuit	(Use independent sets: add "B maximal such that")
7.		$O = \emptyset$, or O is a union of bonds	$|C \cap B| \neq 1$ for all $B \in \mathscr{C}^*$	S meets every bond	For every $x \in I$, some bond intersects I in only x	There is an $x \in D$ such that no bond intersects D in only x	(Use spanning sets: add "B minimal such that")
9.	(Use open sets: add "B is minimal such that" in 9 → 8)	$|O \cap C| \neq 1$ for all $C \in \mathscr{C}^*$ [*Dual set]		For all $A \subseteq E - S$, there is a cycle C such that $C - S = A$	I contains no nonempty cycle	D contains a nonempty cycle	(Use independent sets: add "B maximal such that")
11.	(Use independent sets)	For all $x \in O$, and independent sets I disjoint from O, $I \cup \{x\}$ is independent	For all $x \in C$, there is a maximal independent subset I of C with $x \notin I$	Maximal independent subsets of S are maximal in E [*Dual complement]		D is not independent	B is a maximal independent set
13.	B is minimal with respect to meeting every basis	(Use independent sets)	For all bases B and $x \in C - B$, there is a $y \in C - B$ such that $(B - \{x\}) \cup \{y\}$ is a basis	S contains some basis	I is contained in some basis	D is contained in no basis	[*Dual complement]

5. $\mathscr{H}$: Maximal subsets H of column indices such that the row space of M' contains a nonzero vector $\mathbf{v}$ whose zeros coincide with H. Equivalently, H is a maximal proper subset representable as the kernel of a linear functional $f:F^r \to F$ intersected with E.

→ 6. $\mathscr{C}$: Minimal nonempty subsets C such that $\sum f_i \mathbf{c}_i = \mathbf{0}, f_i \in F - \{0\}, \mathbf{c}_i \in C$.

7. $\mathscr{C}^*$: Minimal supports of nonzero vectors in the row space of M' (see the foregoing item 5).

8. $\mathcal{O}$: Subsets O that index nonzero columns of NM' for some $d \times r$ matrix N. Equivalently, the supports of vectors in the row space of M' over the field $\hat{F}$.

9. $\mathcal{O}^*$: Subsets C such that there exist scalars $\hat{f}_i \in \hat{F} - \{0\}$ such that $\sum_{\mathbf{c}_i \in C} \hat{f}_j \mathbf{c}_i = \mathbf{0}$.

✓10. $\mathscr{L}$: Subsets S such that $\mathbf{x}_j = \sum_{\mathbf{s}_i \in S} f_{ij} \mathbf{s}_i$ for all $\mathbf{x}_j \in E$, and appropriate scalars $f_{ij} \in F$.

✓11. $\mathscr{I}$: Linearly independent subsets (subsets I such that if $\sum_{\mathbf{x}_i \in I} f_i \mathbf{x}_i = 0$, then $f_i = 0$ for all i).

✓12.   Linearly dependent subsets (subsets D such that $\sum_{\mathbf{x}_i \in D} f_i \mathbf{x}_i = \mathbf{0}$, with $f_i \neq 0$ for some i).

→✓13. $\mathscr{B}$: Vector bases (subsets B such that the columns B form a nonsingular matrix).

Affine Matroids

E is a set of distinct points in some affine space G (such as Euclidean space) of dimension $r - 1$ over a field F. Further E spans G in that it lies on no proper affine subspace. Here, E is a combinatorial geometry, and we shall assume that the point x has affine coordinates $\mathbf{x}$.

1. L: L is a supremun subsemilattice of $AG(r - 1, F)$.

→ 2. cl: Affine closure $[x \in cl(A)$ if $\mathbf{x} = \sum_i f_i \mathbf{a}_i, f_i \in F, a_i \in A, \sum_i f_i = 1]$.

3. r: $r(A)$ is the size of a largest simplex contained in A; it is 1 more than the dimension of the smallest affine flat containing A.

✓ 4. $\mathscr{F}$: Affine flats intersected with E.

✓ 5. $\mathscr{H}$: Subsets H such that for a unique $\bar{H}, H = \bar{H} \cap E$, where $\bar{H}$ is a hyperplane of G [i.e., $\bar{H} = f^{-1}(c)$ for $c \in F$, and f a nonzero linear functional].

6. $\mathscr{C}$: Three points on a line, four points on a plane no three on a line, ..., $i + 2$ points in dimension i such that no smaller subset has this property. In Euclidean space they are subsets C with exactly one partition $C = C_1 \cup C_2$ whose parts have intersecting convex hulls. Analytically, they are minimal sets that satisfy the later item 12.

8. $\mathcal{O}$: (If F is large enough) Subsets O of E such that $O = \{x \in E : f(x) \neq c\}$ for some functional f and scalar c.

9. $\mathcal{O}^*$: Subsets C that are not cones (a cone is a subset $\{x\} \cup A$, where the apex x is not in the affine span of A).

10. $\mathscr{S}$: Sets of points that lie on no proper affine flat.
→11. $\mathscr{I}$: Simplices (extreme points of a segment, triangle, tetrahedron, etc.).
→12. $\mathscr{D}$: Sets of points D such that $\sum_{\mathbf{d}_i \in D} f_i \mathbf{d}_i = 0$ for scalars not all zero such that $\sum f_i = 0$.
13. $\mathscr{B}$: Spanning simplices.

Transversal Matroids

The transversal presentation is a relation $R \subseteq E \times Y$, where Y is matched by R^{-1} into E [i.e., $|Y| = r(E)$]. For a subset $A \subseteq E$, $R(A) = \{y \in Y: (x, y) \in R$ for some $x \in A\}$. $A \subseteq E$ is matched by R into Y' if $R \cap [A \times Y']$ contains $|A|$ vertex-disjoint edges.

2. cl: $x \in \mathrm{cl}(A)$ if $x \in A$, or for some subset A' of A, A' is matched onto Y', $R(A') = Y'$, and $R(x) \subseteq Y'$.
3. r: $r(A) = \min_{A' \subseteq A} (|A - A'| + |R(A')|) = \max_{A'' \subseteq A} [|A''|: A''$ is matched into $Y]$.
4. $\mathscr{F}$: Flats that, as subgeometries, have isthmuses are not easily identified, but those that are also cycles are subsets F such that $F = R^{-1}[R(F)] = R^{-1}[R(F - \{x\})]$ for all $x \in F$.
5. $\mathscr{H}$: Cyclic hyperplanes are as in item 4, with the additional restriction that $|R(F)| = |Y| - 1$.
6. $\mathscr{C}$: Subsets C such that $|R(C)| = |C| - 1$, and $|R(C')| \geq |C'|$ for all $C' \subsetneq C$.
9. $\mathcal{O}^*$: See item 4.
10. $\mathscr{S}$: Subsets S that contain the image of a one-to-one function with domain Y contained in R^{-1}.
→11. $\mathscr{I}$: Domains of one-to-one functions contained in R (subsets I matched into Y). Equivalently, subsets I such that $|J| \leq R(J)|$ for all $J \subseteq I$.
12. $\mathscr{D}$: Sets D that contain a subset D' with $|D'| > |R(D')|$.
13. $\mathscr{B}$: Images of one-to-one functions with domain Y contained in R^{-1}. Equivalently, subsets B of size $|Y|$ such that $|B'| \leq |R(B')|$ for all $B' \subseteq B$. B is a system of distinct representatives for Y, where Y is viewed as the family $\{\ldots, R^{-1}(y_i), \ldots\}$ of subsets of E.

Graphs

The graph G with edge set E and vertex set V represents the matroid $M(E)$. Here, G is connected, and $|V| = r + 1$.

1. L: $L(M)$ is a supremum semilattice of the lattice Π_{r+1} of partitions of an $(r + 1)$-element set.
2. cl: $\mathrm{cl}(A) = \{x: x$ joins two vertices connected by a path in $A\}$.
3. r: $r(A) = r + 1 - c(A)$, where $c(A)$ is the number of connected components of V induced by the subgraph A.

4. $\mathscr{F}$: Subsets F such that, for some partition π of V, $x \in F$ whenever x connects two vertices in the same block of π.

5. $\mathscr{H}$: Maximal subsets H of edges such that the subgraph induced by H has two connected components.

→✓ 6. $\mathscr{C}$: Polygons (closed simple paths).

✓ 7. $\mathscr{C}^*$: Minimal cut-sets of edges.

8. $\mathcal{O}$: Minimal sets of edges whose removal partitions the vertex-set into some partition π.

9. $\mathcal{O}^*$: Subsets of edges C such that for all $x \in C$, x is not a cut-set (isthmus) for the subgraph C.

→10. $\mathscr{S}$: Subsets of edges that connect all the vertices.

→11. $\mathscr{I}$: Forests.

→13. $\mathscr{B}$: Spanning trees.

SPECIAL CRYPTOMORPHISMS CHARACTERIZING BINARY MATROIDS

In this section we give, for certain of the axioms in the first section of this appendix, a stronger version that identifies a structure as a binary matroid (a vector matroid over F_2). The first five axioms will be given in strong form:

$A2_B$. "For all n, $p(n)$."

Each such axiom is equivalent to its matroid (nonbinary) counterpart when n is set equal to 1, and, further, is equivalent (using A0 and A1) to the weaker version:

$A2_{B'}$. "$p(1)$ and $p(2)$."

For example, axiom B2′ states the following (to characterize matroid basis families among all clutters): For all B, B', and $x \in B$, if

$$|\{y \in B': (B - x) \cup \{y\} \text{ and } (B' - y) \cup \{x\} \text{ are bases}\}| = c,$$

then $c \neq 0$ (i.e., the foregoing set is nonempty). This is axiom $B2_B$ (which is presented later) when $n = 1$. When $n = 2$, $B2_B$ states (in its weak form) that a matroid M is binary if and only if c, the number of symmetric exchanges, is never equal to 2. The strong version of $B2_B$ states that further, when M is binary, the number of symmetric exchanges is always odd (a consequence of the Laplace expansion formula for determinants over F_2).

A similar scheme holds for the eight binary cryptomorphisms that follow. For example, when $n = 1$, $f^B_{\mathscr{C} \to \mathscr{C}_*}$ is the usual $\mathscr{C} \to \mathscr{C}^*$ cryptomorphism that if $\mathscr{C}$ is the family of circuits of a matroid M, then $\mathscr{C}^*$, the family of minimal nonempty subsets B such that $|B \cap C| \neq 2 \cdot 1 - 1 = 1$ for all $C \in \mathscr{C}$, is the family of bonds of M. The matroid M is binary if and only if $|B \cap C| \neq 2 \cdot 2 - 1 = 3$ for all $B \in \mathscr{C}^*$, $C \in \mathscr{C}$. In addition, when M is binary, bonds and circuits always intersect in an even number of points.[16] Similar interpretations hold for the other cryptomorphisms.

When treating binary matroids, it is natural to define *binary* cycles and open sets. A *binary cycle* (respectively *binary open set*) is a *disjoint* union of circuits (respectively bonds). In the following, we shall use these specialized cycles and open sets.

Binary Axiomatizations

$CL2_B$. [Binary closure] For all subsets $E' \subseteq E$, with $\{x_1, \ldots, x_n\} \subseteq E'$, if $\{y, y'\} \subseteq cl(E') - cl(E' - \{x_i\})$ for $1 \le i \le n$, then $y' \in cl[(E' - \{x_1, \ldots, x_n\}) \cup \{y\}]$.

$H2_B$. If $A \cup \{x\}$ is contained in $n - 1$ hyperplanes, then A is contained in at most 2^{n-1} other hyperplanes.

$C2_B$. If A contains $n - 1$ circuits, then $A \cup \{x\}$ contains at most 2^{n-1} additional circuits.

$B2_B$. [Binary basis exchange] For all B, B', and $x \in B$, $|\{y \in B': (B - x) \cup \{y\}$ and $(B' - y) \cup \{x\}$ are bases$\}| \ne 2n - 2$.

$B2'_B$. [Binary basis subset exchange] For all B, B', and $A \subseteq B$, $|\{A' \subseteq B': (B - A) \cup A'$ and $(B' - A') \cup A$ are bases$\}| \ne 2n - 2$.

Binary Cryptomorphisms

$f^B_{\mathscr{C}, \mathscr{O}^* \to \mathscr{C}^*}$. [Binary circuit-bond or cycle-bond cryptomorphism] Bonds are minimal nonempty subsets B such that $|B \cap C| \ne 2n - 1$ for all circuits (or binary cycles) C.

$f^B_{\mathscr{C}^*, \mathscr{O} \to \mathscr{C}}$. [Binary bond-circuit or open-set-circuit cryptomorphism] Circuits are minimal nonempty subsets C such that $|C \cap B| \ne 2n - 1$ for all bonds (or binary open sets) B.

$f^B_{\mathscr{C}, \mathscr{O}^* \to \mathscr{O}}$. [Binary circuit-open-set or cycle-open-set cryptomorphism] Binary open sets are subsets O such that $|O \cap C| \ne 2n - 1$ for all circuits (or binary cycles) C.

$f^B_{\mathscr{C}^*, \mathscr{O} \to \mathscr{O}^*}$. [Binary bond-cycle or open-set-cycle cryptomorphism] Binary cycles are subsets C such that $|C \cap B| \ne 2n - 1$ for all bonds (or binary open sets) B.

---

[16] Actually, a stronger statement can be made. If $\mathscr{C}$ is a nonempty clutter (i.e., a family of subsets of E that satisfies C0 and C1), we can apply the cryptomorphism $f_{\mathscr{C} \to \mathscr{B}}$, obtaining the family $\mathscr{B}$ of maximal subsets of E that contain no $C \in \mathscr{C}$. From $\mathscr{B}$, we use $f_{\mathscr{B} \to \mathscr{C}_*}$ to get the family $\mathscr{C}^*$ of minimal subsets that meet every member of $\mathscr{B}$. Then $\mathscr{C}$, $\mathscr{B}$, and $\mathscr{C}^*$ are the circuits, bases, and bonds, respectively, of a matroid (respectively binary matroid) if and only if no member of $\mathscr{C}$ intersects a member of $\mathscr{C}^*$ in a singleton (or in a set of odd cardinality, respectively).

Other binary axiomatizations that do not fall into the foregoing scheme but do characterize when a matroid is binary include the following:

$L3_B$. In the geometric lattice L, no interval of height two has greater than five elements.

$R3_B$. [Binary rank function] For all subsets Z and Y such that $|Y| = 4$, if $r(Z \cup Y) = r(Z) + 2$, then $r(Z \cup Y_2) = r(Z) + 1$ for some two-element subset $Y_2 \subseteq Y$.

$F3_B$. No four closed sets, each covering the same set, can be covered by a common closed set.

$H2_B$. [Binary hyperplane inclusion] If H_1 and H_2 are distinct hyperplanes, then $(H_1 \cap H_2) \cup [E - H_1) \cap (E - H_2)]$ is an intersection of hyperplanes: $H'_1 \cap H'_2 \cap \ldots \cap H'_m$ such that for $i \neq j$, $H'_i \cup H'_j = E$.

$C2'_B$. [Binary circuit (or bond) elimination] The symmetric difference of two circuits is the disjoint union of circuits.

$C2''_B$. If $C - C' (\neq C)$ is minimal among all nonempty circuit differences, then the symmetric difference of C and C' is a circuit.

$O2_B$. [Binary open-set (or cycle) elmination] Binary open sets are closed under symmetric difference.

Index

ENCYCLOPEDIA OF MATHEMATICS AND ITS APPLICATIONS
GIAN-CARLO ROTA, *Editor*

Other volumes in preparation